T0181884

Lecture Notes in Artificial Intelligence 12872

Subseries of Lecture Notes in Computer Science

Series Editors

Randy Goebel
University of Alberta, Edmonton, Canada
Yuzuru Tanaka
Hokkaido University, Sapporo, Japan
Wolfgang Wahlster
DFKI and Saarland University, Saarbrücken, Germany

Founding Editor

Jörg Siekmann
DFKI and Saarland University, Saarbrücken, Germany

More information about this subseries at http://www.springer.com/series/1244

Sheela Ramanna · Chris Cornelis ·
Davide Ciucci (Eds.)

Rough Sets

International Joint Conference, IJCRS 2021
Bratislava, Slovakia, September 19–24, 2021
Proceedings

 Springer

Editors
Sheela Ramanna ⓘD
University of Winnipeg
Winnipeg, MB, Canada

Chris Cornelis ⓘD
Ghent University
Gent, Belgium

Davide Ciucci ⓘD
University of Milano-Bicocca
Milan, Italy

ISSN 0302-9743 ISSN 1611-3349 (electronic)
Lecture Notes in Artificial Intelligence
ISBN 978-3-030-87333-2 ISBN 978-3-030-87334-9 (eBook)
https://doi.org/10.1007/978-3-030-87334-9

LNCS Sublibrary: SL7 – Artificial Intelligence

This Springer imprint is published by the registered company Springer Nature Switzerland AG
The registered company address is: Gewerbestrasse 11, 6330 Cham, Switzerland

Preface

The proceedings of the International Joint Conference on Rough Sets (IJCRS 2021) contain the results of the meeting of the International Rough Set Society (IRSS) held at the Hotel Tatra in Bratislava, Slovakia, during September 19–24, 2021. IJCRS 2021 was organized as part of the IFSA-EUSFLAT 2021 multiconference, which also comprised the following conferences: the 19th World Congress of the International Fuzzy Systems Association (IFSA 2021), the 12th Conference of the European Society for Fuzzy Logic and Technology (EUSFLAT 2021), the International Summer School on Aggregation Operators (AGOP 2021), and the International Conference on Flexible Query Answering Systems (FQAS 2021). Owing to the special circumstances created by the COVID-19 pandemic, the multiconference was organized as a hybrid event, facilitating both online and onsite participation.

The topics covered by IJCRS 2021's submissions revolved around three major groups:

- Core Rough Set Models and Methods: covering/neighborhood-based rough set models, decision-theoretic rough set methods, dominance-based rough set methods, rough Bayesian models, rough clustering, rough computing, rough-set-based feature selection, rule-based systems, variable consistency/precision rough sets, logic in different rough set models, handling missing values
- Related Methods and Hybridization: artificial intelligence, machine learning, data mining, pattern recognition, decision support systems, fuzzy sets, uncertain and approximate reasoning, information granulation, formal concept analysis, Petri nets, natural language processing, big data processing
- Areas of Applications: medicine and health, bioinformatics, business intelligence, telecommunications, web mining and text mining, knowledge discovery, knowledge engineering and representation, risk, seismic data

IJCRS 2021 attracted a total of 26 submissions (not including invited contributions), which underwent a rigorous reviewing process. Each submission was evaluated by two to four experts. Based on this process, the Program Committee (PC) chairs accepted 13 contributions as full papers (7–15 pages), and 7 submissions were accepted as short papers (4–6 pages). All of them are included in these proceedings, together with five invited papers written by distinguished scholars in the rough set community. Furthermore, the conference program featured three keynote lectures and three additional presentations. Their abstracts may be found in the IFSA-EUSFLAT Book of Abstracts.

We would like to thank all authors for submitting their papers and the PC members for their valuable contribution to the conference through their anonymous, detailed review reports. We also wish to congratulate those authors whose papers were selected for presentation and publication in the proceedings. IJCRS 2021's success was possible thanks to the dedication and support of many individuals and organizations. First and foremost, we want to thank IRSS, its honorary chairs (Andrzej Skowron and Yiyu Yao) and its Steering Committee members. Also, we want to thank the organizers of the special session "Representing and Managing Uncertainty: different scenarios, different tools." We are also very grateful to our plenary speakers (Didier Dubois, Pradipta Maji, and Jerzy Stefanowski) and our invited paper authors (Jerzy Grzymala-Busse, Marzena Kryszkiewicz, Pradipta Maji, Fan Min, and Jerzy Stefanowski).

Special thanks go to the organizers of the IFSA-EUSFLAT multiconference, without their support it would have been impossible to organize this edition of IJCRS. In particular, we want to thank Martin Štěpnička, Ladislav Šipeki, and Petr Hurtík. We also want to express our gratitude to the developers of the EasyChair conference management software system. Last but not least, we acknowledge the excellent Springer support. Their diligent work was greatly appreciated as they navigated us in a very professional and smooth manner during the compilation and editing of these proceedings.

September 2021

Sheela Ramanna
Chris Cornelis
Davide Ciucci

Organization

Program Committee Chairs

Sheela Ramanna   University of Winnipeg, Canada
Chris Cornelis   Ghent University, Belgium

Advisory Board Member

Davide Ciucci   University of Milano-Bicocca, Italy

Program Committee

Mani A.	Indian Statistical Institute, Kolkata, India
Piotr Artiemjew	University of Warmia and Mazury, Poland
Jaume Baixeries	Universitat Politècnica de Catalunya, Spain
Mohua Banerjee	Indian Institute of Technology Kanpur, India
Jan Bazan	University of Rzeszów, Poland
Rafael Bello	Universidad Central de Las Villas, Cuba
Jerzy Błaszczyński	Poznań University of Technology, Poland
Nizar Bouguila	Concordia University, Canada
Mihir Chakraborty	Jadavpur University, India
Shampa Chakraverty	Netaji Subhas University of Technology, India
Chien-Chung Chan	University of Akron, USA
Mu-Chen Chen	National Chiao Tung University, Taiwan
Costin-Gabriel Chiru	"Politehnica" University of Bucharest, Romania
Davide Ciucci	University of Milano-Bicocca, Italy
Victor Codocedo	Universidad Técnica Federico Santa María, Chile
Chris Cornelis	Ghent University, Belgium
Zoltán Ernő Csajbók	University of Debrecen, Hungary
Jianhua Dai	Hunan Normal University, China
Phuoc Huy Dang	University of Dalat, Vietnam
Dayong Deng	Zhejiang Normal University, China
Thierry Denoeux	Université de Technologie de Compiègne, France
Fernando Diaz	University of Valladolid, Spain
Murat Diker	Hacettepe University, Turkey
Pawel Drozda	University of Warmia and Mazury, Poland
Didier Dubois	IRIT/RPDMP, France
Ivo Düntsch	Brock University, Canada
Soma Dutta	University of Warmia and Mazury, Poland
Zied Elouedi	Institut Supérieur de Gestion de Tunis, Tunisia
Rafael Falcón	University of Ottawa, Canada
Victor Flores	Universidad Católica del Norte, Chile

Wojciech Froelich	University of Silesia, Poland
Brunella Gerla	University of Insubria, Italy
Piotr Gnyś	Polish-Japanese Academy of Information Technology, Poland
Anna Gomolinska	University of Bialystok, Poland
Salvatore Greco	University of Catania, Italy
Rafał Gruszczyński	Nicolaus Copernicus University in Toruń, Poland
Jerzy Grzymala-Busse	University of Kansas, USA
Bineet Gupta	Shri RamSwaroop Memorial University, India
Quang Thuy Ha	Vietnam National University, Vietnam
Christopher Henry	University of Winnipeg, Canada
Christopher Hinde	Loughborough University, UK
Qinghua Hu	Tianjin University, China
Van Nam Huynh	JAIST, Japan
Dmitry Ignatov	National Research University Higher School of Economics, Russia
Masahiro Inuiguchi	Osaka University, Japan
Ryszard Janicki	McMaster University, Canada
Jouni Järvinen	University of Turku, Finland
Richard Jensen	Aberystwyth University, UK
Xiuyi Jia	Nanjing University of Science and Technology, China
Tamás Kádek	University of Debrecen, Hungary
Michal Kepski	University of Rzeszów, Poland
Md. Aquil Khan	Indian Institute of Technology, Indore, India
Yoo-Sung Kim	Inha University, South Korea
Marzena Kryszkiewicz	Warsaw University of Technology, Poland
Yasuo Kudo	Muroran Institute of Technology, Japan
Yoshifumi Kusunoki	Osaka University, Japan
Sergei O. Kuznetsov	National Research University Higher School of Economics, Russia
Xuan Viet Le	Quy Nhon University, Vietnam
Huaxiong Li	Nanjing University, China
Tianrui Li	Southwest Jiaotong University, China
Jiye Liang	Shanxi University, China
Churn-Jung Liau	Academia Sinica, Taiwan
Tsau Young Lin	San Jose State University, USA
Pawan Lingras	Saint Mary's University, Canada
Caihui Liu	Gannan Normal University, China
Guilong Liu	Beijing Language and Culture University, China
Nguyen Long Giang	Vietnam Academy of Science and Technology, Vietnam
Pradipta Maji	Indian Statistical Institute, Kolkata, India
Benedetto Matarazzo	University of Catania, Italy
Jesús Medina Moreno	University of Cádiz, Spain
Ernestina Menasalvas	Universidad Politécnica de Madrid, Spain
Claudio Meneses	Universidad Católica del Norte, Chile

Duoqian Miao	Tongji University, China
Marcin Michalak	Silesian University of Technology, Poland
Tamás Mihálydeák	University of Debrecen, Hungary
Fan Min	Southwest Petroleum University, China
Pabitra Mitra	Indian Institute of Technology, Kharagpur, India
Sadaaki Miyamoto	University of Tsukuba, Japan
Mikhail Moshkov	KAUST, Saudi Arabia
Dávid Nagy	University of Debrecen, Hungary
Michinori Nakata	Josai International University, Japan
Amedeo Napoli	LORIA, France
An Khuong Nguyen	Technical University of the Vietnam National University Ho Chi Minh City, Vietnam
Dinh Thuc Nguyen	University of Science of the Vietnam National University Ho Chi Minh City, Vietnam
Hoang Son Nguyen	Hue University, Vietnam
Hung Son Nguyen	University of Warsaw, Poland
Loan T. T. Nguyen	University of Warsaw, Poland
Sinh Hoa Nguyen	Polish-Japanese Academy of Information Technology, Poland
M. C. Nicoletti	FACCAMP and Federal University of São Carlos, Brazil
Vilem Novak	University of Ostrava, Czech Republic
Agnieszka Nowak-Brzezińska	University of Silesia, Poland
Piero Pagliani	Research Group on Knowledge and Information, Italy
Sankar K. Pal	Indian Statistical Institute, Kolkata, India
Krzysztof Pancerz	University of Rzeszów, Poland
Vladimir Parkhomenko	St. Petersburg State Polytechnical University, Russia
Andrei Paun	University of Bucharest, Romania
Witold Pedrycz	University of Alberta, Canada
Tatiana Penkova	Institute of Computational Modelling, Russian Academy of Sciences, Russia
Georg Peters	Munich University of Applied Sciences, Germany, and Australian Catholic University, Australia
James Peters	University of Manitoba, Canada
Alberto Pettorossi	Università di Roma "Tor Vergata", Italy
Jonas Poelmans	Clarida Technologies, UK
Lech Polkowski	Polish-Japanese Academy of Information Technology, Poland
Henri Prade	IRIT-CNRS, France
Małgorzata Przybyła-Kasperek	University of Silesia, Poland
Mohammed Quafafou	Aix-Marseille University, France
Yuhua Qian	Shanxi University, China
Sándor Radeleczki	University of Miskolc, Hungary
Anna Radzikowska	University of Warsaw, Poland

Contents

Invited Papers

Mining Incomplete Data Using Global and Saturated Probabilistic Approximations Based on Characteristic Sets and Maximal Consistent Blocks

Patrick G. Clark[1], Jerzy W. Grzymala-Busse[1,2(✉)], Zdzislaw S. Hippe[2], and Teresa Mroczek[2]

[1] Department of Electrical Engineering and Computer Science, University of Kansas, Lawrence, KS 66045, USA
jerzy@ku.edu
[2] Department of Artificial Intelligence, University of Information Technology and Management, 35-225 Rzeszow, Poland
{zhippe,tmroczek}@wsiz.rzeszow.pl

Abstract. In this paper we discuss incomplete data sets with missing attribute values interpreted as "do not care" conditions. For data mining, we use two types of probabilistic approximations, global and saturated. Such approximations are constructed from two types of granules, characteristic sets and maximal consistent blocks. We present results of experiments on mining incomplete data sets using four approaches, combining two types of probabilistic approximations, global and saturated, with two types of granules, characteristic sets and maximal consistent blocks. We compare these four approaches, using an error rate computed as the result of ten-fold cross validation. We show that there are significant differences (5% level of significance) between these four approaches to data mining. However, there is no universally best approach. Hence, for an incomplete data set, the best approach to data mining should be chosen by trying all four approaches.

Keywords: Data mining · Rough set theory · Characteristic sets · Maximal consistent blocks · Probabilistic approximations

1 Introduction

Incomplete data sets are affected by missing attribute values. In this paper, we consider an interpretation of missing attribute values called a "do not care" condition. According to this interpretation, a missing attribute value may be replaced by any specified attribute value.

For rule induction we use probabilistic approximations, a generalization of the idea of lower and upper approximations known in rough set theory. A probabilistic approximation of the concept X is associated with a probability α; if

© Springer Nature Switzerland AG 2021
S. Ramanna et al. (Eds.): IJCRS 2021, LNAI 12872, pp. 3–17, 2021.
https://doi.org/10.1007/978-3-030-87334-9_1

$\alpha = 1$, the probabilistic approximation becomes the lower approximation of X; if α is a small positive number, e.g., 0.001, the probabilistic approximation is reduced to the upper approximation of X. Usually, probabilistic approximations are applied to completely specified data sets [18,20–27], such approximations are generalized to incomplete data sets, using characteristic sets, in [13,14], and maximal consistent blocks in [1,2].

Missing attribute values are usually categorized into lost values and "do not care" conditions. A lost value, denoted by "?", is unavailable for the process of data mining, while a 'do not care" condition, denoted by "*", represents any value of the corresponding attribute.

Recently, two new types of approximations were introduced, global probabilistic approximations in [3] and saturated probabilistic approximations in [8]. Results of experiments on an error rate, evaluated by ten-fold cross validation, were presented for characteristic sets in [6–8] and for maximal consistent blocks in [1,2]. In these experiments, global and saturated probabilistic approximations based on characteristic sets were explored using data sets with lost values and "do not care" conditions. Results show that among these four methods there is no universally best method.

The main objective of this paper is a comparison of four approaches to mining data, using two probabilistic approximations, global and saturated, based on two granules, characteristic sets and maximal consistent blocks, in terms of an error rate evaluated by ten-fold cross validation.

Rule induction was conducted using a new version of the Modified Learning from Examples Module, version 2 (MLEM2) [5,12]. The MLEM2 algorithm is a component of the Learning from Examples using Rough Sets (LERS) data mining system [4,11,12].

2   Incomplete Data

We assume that the input data sets are presented in the form of a decision table. An example of the decision table is shown in Table 1. Rows of the decision table represent cases, while columns are labeled by variables. The set of all cases will be denoted by U. In Table 1, $U = \{1, 2, 3, 4, 5, 6, 7, 8\}$. Independent variables are called attributes and a dependent variable is called a decision and is denoted by d. The set of all attributes will be denoted by A. In Table 1, $A = \{$ Temperature, Wind, Humidity$\}$ and d is Trip. The value for a case x and an attribute a will be denoted by $a(x)$. For example, Temperature$(1) = $ normal.

The set X of all cases defined by the same value of the decision d is called a concept. For example, a concept associated with the value yes of the decision Trip is the set $\{1, 2, 3\}$.

A block of the attribute-value pair (a, v), denoted by $[(a, v)]$, is the set $\{x \in U \mid a(x) = v\}$ [10]. For incomplete decision tables, the definition of a block of an attribute-value pair is modified in the following way:

– if for an attribute a and a case x we have $a(x) = ?$, the case x should not be included in any blocks $[(a, v)]$ for all values v of attribute a;

Table 1. A decision table

	Attributes			Decision
Case	Temperature	Wind	Humidity	Trip
1	normal	*	no	yes
2	high	no	?	yes
3	*	?	no	yes
4	normal	*	*	no
5	?	yes	*	no
6	very-high	*	?	no
7	very-high	?	*	no
8	?	?	yes	no

– if for an attribute a and a case x we have $a(x) = *$, the case x should be
included in blocks $[(a, v)]$ for all specified values v of attribute a.

For the data set from Table 1, the blocks of attribute-value pairs are:

[(Temperature, normal)] = {1, 3, 4},   [(Wind, yes)] = {1, 4, 5, 6},
[(Temperature, high)] = {2, 3},   [(Humidity, no)] = {1, 3, 4, 5, 7},
[(Temperature, very-high)] = {3, 6, 7},   [(Humidity, yes)] = {4, 5, 7, 8},
[(Wind, no)] = {1, 2, 4, 6}.

For a case $x \in U$ and $B \subseteq A$, the *characteristic set* $K_B(x)$ is defined as the
intersection of the sets $K(x, a)$, for all $a \in B$, where the set $K(x, a)$ is defined in
the following way:

– if $a(x)$ is specified, then $K(x, a)$ is the block $[(a, a(x))]$ of attribute a and its
value $a(x)$;
– if $a(x) = ?$ or $a(x) = *$, then $K(x, a) = U$.

For Table 1 and $B = A$,
$K_A(1) = \{1, 3, 4\}$, $K_A(5) = \{1, 4, 5, 6\}$,
$K_A(2) = \{2\}$, $K_A(6) = \{3, 6, 7\}$,
$K_A(3) = \{1, 3, 4, 5, 7\}$, $K_A(7) = \{3, 6, 7\}$, and
$K_A(4) = \{1, 3, 4\}$, $K_A(8) = \{4, 5, 7, 8\}$.

A binary relation $R(B)$ on U, defined for $x, y \in U$ in the following way

$$(x, y) \in R(B) \text{ if and only if } y \in K_B(x)$$

will be called the *characteristic relation*. In our example $R(A) = \{(1, 1), (1, 3),$
$(1, 4), (2, 2), (3, 1), (3, 3), (3, 4), (3, 5), (3, 7), (4, 1), (4, 3), (4, 4), (5, 1), (5,$
$4), (5, 5), (5, 6), (6, 3), (6, 6), (6, 7), (7, 3), (7, 6), (7, 7), (8, 4), (8, 5), (8, 7),$
$(8, 8)\}$.
We quote some definitions from [1]. Let X be a subset of U. The set X is
B-consistent if $(x, y) \in R(B)$ for any $x, y \in X$. If there does not exist a B-

consistent subset Y of U such that X is a proper subset of Y, the set X is called a *generalized maximal B-consistent block*. The set of all generalized maximal B-consistent blocks will be denoted by $\mathscr{C}(B)$. In our example, $\mathscr{C}(A) = \{\{1, 3, 4\}, \{2\}, \{3, 7\}, \{5\}, \{6, 7\}, \{8\}\}$.

Let $B \subseteq A$ and $Y \in \mathscr{C}(B)$. The set of all generalized maximal B-consistent blocks which include an element x of the set U, i.e. the set

$$\{Y | Y \in \mathscr{C}(B), x \in Y\}$$

will be denoted by $\mathscr{C}_B(x)$.

For data sets in which all missing attribute values are "do not care" conditions, an idea of a maximal consistent block of B was defined in [19]. Note that in our definition, the generalized maximal consistent blocks of B are defined for arbitrary interpretations of missing attribute values. For Table 1, the generalized maximal A-consistent blocks $\mathscr{C}_A(x)$ are

$\mathscr{C}_A(1) = \{\{1, 3, 4\}\}$, $\mathscr{C}_A(5) = \{\{5\}\}$,
$\mathscr{C}_A(2) = \{\{2\}\}$, $\mathscr{C}_A(6) = \{\{6, 7\}\}$,
$\mathscr{C}_A(3) = \{\{3, 7\}, \{1, 3, 4\}\}$, $\mathscr{C}_A(7) = \{\{3, 7\}, \{6, 7\}\}$, and
$\mathscr{C}_A(4) = \{\{1, 3, 4\}\}$, $\mathscr{C}_A(8) = \{\{8\}\}$.

3   Probabilistic Approximations

In this section, we will discuss two types of probabilistic approximations: global and saturated.

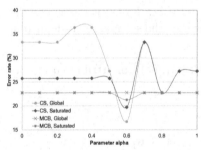

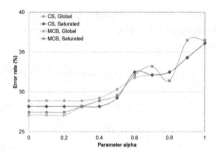

Fig. 1. The *bankruptcy* data set   Fig. 2. The *breast cancer* data set

3.1   Global Probabilistic Approximations Based on Characteristic Sets

An idea of the global probabilistic approximation, restricted to lower and upper approximations, was introduced in [16,17], and presented in a general form in [3]. Let X be a concept, $X \subseteq U$. A *B-global probabilistic approximation* of the

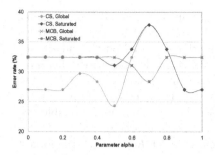

Fig. 3. The *echocardiogram* data set

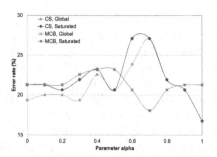

Fig. 4. The *hepatitis* data set

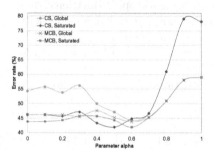

Fig. 5. The *image segmentation* data set

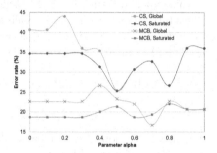

Fig. 6. The *iris* data set

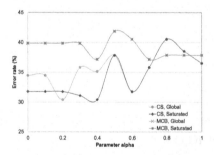

Fig. 7. The *lymphography* data set

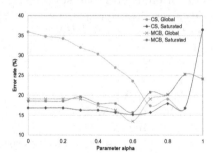

Fig. 8. The *wine recognition* data set

concept X, based on characteristic sets, with the parameter α and denoted by $appr_{\alpha,B}^{global}(X)$ is defined as the following set

$$\bigcup \{K_B(x) \mid \exists\, Y \subseteq U\ \forall x \in Y,\ Pr(X|K_B(x)) \geq \alpha\}. \tag{1}$$

Obviously, for some sets B and X and the parameter α, there exist many B-global probabilistic approximations of X. In addition, the algorithm for computing B-global probabilistic approximations is of exponential computational complexity. Therefore, in our experiments we used a heuristic version of the definition of B-global probabilistic approximation, called a MLEM2 B-global probabilistic approximation of the concept X, associated with a parameter α and denoted by $appr_{\alpha,B}^{mlem2}(X)$ [3]. This definition is based on the rule induction algorithm MLEM2 [12]. The MLEM2 algorithm is used in the Learning from Examples using Rough Sets (LERS) data mining system [4,11,12]. The approximation $appr_{\alpha,B}^{mlem2}(X)$ is constructed from characteristic sets $K_B(y)$, the most relevant to the concept X, i.e., with $|X \cap K_B(y)|$ as large as possible and $Pr(X|K_B(y)) \geq \alpha$, where $y \in U$. If more than one characteristic set $K_B(y)$ satisfies both conditions, we pick the characteristic set $K_B(y)$ with the largest $Pr(X|K_B(y))$. If this criterion ends up with a tie, a characteristic set is picked up heuristically, as the first on the list [3].

In this paper, we study MLEM2 B-global probabilistic approximations based on characteristic sets, with $B = A$. Such approximations are called, for simplicity, *global probabilistic approximations* associated with the parameter α, denoted by $appr_{\alpha}^{global}(X)$. Similarly, for $B = A$, the characteristic set $K_B(X)$ is denoted by $K(x)$.

Let $E_\alpha(X)$ be the set of all eligible characteristic sets defined as follows

$$\{K(x) \mid x \in U, Pr(X|K(x)) \geq \alpha\}. \tag{2}$$

A heuristic version of the global probabilistic approximation based on characteristic sets is presented below.

Global probabilistic approximation
based on characteristic sets algorithm
input: a set X (a concept), a set $E_\alpha(X)$,
output: a set T $(appr_{\alpha}^{global}(X))$
begin

   $G := X$;
   $T := \emptyset$;
   $Y := E_\alpha(X)$;
   while $G \neq \emptyset$ **and** $Y \neq \emptyset$
      begin
         select a characteristic set $K(x) \in Y$
         such that $|K(x) \cap X|$ is maximum;
         if a tie occurs, select $K(x) \in Y$
         with the smallest cardinality;
         if another tie occurs, select the first $K(x)$;
         $T := T \cup K(x)$;
         $G := G - T$;
         $Y := Y - K(x)$
      end

end

For Table 1, all distinct global probabilistic approximations based on characteristic sets are

$$appr_1^{global}(\{1,2,3\}) = \{2\},$$
$$appr_{0.667}^{global}(\{1,2,3\}) = \{1,2,3,4\},$$
$$appr_{0.4}^{global}(\{1,2,3\}) = \{1,2,3,4,5,7\},$$
$$appr_1^{global}(\{4,5,6,7,8\}) = \{4,5,7,8\},$$
$$appr_{0.75}^{global}(\{4,5,6,7,8\}) = \{1,4,5,6,7,8\},$$

3.2 Saturated Probabilistic Approximations Based on Characteristic Sets

Another heuristic version of the probabilistic approximation is based on selection of characteristic sets while giving higher priority to characteristic sets with larger conditional probability $Pr(X|K(x))$. Additionally, if the approximation covers all cases from the concept X, we stop adding characteristic sets.

Let X be a concept and let $x \in U$. Let us compute all conditional probabilities $Pr(X|K(x))$. Then, we sort the set

$$\{Pr(X|K(x)) \mid x \in U\}. \tag{3}$$

Let us denote the sorted list of such conditional probabilities by α_1, α_2,..., α_n, where α_1 is the largest. For any $i = 1, 2,..., n$, the set $E_i(x)$ is defined as follows

$$\{K(x) \mid x \in U, Pr(X|K(x)) = \alpha_i\}. \tag{4}$$

If we want to compute a saturated probabilistic approximation, denoted by $appr_\alpha^{saturated}(X)$, for some α, $0 < \alpha \leq 1$, we need to identify the index m such that

$$\alpha_m \geq \alpha > \alpha_{m+1}, \tag{5}$$

where $m \in \{1, 2, ..., n\}$ and $\alpha_{n+1} = 0$. Then, the saturated probabilistic approximation $appr_{\alpha_m}^{saturated}(X)$ is computed using the following algorithm.

Saturated probabilistic approximation
based on characteristic sets algorithm
input: a set X (a concept), a set $E_i(x)$ for
$i = 1, 2,..., n$ and $x \in U$, index m
output: a set T ($appr_{\alpha_m}^{saturated}(X)$)
begin
    $T := \emptyset$;
    $Y_i(x) := E_i(x)$ for all $i = 1, 2,..., m$ and $x \in U$;
    for $j = 1, 2,..., m$ **do**
        while $Y_j(x) \neq \emptyset$
            begin

> select a characteristic set $K(x) \in Y_j(x)$
> such that $|K(x) \cap X|$ is maximum;
> if a tie occurs, select the first $K(x)$;
> $Y_j(x) := Y_j(x) - K(x)$;
> if $(K(x) - T) \cap X \neq \emptyset$
>   then $T := T \cup K(x)$;
> if $X \subseteq T$ then exit
>
>   end
>
> end

For Table 1, all distinct saturated probabilistic approximations based on characteristic sets are

$$appr_1^{saturated}(\{1,2,3\}) = \{2\},$$
$$appr_{0.667}^{saturated}(\{1,2,3\}) = \{1,2,3,4\},$$
$$appr_1^{saturated}(\{4,5,6,7,8\}) = \{4,5,7,8\},$$
$$appr_{0.75}^{saturated}(\{4,5,6,7,8\}) = \{1,4,5,6,7,8\},$$

3.3 Global Probabilistic Approximations Based on Maximal Consistent Blocks

A special case of the global probabilistic approximation, limited only to lower and upper approximations and to characteristic sets, was introduced in [16,17]. A general definition of the global probabilistic approximation was introduced in [9].

A *B-global probabilistic approximation based on Maximal Consistent Blocks* of the concept X, with the parameter α and denoted by $appr_{\alpha,B}^{global}(X)$ is defined as follows

$$\cup\{Y \mid Y \in \mathscr{C}_x(B),\ x \in X,\ Pr(X|Y) \geq \alpha\}.$$

Obviously, for given sets B and X and the parameter α, there exist many B-global probabilistic approximations of X. Additionally, an algorithm for computing B-global probabilistic approximations is of exponential computational complexity. So, we decided to use a heuristic version of the definition of B-global probabilistic approximation, called the MLEM2 B-global probabilistic approximation of the concept X, associated with a parameter α and denoted by $appr_{\alpha,B}^{mlem2}(X)$ [3]. This definition is based on the rule induction algorithm MLEM2. The approximation $appr_{\alpha,B}^{mlem2}(X)$ is a union of the generalized maximal consistent blocks $Y \in \mathscr{C}(B)$, the most relevant to the concept X, i.e., with $|X \cap Y|$ as large as possible and with $Pr(X|Y) \geq \alpha$. If more than one generalized maximal consistent block Y satisfies both conditions, the generalized maximal consistent block Y with the largest $Pr(X|Y) \geq \alpha$ is selected. If this criterion ends up with a tie, a generalized maximal consistent block Y is picked up heuristically, as the first on the list [3].

Special MLEM2 B-global probabilistic approximations, with $B = A$, are called *global probabilistic approximations* associated with the parameter α, and are denoted by $appr_\alpha^{mlem2}(X)$.

Let $E_\alpha(X)$ be the set of all eligible generalized maximal consistent blocks defined as follows

$$\{Y \mid Y \subseteq \mathscr{C}(A), Pr(X|Y) \geq \alpha\}.$$

A heuristic version of the global probabilistic approximation is computed using the following algorithm

Global probabilistic approximation
based on maximal consistent blocks algorithm
input: a set X (a concept), a set $E_\alpha(X)$,
output: a set T (a global probabilistic approximation $appr_\alpha^{mlem2}(X)$) of X
begin
    $G := X$;
    $T := \emptyset$;
    $\mathscr{Y} := E_\alpha(X)$;
    while $G \neq \emptyset$ and $\mathscr{Y} \neq \emptyset$
        begin
            select a generalized maximal consistent block $Y \in \mathscr{Y}$
            such that $|X \cap Y|$ is maximum;
            if a tie occurs, select $Y \in \mathscr{Y}$
            with the smallest cardinality;
            if another tie occurs, select the first $Y \in \mathscr{Y}$;
            $T := T \cup Y$;
            $G := G - T$;
            $\mathscr{Y} := \mathscr{Y} - Y$
        end
end

For Table 1, all distinct global probabilistic approximations based on maximal consistent blocks are

$$appr_1^{global}(\{1,2,3\}) = \{2\},$$
$$appr_{0.667}^{global}(\{1,2,3\}) = \{1,2,3,4\},$$
$$appr_1^{global}(\{4,5,6,7,8\}) = \{5,6,7,8\},$$
$$appr_{0.5}^{global}(\{4,5,6,7,8\}) = \{3,5,6,7,8\},$$
$$appr_{0.333}^{global}(\{4,5,6,7,8\}) = \{1,3,4,5,6,7,8\},$$

3.4 Saturated Probabilistic Approximations Based on Maximal Consistent Blocks

Saturated probabilistic approximations are unions of generalized maximal consistent blocks while giving higher priority to generalized maximal consistent blocks with larger conditional probability $Pr(X|Y)$. Additionally, if the approximation covers all cases from the concept X, we stop adding generalized maximal consistent blocks.

Let X be a concept and let $x \in U$. Let us compute all conditional probabilities $Pr(X|Z)$, where $Z \in \{Y \mid Y \subseteq \mathscr{C}(A), Pr(X|Y) \geq \alpha\}$. Then we sort the set

$$\{Pr(X|Y) \mid Y \subseteq \mathscr{C}(A)\}$$

in descending order. Let us denote the sorted list of such conditional probabilities by $\alpha_1, \alpha_2, ..., \alpha_n$. For any $i = 1, 2, ..., n$, the set $E_i(X)$ is defined as follows

$$\{Y \mid Y \subseteq \mathscr{C}(A), Pr(X|Y) = \alpha_i\}.$$

If we want to compute a saturated probabilistic approximation, denoted by $appr_\alpha^{saturated}(X)$, for some α, $0 < \alpha \leq 1$, we need to identify the index m such that

$$\alpha_m \geq \alpha > \alpha_{m+1},$$

where $m \in \{1, 2, ..., n\}$ and $\alpha_{n+1} = 0$. The saturated probabilistic approximation $appr_{\alpha_m}^{saturated}(X)$ is computed using the following algorithm

Saturated probabilistic approximation
based on maximal consistent blocks algorithm
input: a set X (a concept), a set $E_i(X)$ for $i = 1, 2, ..., n$, index m
output: a set T (a saturated probabilistic approximation $appr_{\alpha_m}^{saturated}(X)$) of X
begin
    $T := \emptyset$;
    $\mathscr{Y}_i(X) := E_i(X)$ for all $i = 1, 2, ..., m$;
    for $j = 1, 2, ..., m$ **do**
        while $\mathscr{Y}_j(X) \neq \emptyset$
            begin
                select a generalized maximal consistent block $Y \in \mathscr{Y}_j(X)$
                such that $|X \cap Y|$ is maximum;
                if a tie occurs, select the first Y;
                $\mathscr{Y}_j(X) := \mathscr{Y}_j(X) - Y$;
                if $(Y - T) \cap X \neq \emptyset$
                    then $T := T \cup Y$;
                if $X \subseteq T$ **then exit**
            end
end

For Table 1, any saturated probabilistic approximation based on maximal consistent blocks for is the same as corresponding global probabilistic approximation based on maximal consistent blocks for the same concept.

3.5   Rule Induction

Once the global and saturated probabilistic approximations associated with a parameter α are constructed, rule sets are induced using the rule induction algorithm based on another parameter, also interpreted as a probability, and denoted by β. This algorithm also uses the MLEM2 principles [15], and was presented, e.g., in [3].

MLEM2 rule induction algorithm
input: a set Y (an approximation of X) and a parameter β,
output: a set $\mathcal{T}$ (a rule set),
begin

        $G := Y$;
        $D := Y$;
        $\mathcal{T} := \emptyset$;
        $\mathcal{J} := \emptyset$;
        while $G \neq \emptyset$
            begin
                $T := \emptyset$;
                $T_s := \emptyset$;
                $T_n := \emptyset$;
                $T(G) := \{t \mid [t] \cap G \neq \emptyset\}$;
                while $(T = \emptyset$ **or** $[T] \not\subseteq D)$ **and** $T(G) \neq \emptyset$
                    begin
                        select a pair $t = (a_t, v_t) \in T(G)$ with maximum of $|[t] \cap G|$; if a tie occurs, select a pair $t \in T(G)$ with the smallest cardinality of $[t]$; if another tie occurs, select the first pair;
                        $T := T \cup \{t\}$;
                        $G := [t] \cap G$;
                        $T(G) := \{t \mid [t] \cap G \neq \emptyset\}$;
                        if a_t is symbolic $\{$let V_{a_t} be the domain of $a_t\}$
                          then
                              $T_s := T_s \cup \{(a_t, v)|v \in V_{a_t}\}$
                              else $\{a_t$ is numerical, let $t = (a_t, u..v)\}$
                        and $T_n := T_n \cup \{(a_t, x..) \mid$ disjoint $x..y$ and $u..v\} \cup \{(a_t, x..y \mid x..y \supseteq u..v\}$;
                        $T(G) := T(G) - (T_s \cup T_n)$;
                end $\{$while$\}$;
                if $Pr(X \mid [T]) \geq \beta$
                    then

$$\textbf{begin}$$
$$D := D \cup [T];$$
$$\mathcal{T} := \mathcal{T} \cup \{T\};$$
$$\textbf{end } \{\text{then}\}$$
$$\textbf{else } \mathcal{J} := \mathcal{J} \cup \{T\};$$
$$G := D - \cup_{S \in \mathcal{T} \cup \mathcal{J}}[S];$$

  end {while};
for each $T \in \mathcal{T}$ do
  for each numerical attribute a_t with
  $(a_t, u..v) \in T$ do
    while (T contains at least two different
    pairs $(a_t, u..v)$ and $(a_t, x..y)$ with
    the same numerical attribute a_t)
      replace these two pairs with a new pair
      (a_t, common part of $(u..v)$ and $(x..y)$);
    for each $t \in T$ do
      if $[T - \{t\}] \subseteq D$ then $T := T - \{t\}$;
  for each $T \in \mathcal{T}$ do
    if $\cup_{S \in (\mathcal{T} - \{T\})}[S] = \cup_{S \in \mathcal{T}}[S]$ then $\mathcal{T} := \mathcal{T} - \{T\}$;
end {procedure}.

For example, for Table 1 and $\alpha = \beta = 0.5$, using the global probabilistic approximations, the MLEM2 rule induction algorithm induces the following rules:

(Temperature, very-high) & (Headache, yes) $\rightarrow$ (Flu, yes)

(Temperature, high) & (Cough, yes) $\rightarrow$ (Flu, yes)

(Headache, no) & (Cough, no) $\rightarrow$ (Flu, no)

and

(Temperature, normal) $\rightarrow$ (Flu, no)

4 Experiments

For our experiments, we used eight data sets taken from the *Machine Learning Repository* at the University of California at Irvine. For every data set, a new record was created by randomly replacing 35% of existing specified attribute values by *"do not care"* conditions.

In our experiments, the parameter α varied between 0.001 and 1 while the parameter β was equal to 0.5. For any data set, ten-fold cross validation was conducted. Results of our experiments are presented in Figs. 1, 2, 3, 4, 5 6 and 7, where "CS" denotes a characteristic set, "MCB" denotes a generalized maximal consistent block, "Global" denotes a MLEM2 global probabilistic approximation and "Saturated" denotes a saturated probabilistic approximation. In our experiments, four methods for mining incomplete data sets were used, since we combined two types of granules from which approximations are constructed: characteristic sets and generalized maximal consistent blocks with two versions of probabilistic approximations: global and saturated.

These four methods were compared by applying the distribution free Friedman rank sum test and then by the post-hoc test (distribution-free multiple comparisons based on the Friedman rank sums), with a 5% level of significance.

For three data sets: *bankruptcy*, *image segmentation* and *iris*, two methods: global and saturated probabilistic approximations based on maximal consistent blocks are significantly better (error rates evaluated by ten-fold cross validation are smaller) than global probabilistic approximations based on characteristic sets. Additionally, for the *iris* data set saturated probabilistic approximations based on maximal consistent blocks are significantly better than saturated probabilistic approximations based on characteristic sets.

On the other hand, for the data set *lymphography*, saturated global approximations based on characteristic sets are better than both global and saturated probabilistic approximations based on maximal consistent blocks. For the data set *wine recognition*, saturated probabilistic approximations based on characteristic sets are better than both global probabilistic approximations based on characteristic sets and saturated probabilistic approximations based on maximal consistent blocks.

For three data sets, *breast cancer*, *echocardiogram* and *hepatitis*, pairwise differences in an error rate, evaluated by ten-fold cross validation between these four approaches to data mining, are statistically insignificant.

5   Conclusions

We compared four methods for mining incomplete data sets, combining two granules, characteristic sets and generalized maximal consistent blocks with two types of probabilistic approximations, global and saturated. Our criterion of quality was an error rate evaluated by ten-fold cross validation. As follows from our experiments, there are no significant differences between the four methods. The main conclusion is that for data mining all four methods should be applied.

References

1. Clark, P.G., Gao, C., Grzymala-Busse, J.W., Mroczek, T.: Characteristic sets and generalized maximal consistent blocks in mining incomplete data. In: Proceedings of the International Joint Conference on Rough Sets, Part 1, pp. 477–486 (2017)
2. Clark, P.G., Gao, C., Grzymala-Busse, J.W., Mroczek, T.: Characteristic sets and generalized maximal consistent blocks in mining incomplete data. Inf. Sci. **453**, 66–79 (2018)
3. Clark, P.G., Gao, C., Grzymala-Busse, J.W., Mroczek, T., Niemiec, R.: A comparison of concept and global probabilistic approximations based on mining incomplete data. In: Vasiljevienė, G. (ed.) Information and Software Technologies. ICIST 2018. Communications in Computer and Information Science, **920**, pp. 324–335 (2018). https://doi.org/10.1007/978-3-319-99972-2_26

4. Clark, P.G., Grzymala-Busse, J.W.: Experiments on probabilistic approximations. In: Proceedings of the 2011 IEEE International Conference on Granular Computing, pp. 144–149 (2011)
5. Clark, P.G., Grzymala-Busse, J.W.: Experiments on rule induction from incomplete data using three probabilistic approximations. In: Proceedings of the 2012 IEEE International Conference on Granular Computing, pp. 90–95 (2012)
6. Clark, P.G., Grzymala-Busse, J.W., Hippe, Z.S., Mroczek, T., Niemiec, R.: Global and saturated probabilistic approximations based on generalized maximal consistent blocks. In: Proceedings of the 15-th International Conference on Hybrid Artificial Intelligence Systems, pp. 387–396 (2020)
7. Clark, P.G., Grzymala-Busse, J.W., Hippe, Z.S., Mroczek, T., Niemiec, R.: Mining data with many missing attribute values using global and saturated probabilistic approximations. In: Proceedings of the 26-th International Conference on Information and Software Technologies, pp. 72–83 (2020)
8. Clark, P.G., Grzymala-Busse, J.W., Mroczek, T., Niemiec, R.: A comparison of global and saturated probabilistic approximations using characteristic sets in mining incomplete data. In: Proceedings of the Eight International Conference on Intelligent Systems and Applications, pp. 10–15 (2019)
9. Clark, P.G., Grzymala-Busse, J.W., Mroczek, T., Niemiec, R.: Rule set complexity in mining incomplete data using global and saturated probabilistic approximations. In: Proceedings of the 25-th International Conference on Information and Software Technologies, pp. 451–462 (2019)
10. Grzymala-Busse, J.W.: LERS–a system for learning from examples based on rough sets. In: Slowinski, R. (ed.) Intelligent Decision Support. Handbook of Applications and Advances of the Rough Set Theory, pp. 3–18. Kluwer Academic Publishers, Dordrecht, Boston, London (1992)
11. Grzymala-Busse, J.W.: A new version of the rule induction system LERS. Fund. Inf. **31**, 27–39 (1997)
12. Grzymala-Busse, J.W.: MLEM2: a new algorithm for rule induction from imperfect data. In: Proceedings of the 9th International Conference on Information Processing and Management of Uncertainty in Knowledge-Based Systems, pp. 243–250 (2002)
13. Grzymala-Busse, J.W.: Rough set strategies to data with missing attribute values. In: Notes of the Workshop on Foundations and New Directions of Data Mining, in conjunction with the Third International Conference on Data Mining, pp. 56–63 (2003)
14. Grzymala-Busse, J.W.: Generalized parameterized approximations. In: Proceedings of the 6-th International Conference on Rough Sets and Knowledge Technology, pp. 136–145 (2011)
15. Grzymala-Busse, J.W., Clark, P.G., Kuehnhausen, M.: Generalized probabilistic approximations of incomplete data. Int. J. Approx. Reason. **132**, 180–196 (2014)
16. Grzymala-Busse, J.W., Rzasa, W.: Local and global approximations for incomplete data. In: Proceedings of the Fifth International Conference on Rough Sets and Current Trends in Computing. pp. 244–253 (2006)
17. Grzymala-Busse, J.W., Rzasa, W.: Local and global approximations for incomplete data. Trans. Rough Sets **8**, 21–34 (2008)
18. Grzymala-Busse, J.W., Ziarko, W.: Data mining based on rough sets. In: Wang, J. (ed.) Data Mining: Opportunities and Challenges, pp. 142–173. Idea Group Publ., Hershey, PA (2003)
19. Leung, Y., Li, D.: Maximal consistent block technique for rule acquisition in incomplete information systems. Inf. Sci. **153**, 85–106 (2003)

20. Pawlak, Z., Skowron, A.: Rough sets: some extensions. Inf. Sci. **177**, 28–40 (2007)
21. Pawlak, Z., Wong, S.K.M., Ziarko, W.: Rough sets: probabilistic versus deterministic approach. Int. J. Man-Mach. Stud. **29**, 81–95 (1988)
22. Ślęzak, D., Ziarko, W.: The investigation of the Bayesian rough set model. Int. J. Approx. Reason. **40**, 81–91 (2005)
23. Wong, S.K.M., Ziarko, W.: INFER–an adaptive decision support system based on the probabilistic approximate classification. In: Proceedings of the 6-th International Workshop on Expert Systems and their Applications, pp. 713–726 (1986)
24. Yao, Y.Y.: Probabilistic rough set approximations. Int. J. Approx. Reason. **49**, 255–271 (2008)
25. Yao, Y.Y., Wong, S.K.M.: A decision theoretic framework for approximate concepts. Int. J. Man-Mach. Stud. **37**, 793–809 (1992)
26. Ziarko, W.: Variable precision rough set model. J. Comput. Syst. Sci. **46**(1), 39–59 (1993)
27. Ziarko, W.: Probabilistic approach to rough sets. Int. J. Approx. Reason. **49**, 272–284 (2008)

Determining Tanimoto Similarity Neighborhoods of Real-Valued Vectors by Means of the Triangle Inequality and Bounds on Lengths

Marzena Kryszkiewicz[✉]

Institute of Computer Science, Warsaw University of Technology, Nowowiejska 15/19, 00-665 Warsaw, Poland
mkr@ii.pw.edu.pl

Abstract. The Tanimoto similarity is widely used in chemo-informatics, biology, bio-informatics, text mining and information retrieval to determine neighborhoods of sufficiently similar objects or k most similar objects represented by real-valued vectors. For metrics such as the Euclidean distance, the triangle inequality property is often used to efficiently identify vectors that may belong to the sought neighborhood of a given vector. Nevertheless, the Tanimoto similarity as well as the Tanimoto dissimilarity do not fulfill the triangle inequality property for real-valued vectors. In spite of this, in this paper, we show that the problem of looking for a neighborhood with respect to the Tanimoto similarity among real-valued vectors is equivalent to the problem of looking for a neighborhood among normalized forms of these vectors in the Euclidean space. Based on this result, we propose a method that uses the triangle inequality to losslessly identify promising candidates for members of Tanimoto similarity neighborhoods among real-valued vectors. The method requires pre-calculation and storage of the distances from normalized forms of real-valued vectors to so called a reference vector. The normalized forms of vectors themselves do not need to be stored after the pre-calculation of these distances. We also propose two variants of a new combined method which, apart from the triangle inequality, also uses bounds on vector lengths to determine Tanimoto similarity neighborhoods. The usefulness of the new and related methods is illustrated with examples.

Keywords: The Tanimoto similarity · The cosine similarity · The Euclidean distance · Neighborhood · Near duplicates · k nearest neighbors · The triangle inequality · Vector length bounds · Real-valued vector · ZPN-vector · Binary vector

1 Introduction

The Tanimoto similarity is the most popular measure for comparing chemical structures represented by fingerprints. It is also widely used in biology, bio-informatics, text mining and information retrieval for determining neighborhoods of sufficiently similar objects

© Springer Nature Switzerland AG 2021
S. Ramanna et al. (Eds.): IJCRS 2021, LNAI 12872, pp. 18–34, 2021.
https://doi.org/10.1007/978-3-030-87334-9_2

(or near duplicates) or k most similar objects represented by real-valued vectors (please see e.g. [1] for example applications of the Tanimoto similarity measure). The Tanimoto similarity of two vectors is defined in terms of a dot product of both vectors and their lengths. Sometimes, it is convenient to express the Tanimoto similarity of vectors equivalently by means of their cosine similarity and the ratio of their lengths (see e.g. [3]). The discovery of similar objects is challenging when datasets are large or high dimensional. Thus methods that speed up their discovery are of high importance. In [7, 8, 11, 17], we derived bounds on lengths of Tanimoto similar vectors, which can be efficiently used to determine Tanimoto similarity neighborhoods. The vector length bounds derived in [8] were related to so called *ZPN-vectors* whose each dimension takes one of at most three distinct values: a positive value, zero and a negative value. Their tighter version was offered in [17]. The length bounds for Tanimoto similar real-valued vectors were derived in [7]. A tighter version of these bounds for real-valued vectors was obtained in [11]. It used an upper bound on the cosine similarity that was derived in [9]. Also in [1] a tighter version of length bounds from [7] was derived based on upper bounds on the cosine similarity. The solution from [1] was offered for non-negative real-valued vectors.

In the case of metrics, such as the Euclidean distance, the triangle inequality property is often used for efficient non-lossless identification of vectors that are likely to be sufficiently similar to a given vector [2, 3, 12–16, 19, 20, 22, 23]. It was proved in [18] that the Tanimoto dissimilarity satisfies the triangle inequality in the case of weighted binary non-negative vectors. In [5], it was shown how to use this property to efficiently determine cosine similarity neighborhoods among weighted binary non-negative vectors. Nevertheless, the Tanimoto similarity as well as the Tanimoto dissimilarity do not satisfy the triangle inequality property for real-valued vectors. On the other hand, even though the cosine similarity does not fulfill the triangle inequality [3, 6, 10] either, it is possible to transform the search of cosine similarity neighborhoods among real-valued vectors into the search of Euclidean distance neighborhoods among normalized forms of the vectors, and by this, to use the triangle inequality to make the discovery of cosine similar real-valued vectors fast [3, 4, 6, 10, 12]. This result encouraged us to develop a method that may speed up the discovery of Tanimoto similar vectors using the triangle inequality. We obtained this goal by deriving the relationship between the Tanimoto similarity and the Euclidean distance.

The main contribution in the paper is presented in Sects. 5 and 6 and consists in:

- Showing that the problem of looking for Tanimoto similarity neighborhoods among real-valued vectors is equivalent to the problem of looking for neighborhoods among normalized forms of these vectors in the Euclidean space (see Sect. 5).
- Proposing a new method that uses the triangle inequality to losslessly identify promising candidates for members of Tanimoto similarity neighbourhoods among real-valued vectors (see Sect. 5).
- Proposing two variants of a new combined method which, apart from the triangle inequality, also uses the bounds on vector lengths to determine Tanimoto similarity neighborhoods (see Sect. 6).
- Illustration of the usefulness of the new methods against related ones.

Our paper has the following layout. In Sect. 2, we provide definitions of the Euclidean distance, the cosine similarity and the Tanimoto similarity. Next we recall the definitions of similarity and distance neighborhoods. We also recall how to adapt the discovery of neighborhoods to discovery of k nearest neighbors. In Sect. 3, we recall how to calculate the Euclidean and cosine neighborhoods with the triangle inequality property. Then, in Sect. 4, we recall the method for determining Tanimoto similarity neighborhoods using bounds on real-valued vector lengths. In Sect. 5, we derive theoretical results related to the Tanimoto similarity and its relationship with the cosine similarity and the Euclidean distance. Based on these results, we propose then a new method for determining Tanimoto similarity neighborhoods by means of the triangle inequality. In Sect. 6, we propose two variants of a new combined method which use both the triangle inequality and bounds on vector lengths to determine Tanimoto similarity neighborhoods. Section 7 summarizes our contribution.

2 Basic Notions and Properties

2.1 The Euclidean Distance, the Cosine Similarity and the Tanimoto Similarity

In the paper, we consider vectors of the same dimensionality, say n. A vector u will be also denoted as $[u_1, \ldots, u_n]$, where u_i is the value of the i-th dimension of u, $i = 1..n$. A vector will be called a *zero vector* if all its dimensions are equal to zero. Otherwise, the vector will be called *non-zero*.

Vectors' similarity and dissimilarity can be defined in many ways. An important class of dissimilarity measures are distance metrics, which preserve the triangle inequality. We say that measure μ *preserves the triangle inequality* if for any vectors u, v, and r, $\mu(u, r) \leq \mu(u, v) + \mu(v, r)$ or, alternatively $\mu(u, r) \geq \mu(u, v) - \mu(v, r)$.

The most popular distance metric is the *Euclidean distance*, which takes only non-negative values. The *Euclidean distance* between vectors u and v is denoted by *Euclidean(u, v)* and is defined as follows:

$$Euclidean(u, v) = \sqrt{\sum_{i=1..n} (u_i - v_i)^2}.$$

Among most popular similarity measures are the *cosine similarity* and the *Tanimoto similarity*, which are defined in terms of *dot products of vectors and vector lengths* as follows:

$$cosSim(u, v) = \frac{u \cdot v}{|u| \cdot |v|},$$

$$TSim(u, v) = \frac{u \cdot v}{u \cdot u + v \cdot v - u \cdot v} = \frac{u \cdot v}{|u|^2 + |v|^2 - u \cdot v},$$

where

- $u \cdot v$ is the *dot product of vectors u and v* and equals $\sum_{i=1..n} u_i \cdot v_i$;
- $|u|$ is *the length of vector u* and equals $\sqrt{u \cdot u}$.

The Tanimoto similarity takes values from interval [−1/3, 1] [21], while the cosine similarity takes values from interval [−1, 1].

Both similarity measures as well as their dissimilarity variants: $1 - cosSim$ and $1 - TSim$ do not satisfy the triangle inequality property in general for real valued vectors. Nevertheless, $1 - TSim$ fulfills the triangle inequality property for binary vectors and weighted binary vectors each dimension of which may take either 0 or one positive value [18]. In our paper, however, we focus on real-valued vectors, so we cannot use directly the triangle inequality to determine $1 - TSim$ dissimilarity of vectors, and in consequence their Tanimoto similarity. In spite of this, we will derive the theoretical results that will enable us to propose methods using the triangle inequality for determining Tanimoto similarity neighborhoods.

2.2 ε-Neighborhoods and k Nearest Neighbors

Let sim denote a similarity measure (such as $TSim$ and $cosSim$) and $disSim$ denote a dissimilarity measure (such as $Euclidean$, 1-$TSim$ and 1-$cosSim$). $disSim$ ε-neighborhood of vector u in D is denoted by ε-$NB_{disSim}{}^{D}(u)$ and is defined as the set of all vectors in dataset D whose dissimilarity from u in terms of the $disSim$ measure is not greater than ε; that is,

$$\varepsilon\text{-}NB_{disSim}{}^{D}(u) = \{v \in D \mid disSim(u, v) \leq \varepsilon\}.$$

sim ε-neighborhood of vector u in D is denoted by ε-$SNB_{sim}{}^{D}(u)$ and is defined as the set of all vectors in dataset D whose similarity to u in terms of the sim measure is at least ε; that is,

$$\varepsilon\text{-}SNB_{sim}{}^{D}(u) = \{v \in D \mid sim(u, v) \geq \varepsilon\}.$$

Instead of looking for a (dis)similarity ε-neighborhood, one may be interested in determining k $(dis)Sim$ $nearest$ $neighbors$. The task of searching k $disSim$ nearest neighbors of vector u still can be considered as searching an ε-Euclidean neighborhood for some ε value (possibly different for different vectors and adjusted dynamically) as follows: Let K be a set containing any k vectors from $D \setminus \{u\}$ and $\varepsilon = \max\{disSim(u, v) \mid v \in K\}$. Then, k $disSim$ nearest neighbors are guaranteed to be found within ε radius from vector u; namely, they are contained in ε-$NB_{disSim}{}^{D}(u) \setminus \{u\}$. In practice, one may apply some heuristics to determine possibly best value (that is, as little as possible) of ε within which k-nearest neighbors of u are guaranteed to be found and the value of ε can be re-estimated (and thus possibly narrowed) when calculating the $disSim$ dissimilarity between u and the next vectors from $D \setminus (K \cup \{u\})$ [12, 15, 16]. The above approach to searching k $disSim$ nearest neighbors can be easily adapted to searching k sim nearest neighbors. At the beginning, ε should be assigned $\min\{sim(u, v) \mid v \in K\}$ and then could be re-estimated (and thus possibly increased) when calculating the similarity sim between u and the next vectors from $D \setminus (K \cup \{u\})$.

3 Using the Triangle Inequality Property to Calculate Euclidean and Cosine ε-Neighborhoods

3.1 Using the Triangle Inequality to Calculate Euclidean ε-Neighborhoods

In this subsection, we recall the method of determining ε-Euclidean neighborhoods using the triangle inequality property, as proposed in [13, 14]. The method is based on the following observation: As the Euclidean distance fulfills the triangle property, then for any triple of vectors u, v and r; $Euclidean(u, v) \geq Euclidean(u, r) - Euclidean(v, r)$. Thus, the difference $Euclidean(u, r) - Euclidean(v, r)$ (as well as $Euclidean(v, r) - Euclidean(u, r)$) is a *pessimistic estimation* of the real Euclidean distance between u and v. Clearly, if this pessimistic estimation is greater than ε, then the real Euclidean distance between the two vectors is also greater than ε.

Property 1. Let r, u and v be vectors and $\varepsilon \geq 0$.

a) If $Euclidean(u, r) - Euclidean(v, r) > \varepsilon$, then $Euclidean(u, v) > \varepsilon$.
b) If $Euclidean(u, v) \leq \varepsilon$, then $Euclidean(u, r) - Euclidean(v, r) \leq \varepsilon$.

Let us assume that we would like to identify promising candidates for members of ε-Euclidean neighborhood of vector u. Property 1 allows deducing that only vectors v for which $Euclidean(v, r) \in [Euclidean(u, r) - \varepsilon, Euclidean(u, r) + \varepsilon]$ are likely to belong to ε-Euclidean neighborhood of u. In the remainder of the paper, checking if the Euclidean distance of a vector from a reference vector is within a given interval will be called *the Euclidean triangle inequality condition with two bounds*.

Corollary 1. (Using the Euclidean triangle inequality condition with two bounds). Let r, u and v be vectors, the Euclidean distance threshold $\varepsilon \geq 0$ and corresponding Euclidean bounds $\varepsilon_1 = Euclidean(u, r) - \varepsilon$ and $\varepsilon_2 = Euclidean(u, r) + \varepsilon$.

a) If $Euclidean(v, r) \notin [\varepsilon_1, \varepsilon_2]$, then $Euclidean(u, v) > \varepsilon$.
b) If $Euclidean(u, v) \leq \varepsilon$, then $Euclidean(v, r) \in [\varepsilon_1, \varepsilon_2]$.

The method proposed in [13, 14] assumes that the Euclidean distances from all vectors in dataset D to so called *reference vector r* are pre-calculated and sorted with respect to these distances. The reference vector is a parameter of the method. It is suggested to build a reference vector from extreme (minimal and/or maximal) dimension values of vectors in D. The first step in determining ε-$NB_{Euclidean}{}^{D}(u)$ is to identify set R of all vectors v in D whose pessimistic estimation of the Euclidean distance to u does not exceed ε; that is, the vectors v in D for which $Euclidean(v, r) \in [\varepsilon_1, \varepsilon_2]$, where $\varepsilon_1 = Euclidean(u, r) - \varepsilon$ and $\varepsilon_2 = Euclidean(u, r) + \varepsilon$. Thanks to pre-sorting of vectors in D, the number of vectors v for which it is checked whether $Euclidean(v, r) \in [\varepsilon_1, \varepsilon_2]$ is restricted to at most $|R| + 2$. By Corollary 1, the vectors in $D \setminus R$ do not belong to ε-$NB_{Euclidean}{}^{D}(u)$. Hence, ε-$NB_{Euclidean}{}^{D}(u) \subseteq R$. Thus, in order to determine $NB_{Euclidean}{}^{D}(u)$, it suffices to calculate the Euclidean distances to u only from vectors in R.

Example 1. (Determining Euclidean ε-neighborhood using the Euclidean triangle inequality condition with two bounds). Table 1 shows example set D of 10 vectors of dimensionality $n = 9$, which will be used throughout the paper. We assume that the Euclidean distances of all vectors in D to reference vector $r = [-3, -2, 6, 4, -3, 0, 0, 6, 0]$ are pre-calculated and the vectors in D are sorted with respect to these distances. The reference vector was created from extreme dimension values of vectors in D (see Table 1). Our task is to calculate $\varepsilon\text{-}NB_{cosSim}{}^{D}(v5)$ for $\varepsilon = 2$. Let R be the set of vectors in D whose pessimistic estimations of the Euclidean distances to $v5$ do not exceed ε. Thus, R contains only those vectors in D whose Euclidean distances to r differ from $Euclidean(v5, r)$ by at most ε; that is, vectors $v \in D$ for which $Euclidean(v, r) \in [\varepsilon_1, \varepsilon_2] = [9.70 -2, 9.70 +2] = [7.70, 11.70]$. Hence, $R = \{v5, v2\}$ and R is a superset of $NB_{Euclidean}{}^{D}(v5)$. Thus, the determination of $NB_{Euclidean}{}^{D}(v5)$ requires 2 calculations of the Euclidean distances to vector $v5$ instead of 10 calculations. Now, one may wonder for how many vectors v in D it was necessary to check if $Euclidean(v, r) \in [7.70, 11.70]$. As the vectors in D are pre-sorted with respect to their Euclidean distances to the reference vector, then this check was carried out only for the vectors in R plus for vectors $v3$ and $v8$. Vector $v3$ is the first vector preceding $v5$ whose Euclidean distance to r is less than 7.70, while vector $v8$ is the first vector following $v5$ whose Euclidean distance to r is greater than 11.70. Thanks to pre-sorting, it is known in advance that the Euclidean distance to r from each vector preceding $v3$ is less than 7.70 and that the Euclidean distance to r from each vector following $v8$ is greater than 11.70. Hence, $v3$ and vectors preceding it in D as well as $v8$ and vectors following it in D cannot belong to $\varepsilon\text{-}NB_{cosSim}{}^{D}(v5)$. $\square$

Table 1. Example set of vectors D sorted with respect to their Euclidean distances to reference vector $r = [-3, -2, 6, 4, -3, 0, 0, 6, 0]$. By Corollary 1, $\varepsilon\text{-}NB_{Euclidean}{}^{D}(v5)$ is guaranteed to be a subset of $R = \{v5, v2\}$ for $\varepsilon = 2$. The Euclidean triangle inequality condition check with two bounds was carried out for all vectors in R and for vectors $v3$ and $v8$.

$v\# \in D$	$v\#_1$	$v\#_2$	$v\#_3$	$v\#_4$	$v\#_5$	$v\#_6$	$v\#_7$	$v\#_8$	$v\#_9$	$Euclidean\,(v\#,r)$
$v7$	0	0	6	4	0	0	0	0	0	7.62
$v3$	0	0	5	4	0	0	0	0	0	7.68
$v5$	0	0	0	4	-3	0	3	0	0	9.70
$v2$	3	-2	0	0	0	5	0	6	0	11.05
$v8$	0	4	0	4	0	0	0	0	5	12.29
$v1$	-3	4	0	0	3	5	3	6	0	12.57
$v9$	0	0	0	-2	3	0	3	0	5	13.82
$v4$	0	-2	0	4	0	5	0	-5	0	14.14
$v10$	0	-2	-9	0	0	0	0	0	0	17.18
$v6$	0	0	-9	4	0	0	0	0	5	17.55

Please note that the calculation of the Euclidean distance between two n-dimensional vectors requires n subtractions, n multiplications, $n-1$ additions and 1 calculation of the square root. Thus, it is costly, especially in the case of dense high dimensional data. The method offered in [13, 14] for calculation of Euclidean ε-neighborhoods enables considerable elimination of the costly calculation of the Euclidean distances between vectors. In fact, one may use more than one reference vector to further restrict the number of calculations of the Euclidean distances [13, 14].

3.2 Calculating Cosine ε-Neighborhoods by Means of the Triangle Inequality

Let us start with recalling the relationship between the cosine similarity and the Euclidean distance.

Lemma 1 [3, 6] . Let u,v be non-zero vectors. Then:

$$cosSim(u, v) = \frac{|u|^2 + |v|^2 - Euclidean^2(u, v)}{2|u| \cdot |v|}$$

Clearly, the cosine similarity between any vectors u and v depends solely on the angle between the vectors and does not depend on their lengths, hence the calculation of the $cosSim(u, v)$ may be carried out on *normalized forms* of vectors as follows:

A *normalized form of a vector* u is denoted by $NF(u)$ and is defined as the ratio of u to its length $|u|$. A vector u is defined as a *normalized vector* if $u = NF(u)$. Obviously, the length of a normalized vector equals 1. Table 2 presents the normalized forms of vectors from Table 1.

Table 2. Normalized forms of vectors from example vector set D

$NF(v\#)$	$NF(v\#)_1$	$NF(v\#)_2$	$NF(v\#)_3$	$NF(v\#)_4$	$NF(v\#)_5$	$NF(v\#)_6$	$NF(v\#)_7$	$NF(v\#)_8$	$NF(v\#)_9$
$NF(v1)$	-0.29	0.39	0.00	0.00	0.29	0.49	0.29	0.59	0.00
$NF(v2)$	0.35	-0.23	0.00	0.00	0.00	0.58	0.00	0.70	0.00
$NF(v3)$	0.00	0.00	0.78	0.62	0.00	0.00	0.00	0.00	0.00
$NF(v4)$	0.00	-0.24	0.00	0.48	0.00	0.60	0.00	-0.60	0.00
$NF(v5)$	0.00	0.00	0.00	0.69	-0.51	0.00	0.51	0.00	0.00
$NF(v6)$	0.00	0.00	-0.81	0.36	0.00	0.00	0.00	0.00	0.45
$NF(v7)$	0.00	0.00	0.83	0.55	0.00	0.00	0.00	0.00	0.00
$NF(v8)$	0.00	0.53	0.00	0.53	0.00	0.00	0.00	0.00	0.66
$NF(v9)$	0.00	0.00	0.00	-0.29	0.44	0.00	0.44	0.00	0.73
$NF(v10)$	0.00	-0.22	-0.98	0.00	0.00	0.00	0.00	0.00	0.00

Theorem 1 [3, 6] . Let u and v be non-zero vectors. Then:

$$cosSim(u, v) = cosSim(NF(u), NF(v)) = \frac{2 - Euclidean^2(NF(u), NF(v))}{2}.$$

Theorem 1 allows deducing that checking whether the cosine similarity between any two vectors does not exceed the ε threshold, where $\in \varepsilon[-1, 1]$, can be carried out as checking if the Euclidean distance between the normalized forms of the vectors does not exceed the modified threshold $\varepsilon'(\varepsilon) = \sqrt{2 - 2\varepsilon}$:

Corollary 2 [3, 6] . Let u, v be non-zero vectors, $\varepsilon \in [-1, 1]$ and $\varepsilon'(\varepsilon) = \sqrt{2 - 2\varepsilon}$. Then:

$$cosSim(u, v) \geq \varepsilon \Leftrightarrow Euclidean(NF(u), NF(v)) \leq \varepsilon'(\varepsilon).$$

Now, if for reference vector r, $Euclidean(NF(v), r)$ - $Euclidean(NF(u), r) > \varepsilon'(\varepsilon)$, then $Euclidean(NF(u), NF(v)) > \varepsilon'(\varepsilon)$ (by Property 1), and $cosSim(u, v) < \varepsilon$ (by Corollary 2). Corollary 3 follows immediately from Corollary 2 and Corollary 1.

Corollary 3. (Using the Euclidean triangle inequality condition with two bounds). Let u and v be non-zero vectors, r be a vector, the Tanimoto similarity threshold $\varepsilon \in [-1, 1]$, the corresponding Euclidean threshold $\varepsilon'(\varepsilon) = \sqrt{2 - 2\varepsilon}$ and bounds $\varepsilon_1'(\varepsilon) = Euclidean(NF(u), r) - \varepsilon'(\varepsilon)$ and $\varepsilon_2'(\varepsilon) = Euclidean(NF(u), r) + \varepsilon'(\varepsilon)$.

a) If $Euclidean(NF(v), r) \notin [\varepsilon_1'(\varepsilon), \varepsilon_2'(\varepsilon)]$, then $cosSim(u, v) < \varepsilon$.
b) If $cosSim(u, v) \geq \varepsilon$, then $Euclidean(NF(v), r) \in [\varepsilon_1'(\varepsilon), \varepsilon_2'(\varepsilon)]$.

This observation enables efficient determination of promising candidates for the members of cosine similarity ε-neigborhood based on the Euclidean distances of normalized forms of non-zero vectors to some reference vector r. Note that the reference vector itself does not need to be normalized.

Example 2. (Determining cosine ε-neighborhood using the Euclidean triangle inequality condition with two bounds). Let $\varepsilon = 0.95$. We are to find promising candidates for ε-$SNB_{cosSim}{}^D(v5)$ in example set D of vectors from Table 1, which was used in Example 1. Nevertheless, this time the Euclidean distances of normalized forms of all vectors in D to reference vector $r = [-0.29, -0.24, 0.83, 0.69, -0.51, 0, 0, 0.70, 0]$ were pre-calculated and the vectors in D were sorted with respect to these distances. The reference vector was created from extreme dimension values of normalized forms of vectors in D (the chosen extreme dimension values are bolded in Table 2). Let R be the set of vectors in D whose pessimistic estimations of the Euclidean distances of their normalized forms to $NF(v5)$ do not exceed $\varepsilon'(\varepsilon) = \sqrt{2 - 2\varepsilon} \approx 0.32$; that is, vectors $v \in D$ for which $Euclidean(NF(v), r) \in [\varepsilon_1'(\varepsilon), \varepsilon_2'(\varepsilon)] = [Euclidean(NF(v5), r) - \varepsilon'(\varepsilon), Euclidean(NF(v5), r) + \varepsilon'(\varepsilon)] = [1.26 - 0.32, 1.26 + 0.32] = [0.94, 1.58]$. Hence, $R = \{v3, v7 \ v5, v2\}$ (please see Table 3) and R is a superset of $SNB_{cosSim}{}^D(v5)$. Thus, the determination of $SNB_{cosSim}{}^D(v5)$ requires 4 calculations of the cosine similarity to vector $v5$ instead of 10 calculations. As the vectors in D were pre-sorted with respect to the Euclidean distances of their normalized forms to the reference vector, it was sufficient to check for 5 vectors v in D whether $Euclidean(NF(v), r) \in [\varepsilon_1'(\varepsilon), \varepsilon_2'(\varepsilon)]$; namely for all for vectors in R and for $v1$. $\square$

Table 3. Set of vectors in D sorted with respect to the Euclidean distances of their normalized forms (from Table 2) to reference vector $r = [-0.29, -0.24, 0.83, 0.69, -0.51, 0, 0, 0.70, 0]$. By Corollary 3, ε-$SNB_{cosSim}{}^{D}(v5)$ is guaranteed to be a subset of $R = \{v3, v7, v5, v2\}$ for $\varepsilon = 0.95$. The Euclidean triangle inequality condition check with two bounds was carried out for all vectors in R and for vector $v1$.

$NF(v\#)$	$NF(v\#)_1$	$NF(v\#)_2$	$NF(v\#)_3$	$NF(v\#)_4$	$NF(v\#)_5$	$NF(v\#)_6$	$NF(v\#)_7$	$NF(v\#)_8$	$NF(v\#)_9$	$Euclidean\,(NF(v\#),r)$
$NF(v3)$	0.00	0.00	0.78	0.62	0.00	0.00	0.00	0.00	0.00	0.95
$NF(v7)$	0.00	0.00	0.83	0.55	0.00	0.00	0.00	0.00	0.00	0.96
$NF(v5)$	0.00	0.00	0.00	0.69	-0.51	0.00	0.51	0.00	0.00	1.26
$NF(v2)$	0.35	-0.23	0.00	0.00	0.00	0.58	0.00	0.70	0.00	1.48
$NF(v1)$	-0.29	0.39	0.00	0.00	0.29	0.49	0.29	0.59	0.00	1.60
$NF(v8)$	0.00	0.53	0.00	0.53	0.00	0.00	0.00	0.00	0.66	1.61
$NF(v4)$	0.00	-0.24	0.00	0.48	0.00	0.60	0.00	-0.60	0.00	1.77
$NF(v9)$	0.00	0.00	0.00	-0.29	0.44	0.00	0.44	0.00	0.73	1.98
$NF(v6)$	0.00	0.00	-0.81	0.36	0.00	0.00	0.00	0.00	0.45	1.98
$NF(v10)$	0.00	-0.22	-0.98	0.00	0.00	0.00	0.00	0.00	0.00	2.14

4 Using Bounds on Vector Lengths to Calculate Tanimoto Similarity ε-Neighborhoods

In this section, we recall the bounds on lengths of Tanimoto ε-similar vectors after [7] and illustrate their usefulness with an example.

Theorem 2 [7]. Let u and v be non-zero vectors, $\alpha = \frac{1}{2}\left(\frac{1+\varepsilon}{\varepsilon} + \sqrt{\left(\frac{1+\varepsilon}{\varepsilon}\right)^2 - 4}\right)$ and $\varepsilon \in (0, 1]$. If $T(u, v) \geq \varepsilon$, then $|v| \in \left[\frac{1}{\alpha}|u|, \alpha|u|\right]$.

Example 3. (Determining Tanimoto ε-neighborhood with bounds on vector lengths). Let $\varepsilon = 0.95$ and $v5$ be the vector in example dataset D for which we wish to calculate its Tanimoto similarity ε-neighborhood. We will use the bounds on vector lengths in order to reduce the number of candidates for members of ε-$SNB_{TSim}{}^{D}(v5)$. By Theorem 2, $\alpha \approx 1.26$ and only vectors the lengths of which belong to the interval $\left[\frac{1}{\alpha} \cdot |v5|, \alpha \cdot |v5|\right]$ $\approx \left[\frac{1}{1.26} \cdot 5.83, 1.26 \cdot 5.83\right] \approx [4.62, 7.35]$ are likely to be sufficiently Tanimoto similar to $v5$. Let R be the set of such vectors. As follows from Table 4, $R = \{v5, v3, v9, v7\}$. Thus, only 4 out of 10 vectors in D may belong to ε-$SNB_{TSim}{}^{D}(v5)$. If vectors in dataset D are sorted with respect to their lengths, then the vector length condition is checked only for the vectors in R and for vector $v8$, which is the first vector following $v5$ in the sorted dataset that does not fulfill the vector length condition. □

Table 4. The *TSimVectorLengthCondition* approach: Example set of vectors D is sorted with respect to their lengths. By Theorem 2, ε-$SNB_{TSim}{}^D(v5)$ is guaranteed to be a subset of $R = \{v5, v3, v9, v7\}$ for $\varepsilon = 0.95$. The vector length condition was carried out for all vectors in R and for vector $v8$.

$v\# \in D$	$v\#_1$	$v\#_2$	$v\#_3$	$v\#_4$	$v\#_5$	$v\#_6$	$v\#_7$	$v\#_8$	$v\#_9$	$v\#$ length
$v5$	0	0	0	4	-3	0	3	0	0	5.83
$v3$	0	0	5	4	0	0	0	0	0	6.40
$v9$	0	0	0	-2	3	0	3	0	5	6.86
$v7$	0	0	6	4	0	0	0	0	0	7.21
$v8$	0	4	0	4	0	0	0	0	5	7.55
$v4$	0	-2	0	4	0	5	0	-5	0	8.37
$v2$	3	-2	0	0	0	5	0	6	0	8.60
$v10$	0	-2	-9	0	0	0	0	0	0	9.22
$v1$	-3	4	0	0	3	5	3	6	0	10.20
$v6$	0	0	-9	4	0	0	0	0	5	11.05

In the remainder of the paper, the approach to determination of promising candidates for Tanimoto ε-similarity neighbourhood that was presented in this section will be called the *TSimVectorLengthCondition* approach.

5 Calculating Tanimoto ε-Neighborhoods by Means of the Triangle Inequality

In this section, we precede the examination of the relation of the Tanimoto similarity with the Euclidean distance by the examination of its relation with the cosine similarity. Let us start with recalling Property 1.

Property 2 [8] . For any non-zero vectors u and v, $TSim(u, v) = \dfrac{cosSim(u,v)}{\frac{|u|}{|v|}+\frac{|v|}{|u|}-cosSim(u,v)}$.

Proof. $TSim(u, v) = \dfrac{u \cdot v}{|u|^2+|v|^2-u \cdot v}$. After dividing the numerator and denominator of the expression by $|u| \cdot |v|$, one obtains: $TSim(u, v) = \dfrac{cosSim(u,v)}{\frac{|u|}{|v|}+\frac{|v|}{|u|}-cosSim(u,v)}$. □

By Property 2, the Tanimoto similarity between two vectors depends on the cosine of the angle between the vectors as well as on the ratio of the lengths of both vectors. One may observe that among pairs of vectors with the same angle (and by this, with the same cosine similarity), the pair, say (u, v), with the least value of $\frac{|u|}{|v|} + \frac{|v|}{|u|}$ has the greatest Tanimoto similarity. One may easily derive that $\frac{|u|}{|v|} + \frac{|v|}{|u|} \geq 2$ from the inequation $(|u| - |v|)^2 \geq 0$. Nevertheless, in Lemma 2, we also claim that 2 is the tight lower bound for $\frac{|u|}{|v|} + \frac{|v|}{|u|}$ and prove this statement.

Lemma 2. Let u and v be non-zero vectors. Then 2 is the tight lower bound for $\frac{|u|}{|v|} + \frac{|v|}{|u|}$.

Proof. Let us assume that $\frac{|u|}{|v|} + \frac{|v|}{|u|} < 2$ (*) for any non-zero vectors u and v. We will prove by contradiction that this assumption is wrong. Let $k = \frac{|u|}{|v|}$. Then, $k > 0$ and $k + \frac{1}{k} - 2 < 0$ (by *). Hence, $k^2 - 2k + 1 < 0$ (**). On the other hand, $k^2 - 2k + 1 =$

$(k-1)^2 \geq 0$, which contradicts (**). Thus, there are no vectors u and v for which $\frac{|u|}{|v|} + \frac{|v|}{|u|} < 2$ (***).

We also note that $\frac{|u|}{|v|} + \frac{|v|}{|u|} = 2$ for $|u| = |v|$ (****). By (***) and (****), 2 is the tight lower bound for $\frac{|u|}{|v|} + \frac{|v|}{|u|}$. $\qquad\square$

Lemma 3. (Relationship between the Tanimoto similarity and the cosine similarity). Let u and v be non-zero vectors, $\varepsilon \in (0, 1]$, $\varepsilon''(\varepsilon, |u|, |v|) = \frac{\varepsilon}{1+\varepsilon}\left(\frac{|u|}{|v|} + \frac{|v|}{|u|}\right)$ and $\varepsilon''(\varepsilon) = \frac{2\varepsilon}{1+\varepsilon}$. Then:

a) $TSim(u,v) \geq \varepsilon \Leftrightarrow cosSim(u,v) \geq \varepsilon''(\varepsilon, |u|, |v|)$.
b) $\varepsilon''(\varepsilon, |u|, |v|) > 0$.
c) $\varepsilon''(\varepsilon)$ is the tight lower bound on $\varepsilon''(\varepsilon, |u|, |v|)$.
d) $\varepsilon''(\varepsilon) \in (0, 1]$.
e) If $TSim(u,v) \geq \varepsilon$, then $cosSim(u,v) \geq \varepsilon''(\varepsilon)$.

Proof. Ad a) Follows from Property 2.
　　Ad b) Follows from the fact that $\varepsilon \in (0, 1]$.
　　Ad c) Follows from Lemma 2.
　　Ad d) Follows from the fact that $\varepsilon \in (0, 1]$.
　　Ad e) Follows from Lemma 3a) and 3c). $\qquad\square$

We already know that checking if $TSim(u,v) \geq \varepsilon$ is equivalent to checking whether $cosSim(u, v) \geq \varepsilon''(\varepsilon, |u|, |v|)$ (by Lemma 3a). By Corollary 2, the latter condition is equivalent to the following condition expressed in terms of the Euclidean distance between normalized forms of vectors: $Euclidean(NF(u), NF(v)) \leq \sqrt{2 - 2\varepsilon''(\varepsilon, |u|, |v|)}$.

Theorem 3. (Relationship between the Tanimoto similarity and Euclidean distance). Let u and v be non-zero vectors, r be a vector, $\varepsilon \in (0, 1]$, $\varepsilon''(\varepsilon, |u|, |v|) = = \frac{\varepsilon}{1+\varepsilon}\left(\frac{|u|}{|v|} + \frac{|v|}{|u|}\right)$, $\xi(\varepsilon, |u|, |v|) = \sqrt{2 - 2\varepsilon''(\varepsilon, |u|, |v|)}$, $\varepsilon''(\varepsilon) = \frac{2}{1+\varepsilon}$ and $\xi(\varepsilon) = \sqrt{2 - 2\varepsilon''(\varepsilon)}$. Then:

a) $TSim(u,v) \geq \varepsilon \Leftrightarrow Euclidean(NF(u), NF(v)) \leq \xi(\varepsilon, |u|, |v|)$.
b) $\xi(\varepsilon)$ is the tight upper bound on $\xi(\varepsilon, |u|, |v|)$.
c) $\xi(\varepsilon) \in [0, 1)$.
d) If $TSim(u,v) \geq \varepsilon$, then $Euclidean(NF(u), NF(v)) \leq \xi(\varepsilon)$.
e) If $TSim(u,v) \geq \varepsilon$, then $Euclidean(NF(u), r) - Euclidean(NF(u), r) \leq \xi(\varepsilon)$.
f) If $Euclidean(NF(u), r) - Euclidean(NF(u), r) > \xi(\varepsilon)$, then $TSim(u,v) < \varepsilon$.
g) $\xi(\varepsilon, |u|, |v|) = \sqrt{2 - \frac{2\varepsilon}{1+\varepsilon}\left(\frac{|u|}{|v|} + \frac{|v|}{|u|}\right)}$.
h) $\xi(\varepsilon) = \sqrt{2 - \frac{4\varepsilon}{1+\varepsilon}}$.

Proof. Ad a) Follows from Lemma 3a) and Corollary 2.
　　Ad b) Follows from Lemma 3c).

Ad c) Follows from Lemma 3d).
Ad d) Follows from Theorem 3a-b).
Ad e) Follows from Theorem 3d) and Property 1b).
Ad f) Follows from Theorem 3e).
Ad g-h) Trivial. □

Corollary 4. (Using the Euclidean triangle inequality condition with two bounds). Let u and v be non-zero vectors, r be a vector, the Tanimoto similarity threshold $\varepsilon \in (0, 1]$, the corresponding Euclidean threshold $\xi(\varepsilon) = \sqrt{2 - \frac{4\varepsilon}{1+\varepsilon}}$ and bounds $\xi_1(\varepsilon) = Euclidean(NF(u), r) - \xi(\varepsilon)$ and $\xi_2(\varepsilon) = Euclidean(NF(u), r) + \xi(\varepsilon)$.

a) If $TSim(u, v) \geq \varepsilon$, then $Euclidean(NF(v), r) \in [\xi_1(\varepsilon), \xi_2(\varepsilon)]$.
b) If $Euclidean(NF(v), r) \notin [\xi_1(\varepsilon), \xi_2(\varepsilon)]$, then $TSim(u, v) < \varepsilon$.

Proof. Ad a) Follows from Theorem 3d), Theorem 3h) and Corollary 1b).
Ad b) Follows from Corollary 4a). □

Theorem 3 and Corollary 4 allow us to propose a new approach to determination of the Tanimoto similarity by means of the triangle inequality. In this approach, we assume that the Euclidean distances of normalized forms of all vectors in D to a reference vector r are pre-calculated and all vectors are sorted with respect to these distances. For a given vector u, set R of vectors v in D whose pessimistic estimations of the Euclidean distances from $NF(v)$ to $NF(u)$ do not exceed $\xi(\varepsilon)$; that is, those vectors v in D for which $Euclidean(NF(v), r) \in [\xi_1(\varepsilon), \xi_2(\varepsilon)]$ are determined. By Corollary 4, R is a superset of $\varepsilon\text{-}SNB_{TSim}{}^D(u)$.

Example 4. (Determining Tanimoto ε-neighborhood using the Euclidean triangle inequality condition with two bounds). Let $\varepsilon = 0.95$. Our task is to reduce the number of promising candidates for members of $SNB_{TSim}{}^D(v5)$ by means of the triangle inequality. To this end, we will use Corollary 4. It is assumed that the Euclidean distances of normalized forms of all vectors in D to reference vector $r = [-0.29, -0.24, 0.83, 0.69, -0.51, 0, 0, 0.70, 0]$ (the same as in Example 2) are pre-calculated and the vectors in D are sorted with respect to these distances (see Table 5). Let R be the set of vectors whose pessimistic estimations of the Euclidean distance of normalized forms of vectors in D to $NF(v5)$ do not exceed $\xi(\varepsilon) = \sqrt{2 - \frac{4\varepsilon}{1+\varepsilon}} \approx 0.23$; that is, R contains those vectors v in D for which $Euclidean(NF(v), r) \in [\xi_1(\varepsilon), \xi_2(\varepsilon)] = [Euclidean(NF(v5), r) - \xi(\varepsilon), Euclidean(NF(v5), r) + \xi(\varepsilon)] = [1.26 - 0.23, 1.26 + 0.23] = [1.03, 1.49]$. Hence, $R = \{v5, v2\}$ and R is a superset of $SNB_{TSim}{}^D(v5)$. Thus, the determination of $SNB_{TSim}{}^D(v5)$ requires 2 calculations of the cosine similarity to vector $v5$ instead of 10 calculations. As the vectors in D are pre-sorted with respect to the Euclidean distances of their normalized forms to the reference vector, it is sufficient (by Corollary 4) to carry out the triangle inequality check only for the vectors in R and for vectors: $v7$ and $v1$. □

In the remainder of the paper, the proposed approach to determination of promising candidates for Tanimoto ε-neighbourhood will be called the *TSimTriangleInequality-Condition* approach.

Table 5. The *TSimTriangleInequalityCondition* approach: Set of vectors D sorted with respect to the Euclidean distances of their normalized forms to reference vector $r = [-0.29, -0.24, 0.83, 0.69, -0.51, 0, 0, 0.70, 0]$. By Corollary 4, $\varepsilon\text{-}SNB_{TSim}^{D}(v5)$ is guaranteed to be a subset of $R = \{v5, v2\}$ for $\varepsilon = 0.95$. The Euclidean triangle inequality condition check with two bounds was carried out for all vectors in R and for vectors $v7$ and $v1$.

$v\# \in D$	$v\#_1$	$v\#_2$	$v\#_3$	$v\#_4$	$v\#_5$	$v\#_6$	$v\#_7$	$v\#_8$	$v\#_9$	Euclidean $(NF(v\#), r)$
$v3$	0	0	5	4	0	0	0	0	0	0.95
$v7$	0	0	6	4	0	0	0	0	0	0.96
$v5$	0	0	0	4	-3	0	3	0	0	1.26
$v2$	3	-2	0	0	0	5	0	6	0	1.48
$v1$	-3	4	0	0	3	5	3	6	0	1.60
$v8$	0	4	0	4	0	0	0	0	5	1.61
$v4$	0	-2	0	4	0	5	0	-5	0	1.77
$v9$	0	0	0	-2	3	0	3	0	5	1.98
$v6$	0	0	-9	4	0	0	0	0	5	1.98
$v10$	0	-2	-9	0	0	0	0	0	0	2.14

6 Calculating Tanimoto ε-Neighborhoods by Means of the Triangle Inequality and Lengths of Vectors

In this section, we consider the use of both the triangle inequality and bounds on vector lengths to reduce the number of candidates for Tanimoto similarity ε-neighborhood. Depending on which condition is applied as first, the combined approach will be called as either *TSimTriangleInequalityConditionAsFirst* or *TSimVectorLengthConditionAsFirst*. In both cases, we assume that the lengths of vectors in D and the distances of normalized forms of these vectors to a reference vector are pre-calculated.

In the case of the *TSimTriangleInequalityConditionAsFirst* approach, it is assumed that all vectors are sorted with respect to the Euclidean distances of their normalized forms to a reference vector. For a given vector u, set R is determined according the *TSimTriangleInequalityCondition* approach (see Corollary 4), which was proposed in Sect. 5. Next, R is reduced to *RedR* containing only those vectors from R that fulfill the condition on vector lengths (see Theorem 2). Clearly, $R \supseteq RedR \supseteq \varepsilon\text{-}SNB_{TSim}^{D}(u)$. Please see Table 6 for the results obtained with the *TSimTriangleInequalityConditionAsFirst* approach when looking for promising candidates for $\varepsilon\text{-}SNB_{TSim}^{D}(v5)$ given $\varepsilon = 0.95$.

In the case of the *TSimVectorLengthConditionAsFirst* approach, we assume that all vectors are sorted with respect to their lengths. For a given vector u, set R is determined according the *TSimVectorLengthCondition* approach (see Theorem 2), which was described in Sect. 4. Next, R is reduced to *RedR* containing only those vectors from R whose pessimistic estimations of the Euclidean distances of their normalized forms to $NF(u)$ do not exceed $\xi(\varepsilon)$ (see Theorem 3 or Corollary 4). Clearly, $R \supseteq RedR \supseteq \varepsilon\text{-}SNB_{TSim}^{D}(u)$. Table 7 presents the results obtained with *TSimVectorLengthConditionAsFirst* approach when looking for promising candidates for $\varepsilon\text{-}SNB_{TSim}^{D}(v5)$ given $\varepsilon = 0.95$.

In Table 8, we summarize the results obtained when looking for promising candidates for $\varepsilon\text{-}SNB_{TSim}^{D}(v5)$ given $\varepsilon = 0.95$ for five methods: (i) without any optimization, (ii) based on vector lengths (*TSimVectorLengthCondition*), (iii) for the new method using the

Table 6. The *TSimTriangleInequalityConditionAsFirst* approach: Set of vectors D is sorted with respect to the Euclidean distances of their normalized forms to reference vector $r = [-0.29, -0.24, 0.83, 0.69, -0.51, 0, 0, 0.70, 0]$. By Corollary 4, ε-$SNB_{TSim}^{D}(v5)$ is guaranteed to be a subset of $R = \{v5, v2\}$ for $\varepsilon = 0.95$. After checking $v5$ and $v2$ w.r.t. the vector length condition (see Theorem 2), only $v5$ remains in reduced set $RedR$ of promising candidates for the members of $SNB_{TSim}^{D}(v5)$.

$v\# \in D$	$v\#_1$	$v\#_2$	$v\#_3$	$v\#_4$	$v\#_5$	$v\#_6$	$v\#_7$	$v\#_8$	$v\#_9$	Euclidean $(NF(v\#), r)$	$v\#$ length
$v3$	0	0	5	4	0	0	0	0	0	0.95	6.40
$v7$	0	0	6	4	0	0	0	0	0	0.96	7.21
$v5$	0	0	0	4	-3	0	3	0	0	1.26	5.83
$v2$	3	-2	0	0	0	5	0	6	0	1.48	8.60
$v1$	-3	4	0	0	3	5	3	6	0	1.60	10.20
$v8$	0	4	0	4	0	0	0	0	5	1.61	7.55
$v4$	0	-2	0	4	0	5	0	-5	0	1.77	8.37
$v9$	0	0	0	-2	3	0	3	0	5	1.98	6.86
$v6$	0	0	-9	4	0	0	0	0	5	1.98	11.05
$v10$	0	-2	-9	0	0	0	0	0	0	2.14	9.22

Table 7. The *TSimVectorLengthConditionAsFirst* approach: Example set of vectors D is sorted with respect to their lengths. By Theorem 2, ε-$SNB_{TSim}^{D}(v5)$ is guaranteed to be a subset of $R = \{v5, v3, v9, v7\}$ for $\varepsilon = 0.95$. After checking $v5$, $v3$, $v9$ and $v7$ w.r.t. the triangle inequality condition (see Corollary 4), only $v5$ remains in reduced set $RedR$ of promising candidates for the members of $SNB_{TSim}^{D}(v5)$.

$v\# \in D$	$v\#_1$	$v\#_2$	$v\#_3$	$v\#_4$	$v\#_5$	$v\#_6$	$v\#_7$	$v\#_8$	$v\#_9$	$v\#$ length	Euclidean $(NF(v\#), r)$
$v5$	0	0	0	4	-3	0	3	0	0	5.83	1.26
$v3$	0	0	5	4	0	0	0	0	0	6.40	0.95
$v9$	0	0	0	-2	3	0	3	0	5	6.86	1.98
$v7$	0	0	6	4	0	0	0	0	0	7.21	0.96
$v8$	0	4	0	4	0	0	0	0	5	7.55	1.61
$v4$	0	-2	0	4	0	5	0	-5	0	8.37	1.77
$v2$	3	-2	0	0	0	5	0	6	0	8.60	1.48
$v10$	0	-2	-9	0	0	0	0	0	0	9.22	2.14
$v1$	-3	4	0	0	3	5	3	6	0	10.20	1.60
$v6$	0	0	-9	4	0	0	0	0	5	11.05	1.98

triangle inequality property (*TSimTriangleInequalityCondition*) and (iv) for two variants (*TSimTriangleInequalityConditionAsFirst* and *TSimVectorLengthConditionAsFirst*) of the new combined method using both vector length condition and the triangle inequality property. As follows from Table 8, the new approaches are likely to considerably decrease the number of vectors in D for which it is necessary to carry out the costly Tanimoto similarity calculations. It also suggests that the combined approach is likely to be more efficient than the approaches using one of the conditions.

Table 8. The summary of the results obtained when looking for promising candidates for $\varepsilon\text{-}SNB_{TSim}{}^{D}(v5)$ given $\varepsilon = 0.95$ for five methods: (i) without any optimization, (ii) based on vector lengths, (iii) the new method using the triangle inequality property and (iv) 2 variants of the combined method using both bounds on vector length and the triangle inequality.

Approach	Set R of promising candidates	Reduced set $RedR$ of promising candidates	# of computationally inexpensive calculations of conditions	# of computationally expensive $TSim$ calculations
No optimization	D	N/A	0	10
$TSimVectorLengthCondition$	$\{v5, v3, v9, v7\}$	N/A	5 vector length conditions	4
$TSimVectorLengthConditionAsFirst$	$\{v5, v3, v9, v7\}$	$\{v5\}$	5 vector length conditions + 4 triangle iinequation conditions	1
$TSimTriangleInequalityCondition$	$\{v5, v2\}$	N/A	4 triangle inequation conditions	2
$TSimTriangleInequalityConditionAsFirst$	$\{v5, v2\}$	$\{v5\}$	4 triangle inequation conditions + 2 vector length conditions	1

7 Summary

In the paper, we showed that the problem of looking for Tanimoto similarity neighborhoods among real-valued vectors is equivalent to the problem of looking for neighborhoods among normalized forms of these vectors in the Euclidean space. Based on the obtained theoretical results, we proposed a new method that uses the triangle inequality to losslessly identify promising candidates for members of the Tanimoto similarity neighbourhoods among real-valued vectors. Also, we proposed two variants of a new combined method which, apart from the triangle inequality, also uses the bounds on vector lengths to determine Tanimoto similarity neighborhoods. By means of examples we showed that the method based solely on bounds on vector lengths as well as the new method that uses the triangle inequality may result in significant reduction of the number of costly calculations of the Tanimoto similarities of vectors as well as we showed that both variants of the combined method may further considerably reduce the number of these calculations.

References

1. Anastasiu, D.C., Karypis, G.: Efficient identification of Tanimoto nearest neighbors. Int. J. Data Sci. Anal. **4**(3), 153–172 (2017). https://doi.org/10.1007/s41060-017-0064-z
2. Elkan, C.: Using the triangle inequality to accelerate k-means. In: ICML'03, pp. 147–153, Washington (2003)
3. Kryszkiewicz, M.: Efficient determination of neighborhoods defined in terms of cosine similarity measure, ICS Research Report 4/2011, Warsaw University of Technology (2011)
4. Kryszkiewicz, M.: The triangle inequality versus projection onto a dimension in determining cosine similarity neighborhoods of non-negative vectors. In: Yao, JingTao, et al. (eds.) RSCTC 2012. LNCS (LNAI), vol. 7413, pp. 229–236. Springer, Heidelberg (2012). https://doi.org/10.1007/978-3-642-32115-3_27
5. Kryszkiewicz, M.: Efficient determination of binary non-negative vector neighbors with regard to cosine similarity. In: Jiang, H., Ding, W., Ali, M., Wu, X. (eds.) IEA/AIE 2012. LNCS (LNAI), vol. 7345, pp. 48–57. Springer, Heidelberg (2012). https://doi.org/10.1007/978-3-642-31087-4_6
6. Kryszkiewicz, M.: Determining cosine similarity neighborhoods by means of the euclidean distance. In: Skowron, A., Suraj, Z. (eds.) Rough Sets and Intelligent Systems, Intelligent Systems Reference Library, vol. 43, pp. 323–345. Springer, Heidelberg (2013). https://doi.org/10.1007/978-3-642-30341-8_17
7. Kryszkiewicz, M.: Bounds on lengths of real valued vectors similar with regard to the Tanimoto similarity. In: Selamat, A., Nguyen, N.T., Haron, H. (eds.) ACIIDS 2013. LNCS (LNAI), vol. 7802, pp. 445–454. Springer, Heidelberg (2013). https://doi.org/10.1007/978-3-642-36546-1_46
8. Kryszkiewicz, M.: On cosine and Tanimoto near duplicates search among vectors with domains consisting of zero, a positive number and a negative number. In: Larsen, H.L., Martin-Bautista, M.J., Vila, M.A., Andreasen, T., Christiansen, H. (eds.) FQAS 2013. LNCS (LNAI), vol. 8132, pp. 531–542. Springer, Heidelberg (2013). https://doi.org/10.1007/978-3-642-40769-7_46
9. Kryszkiewicz, M.: Using non-zero dimensions for the cosine and Tanimoto similarity search among real valued vectors. Fund. Inform. **127**(1–4), 307–323 (2013)
10. Kryszkiewicz, M.: The cosine similarity in terms of the Euclidean distance. In: Encyclopedia of Business Analytics and Optimization (2014)

11. Kryszkiewicz, M.: Using non-zero dimensions and lengths of vectors for the Tanimoto similarity search among real valued vectors. In: Nguyen, N.T., Attachoo, B., Trawiński, B., Somboonviwat, K. (eds.) ACIIDS 2014. LNCS (LNAI), vol. 8397, pp. 173–182. Springer, Cham (2014). https://doi.org/10.1007/978-3-319-05476-6_18

12. Kryszkiewicz, M., Jańczak, B.: Basic triangle inequality approach versus metric VP-tree and projection in determining Euclidean and cosine neighbors. In: Bembenik, R., Skonieczny, Ł, Rybiński, H., Kryszkiewicz, M., Niezgódka, M. (eds.) Intelligent Tools for Building a Scientific Information Platform: From Research to Implementation. SCI, vol. 541, pp. 27–49. Springer, Cham (2014). https://doi.org/10.1007/978-3-319-04714-0_3

13. Kryszkiewicz, M., Lasek, P.: TI-DBSCAN: clustering with DBSCAN by means of the triangle inequality, ICS Research Report 3/2010, Warsaw University of Technology (2010)

14. Kryszkiewicz, M., Lasek, P.: TI-DBSCAN: clustering with DBSCAN by means of the triangle inequality. In: Szczuka, M., Kryszkiewicz, M., Ramanna, S., Jensen, R., Hu, Q. (eds.) RSCTC 2010. LN CS, vol. 6086, pp. 60–69. Springer, Heidelberg (2010). https://doi.org/10.1007/978-3-642-13529-3_8

15. Kryszkiewicz, M., Lasek, P.: A neighborhood-based clustering by means of the triangle inequality. In: Fyfe, C., Tino, P., Charles, D., Garcia-Osorio, C., Yin, H. (eds.) IDEAL 2010. LNCS, vol. 6283, pp. 284–291. Springer, Heidelberg (2010). https://doi.org/10.1007/978-3-642-15381-5_35

16. Kryszkiewicz, M., Lasek, P.: A neighborhood-based clustering by means of the triangle inequality and reference points, ICS Research Report 3/2011, Warsaw University of Technology (2011)

17. Kryszkiewicz, M., Podsiadly, P.: Efficient search of cosine and Tanimoto near duplicates among vectors with domains consisting of zero, a positive number and a negative number. IEA/AIE (2), 160–170 (2014)

18. Lipkus, A.H.: A proof of the triangle inequality for the Tanimoto distance. J. Math. Chem. **26**, 263–265 (1999)

19. Moore, A.W.: The anchors hierarchy: using the triangle inequality to survive high dimensional data. In: Proceedings of UAI, Stanford, pp. 397–405 (2000)

20. Uhlmann, J.K.: Satisfying general proximity/similarity queries with metric trees. Inf. Process. Lett. **40**(4), 175–179 (1991)

21. Willett, P., Barnard, J.M., Downs, G.M.: Chemical similarity searching. J. Chem. Inf. Comput. Sci. **38**(6), 983–996 (1998)

22. Yanilos, P.N.: Data structures and algorithms of nearest neighbor search in general metric spaces. In: Proceedings of 4th ACM-SIAM Symposium on Discrete Algorithms, pp. 311–321 (1993)

23. Zezula, P., Amato, G., Dohnal, V., Bratko, M.: Similarity Search: The Metric Space Approach. Springer, Heidelberg (2006). https://doi.org/10.1007/0-387-29151-2

Rough-Fuzzy Segmentation of Brain MR Volumes: Applications in Tumor Detection and Malignancy Assessment

Pradipta Maji[1]([envelope]) [ORCID] and Shaswati Roy[2]

[1] Biomedical Imaging and Bioinformatics Lab, Machine Intelligence Unit,
Indian Statistical Institute, Kolkata, India
`pmaji@isical.ac.in`
[2] Department of Information Technology, RCC Institute of Information Technology,
Kolkata, India

Abstract. An important diagnostic technique for providing accurate information about the spatial distribution of brain soft tissues non-invasively is magnetic resonance (MR) imaging. In MR images, different imaging artifacts give rise to uncertainties in brain volume segmentation into major soft tissue classes; as well as in extracting brain tumor and evaluating its malignancy state. Among various soft computing techniques, rough sets provide a powerful tool to handle uncertainties and incompleteness associated with data, while fuzzy set serves as an analytical tool for dealing with uncertainty that arises due to the overlapping characteristics in the data. In this regard, the paper presents a brief review on the recent advances of rough-fuzzy hybridized approaches for brain MR volume segmentation, brain tumor detection and gradation.

Keywords: Medical image analysis · Segmentation · Gradation · Rough sets · Fuzzy set · Hybrid intelligent systems

1 Introduction

Magnetic resonance (MR) imaging is an important diagnostic technique for providing accurate information about the spatial distribution of brain soft tissues non-invasively. In the last few decades, MR imaging has evolved into the most powerful and versatile medical imaging tool for diagnosis of neurological diseases and brain cancer. The analysis of non-invasive brain MR scans enables to impart enormous potential value for improved diagnosis, treatment planning, and follow-up of individual patients.

However, various imaging artifacts like intensity inhomogeneity or bias field and noise may give rise to uncertainties in segmentation of brain MR volumes into three major soft tissue classes, namely, gray matter, white matter, and cerebro-spinal fluid; as well as in extracting brain tumor and evaluating its malignancy state. Three main factors, namely, vagueness and incompleteness in class

© Springer Nature Switzerland AG 2021
S. Ramanna et al. (Eds.): IJCRS 2021, LNAI 12872, pp. 35–43, 2021.
https://doi.org/10.1007/978-3-030-87334-9_3

definitions, imprecision in computations, and overlapping characteristics among class boundaries, are some of the sources of this uncertainty. In this regard, the theories of rough sets [6] and fuzzy sets [12] have been integrated to model and handle the uncertainties in brain MR image segmentation and tumor classification tasks. Both fuzzy sets and rough sets provide a mathematical framework to capture the uncertainties associated with the real-life data. The rough set theory comes up with a successful tool for data analysis by synthesizing approximations of concepts or sets from the given data [6]. There are generally fuzzy information in real-life applications and the data is real valued. Combining rough sets and fuzzy sets, therefore, provides an important direction in reasoning with uncertainty for real valued data.

The theory of fuzzy set is based on the notion of a membership function on the domain of discourse, assigning a grade of belongingness to each object in order to model an overlapping or imprecise concept. On the other hand, the theory of rough sets is based on the ambiguity caused by limited discernibility of the objects in the domain of discourse. The main idea here is to approximate any crisp subset or concept by a pair of exact sets, called the upper and lower approximations. However, in granular universe, the concepts may well be imprecise, since the crisp subsets may not represent these concepts. This leads to an important direction, in which the notions of fuzzy sets and rough sets can be integrated, to develop a model of uncertainty, stronger than either, under the umbrella called rough-fuzzy computing [2,5].

Rough-fuzzy techniques are efficient hybrid techniques based on judicious integration of principles of rough sets and fuzzy sets. While the membership functions of fuzzy sets enable efficient handling of overlapping classes, the concept of lower and upper approximations of rough sets deals with uncertainty, vagueness, and incompleteness in class definitions. Since the rough-fuzzy approach has the capability of providing a stronger paradigm than either of the rough sets or fuzzy sets for uncertainty handling, it has greater promise in application domains of medical image analysis. In this regard, the paper presents some new algorithms, based on the theory of rough-fuzzy computing. These algorithms have been introduced in recent past in [7], for the analysis of brain MR image data, in the presence of intensity inhomogeneity and noise.

2   Segmentation of Brain MR Images

Segmentation is the process of partitioning an image domain into a number of non-overlapping, meaningful, and homogeneous regions. Accurate and reliable segmentation of brain MR images into three major soft tissue classes (Fig. 1), namely, gray matter (GM), white matter (WM), and cerebro-spinal fluid (CSF), is of great importance to identify the disease-specific morphological differences. Manual segmentation of brain MR images has been a common practice in clinics. But, it is labor intensive, time-consuming, and varies according to the expert's perception. It thus makes an automated, consistent, and accurate brain MR image segmentation method desirable for clinical use.

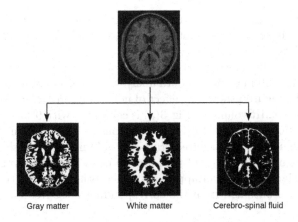

Gray matter   White matter   Cerebro-spinal fluid

Fig. 1. Example of brain MR image segmentation

However, during segmentation of brain tissues, namely, GM, WM, and CSF, some non-brain tissues may seem to be indiscernible with respect to the brain tissues. Therefore, it may produce high classification error during segmentation of brain tissues or lesion. Since the region of interest for segmentation is the area containing only brain tissues, the proper surface representation of the human cerebral cortex is thus highly desirable prior to the segmentation. So, one of the important tasks of brain image analysis is the extraction of brain region from non-brain region, which is known as skull stripping. The process of brain extraction or skull stripping (Fig. 2) includes the removal of non-brain areas, like eyes, dura, scalp, skull, etc., from brain MR volumes. It acts as an important pre-processing step, which not only increases the segmentation accuracy, but also minimizes the execution time of the segmentation in manifold.

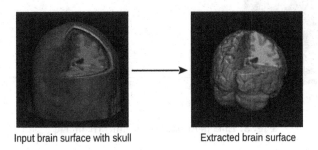

Input brain surface with skull   Extracted brain surface

Fig. 2. Skull stripping in brain MR image

Even though the problem of brain tissue segmentation has been evolved for many years in medical research, the design of robust and efficient segmentation algorithms is still a very challenging research topic, due to the variety and complexity of the brain MR images. The inherent limitations of spatial and temporal resolutions of imaging devices and material heterogeneity lead to different imaging artifacts in images obtained from various modalities. These imaging artifacts include mainly noise and intensity inhomogeneity.

These artifacts introduce misclassification of brain tissues during segmentation of brain MR volumes. In real data analysis, the effects of noise and outliers are unavoidable. Figure 3 shows the effect of noise in brain MR images, where the example noise-free MR volume of Fig. 3(a) is corrupted with 9% noise (Fig. 3(c)). The histograms of different tissue classes, namely, GM, WM, and CSF, generated from these noisy and noise-free brain MR volumes are also shown in Fig. 3. The histograms of GM, WM, and CSF for the noisy brain MR volume indicate that they are significantly overlapped with each other (Fig. 3(d)) as compared to the histograms obtained from the corresponding noise-free brain MR image (Fig. 3(b)).

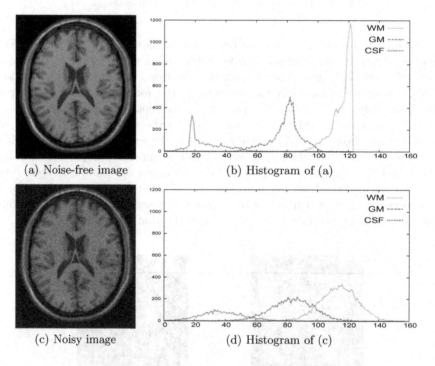

(a) Noise-free image   (b) Histogram of (a)

(c) Noisy image   (d) Histogram of (c)

Fig. 3. Effects of noise in brain MR image

Another major artifact in brain MR volumes is the bias field artifact, also known as intensity inhomogeneity. Intensity inhomogeneity in MR images generally occurs due to the faulty image acquisition process and also from imaged object. Although this artifact is hardly observable to a person, it degrades the quality of segmentation drastically. Bias field artifact is conventionally defined as a smooth, but spurious, variation of intensities within same tissue class across the

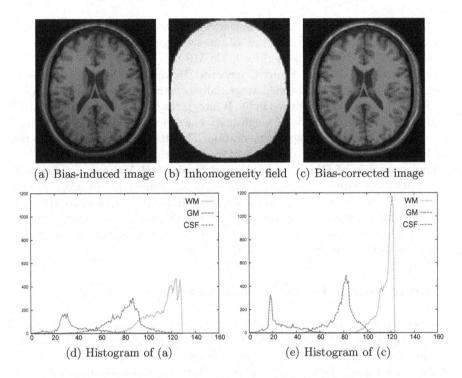

(a) Bias-induced image (b) Inhomogeneity field (c) Bias-corrected image

(d) Histogram of (a) (e) Histogram of (c)

Fig. 4. Effects of intensity inhomogeneity in brain MR image

image. Figure 4 presents the effects of bias field artifact on brain MR images. The intensity non-uniformity level of bias-induced image (Fig. 4) is 80%, while both bias-corrected and bias-induced images are noise-free in this example. Because of the smooth spatially varying nature of this artifact, the intensity values of a tissue class vary according to their spatial location within the image domain. This results in modification of mean and increase in within-class variation of each tissue class, such as GM, WM, and CSF, as shown in Fig. 4. In effect, the overlapping region between the histograms of two tissue classes is large. In other words, the bias field artifact increases the overlaps between the intensity distribution of different soft tissues significantly.

Hence, it makes the segmentation process highly sensitive to this spurious intensity variations. In order to apply any automatic image analysis tool to a brain MR volume, a noise reduction technique as well as an inhomogeneity correction step are thus indispensable. In this regard, the following algorithms have been introduced recently, which are developed based on rough-fuzzy computing.

1. **Skull Stripping of Brain MR Volumes:** Recently, Roy and Maji [8] introduced an accurate and robust skull stripping algorithm, termed as ARoSi. It is based on a novel concept, called rough-fuzzy connectedness. The connectedness of a voxel to the brain depends on its degree of belongingness to the brain region as well as the degree of adjacency to the brain. The performance

of ARoSi on several healthy and diseased real-life 3-D brain MR images, along with a comparison with other state-of-the-art algorithms, establishes it as an effective skull stripping method for brain MR volumes.

2. **Segmentation of Bias Field Corrected Brain MR Volumes:** A novel segmentation algorithm, termed as spatially constrained rough-fuzzy c-means (sRFCM), has been introduced in [9]. It integrates judiciously local contextual information and the merits of rough-fuzzy clustering [1], which are very much effective for brain MR image segmentation. The sRFCM algorithm assigns the label of a voxel based on the labels of its local neighbors.

3. **Segmentation of Bias Field Induced Brain MR Volumes:** A new segmentation algorithm, termed as coherent local intensity rough segmentation (CoLoRS), has been introduced in [10], for brain MR volumes corrupted with bias field artifact. It judiciously integrates the theory of rough sets and the merits of coherent local intensity clustering for simultaneous segmentation and bias field correction of brain MR volumes.

3   Brain Tumor Detection and Gradation

Treatment strategy differs significantly from one type of brain tumor to another, ranging in grade from low to high. Patients with low-grade brain tumor may live for a long time, giving median survival rate more than five years. In contrast, patients with high-grade tumors have worse prognosis. Glioblastoma multiforme is the most frequent and most malignant nervous system tumor. Glioblastomas, shown in Fig. 5(a), develop very rapidly, and thus, yield poor median survival rate of about 14 months. Meningiomas, shown in Fig. 5(b), for example, are generally benign or low grade and slowly growing intracranial tumors, attached to the dura matter.

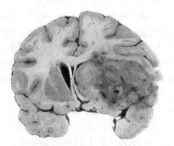   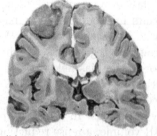

(a) Glioblastoma (high grade)   (b) Meningioma (low grade)

Fig. 5. Examples of high grade and low grade brain tumors

The prognosis for high grade brain tumors remains poor, even with aggressive treatment strategies, including surgical resection, along with recent advances in radiotherapy and chemotherapy. Surgery is the cornerstone of treatment for the majority of low-grade tumors, while post-operative radiotherapy has been shown to lengthen the progression-free period without significantly affecting the overall survival. Therefore, the diagnosis of brain tumors is critical for planning therapeutic strategies, assessing prognosis, and monitoring response to therapy. In current clinical studies, the acquired images are evaluated based on qualitative criteria. An analysis of the characteristics of hyper-intense tissue in contrast-enhanced T1-weighted MR image is often relied to assess the tumor malignancy.

The characteristics of brain tumors can be quantified using a variety of image processing routines, which make the assessment highly accurate and reproducible. Moreover, these automated methods of analyzing brain tumors enable treatment planning, monitoring of therapy, examining efficacy of radiation and drug treatments. In order to evaluate the degree of malignancy of brain tumors, several properties or features can be extracted from tumor regions and tumors are then classified according to these extracted features. Therefore, the necessity of brain tumor classification, in turn, compels the accurate identification of the tumor region.

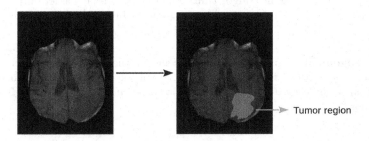

Fig. 6. Example of tumor detection from brain MR image

For brain tumor studies, tumor segmentation is, therefore, crucial for monitoring tumor growth or shrinkage in patients during therapy and studying the differences of healthy subjects and subjects with tumor. The brain tumor detection method (Fig. 6) aims at segmentation of healthy and pathologic brain tissues from images. Conventionally, the brain tumor from MR images is interpreted visually and qualitatively by radiologists. Manual segmentation of brain tumor from MR images is a time-consuming task and subject to considerable variation in intra- and inter-operator performance. In addition, any manual segmentation method suffers from lack of reliability and reproducibility. Therefore, a consistent, accurate, automated segmentation method for clinical brain tumor segmentation and volume measurement is much needed (Fig. 6). In this regard, the following algorithms, based on rough-fuzzy computing, have been introduced.

1. **Brain Tumor Detection from 3-D MR Images:** The brain tumor detection from 2-D MR images using rough-fuzzy technique has been presented in [3,4]. A new method has been introduced in [7], for segmentation of brain tumor from MR volumes. To address the problems of uncertainty and bias

field artifact of brain MR image segmentation, the new method uses simultaneous segmentation and inhomogeneity correction algorithm, described in [10]. One of the major issues of the unsupervised brain tumor segmentation method is how to extract brain tumor accurately, since tumors may not have clearly defined intensity or textural boundaries. Therefore, a new postprocessing method has been presented in [7], for clustering based brain tumor detection. It combines the merits of mathematical morphology and the concept of rough set based region growing approach to refine the result obtained after segmentation, thereby, ensuring the accurateness of brain tumor segmentation application.

2. **Gradation of Brain Tumor Malignancy Using 3-D MR Images:** A new algorithm has been introduced in [11], using conventional MR sequences, namely, T1, T2, T1 with contrast enhancement and FLAIR, for the assessment of tumor grades.

The effectiveness of the rough-fuzzy computing based brain tumor detection and gradation methods has been demonstrated on several real brain MR volume databases. However, the new algorithms, mentioned here, consider only intensity feature for each voxel. In near future, these algorithms will also be extended to incorporate the textural properties as well to describe a voxel.

Acknowledgement. This work is an outcome of the R&D work undertaken in the project under the Visvesvaraya Ph.D. Scheme of Ministry of Electronics and Information Technology, Government of India, being implemented by Digital India Corporation.

References

1. Maji, P., Pal, S.K.: Rough set based generalized fuzzy C-means algorithm and quantitative indices. IEEE Trans. Syst. Man Cybern. Part B: Cybern. **37**(6), 1529–1540 (2007)
2. Maji, P., Pal, S.K.: Rough-Fuzzy Pattern Recognition: Applications in Bioinformatics and Medical Imaging. Wiley-IEEE Computer Society Press, New Jersey (2012)
3. Maji, P., Roy, S.: Rough-fuzzy clustering and unsupervised feature selection for wavelet based MR image segmentation. PLoS ONE **10**(4), e0123677 (2015). https://doi.org/10.1371/journal.pone.0123677
4. Maji, P., Roy, S.: SoBT-RFW: Rough-fuzzy computing and wavelet analysis based automatic brain tumor detection method from MR image. Fund. Inf. **142**, 237–267 (2015)
5. Pal, S.K., Skowron, A. (eds.): Rough-Fuzzy Hybridization: A New Trend in Decision Making. Springer-Verlag, Singapore (1999)
6. Pawlak, Z.: Rough sets. Int. J. Comput. Inf. Sci. **11**, 341–356 (1982)
7. Roy, S.: Rough-Fuzzy Segmentation of Brain MR Volumes and Its Applications in Tumor Detection and Gradation. Ph.D. thesis, University of Calcutta, West Bengal, India (2021)
8. Roy, S., Maji, P.: An accurate and robust skull stripping method for 3-D magnetic resonance brain images. Magn. Reson. Imaging **4**, 46–57 (2018)

9. Roy, S., Maji, P.: Medical image segmentation by partitioning spatially constrained fuzzy approximation spaces. IEEE Trans. Fuzzy Syst. **28**(5), 965–977 (2020)
10. Roy, S., Maji, P.: Rough segmentation of coherent local intensity for bias induced 3-D MR brain images. Pattern Recogn. **97**, 106997 (2020)
11. Roy, S., Maji, P.: Multispectral co-occurrence of wavelet coefficients for malignancy assessment of brain tumors. PLoS ONE **16**(6), e0250964 (2021). https://doi.org/10.1371/journal.pone.0250964
12. Zadeh, L.A.: Fuzzy sets. Inf. Control **8**, 338–353 (1965)

DDAE-GAN: Seismic Data Denoising by Integrating Autoencoder and Generative Adversarial Network

Fan Min[1(✉)][iD], Lin-Rong Wang[1], Shu-Lin Pan[2], and Guo-Jie Song[3]

[1] School of Computer Science, Southwest Petroleum University,
Chengdu 610500, China
minfan@swpu.edu.cn
[2] School of Earth Science and Technology, Southwest Petroleum University,
Chengdu 610500, China
[3] School of Science, Southwest Petroleum University,
Chengdu 610500, China

Abstract. Machine learning methods face two main challenges in denoising tasks. One is the lack of supervised training data, and the other is the limited knowledge of complex unknown noise. In this paper, for seismic denoising, we propose a new method with three techniques to handle them effectively. First, a Generative Adversarial Network (GAN) is employed to generate a large number of paired clean-noisy data using real noise. Second, a deep denoising autoencoder (DDAE) is pre-trained using these data. Third, a transfer learning technique is used to train the DDAE further on a few field data. We have assessed the proposed method based on qualitative and quantitative analysis. Results show that the method can suppress seismic data noise well.

Keywords: DDAE · Denoise · GAN · Seismic data

1 Introduction

Machine learning methods, especially discriminative learning, have achieved great success in denoising tasks for various applications. These include convolutional neural network (CNN) with batch normalization and residual learning for image denoising [1], frame-to-frame training DnCNN [2] for video denoising [3], branch-based CNN for seismic data denoising [4], etc. They usually require a large amount of supervised training data, i.e., clean-noisy data pairs. However, such data are difficult to obtain in the field of seismic data processing. Therefore, it is a challenging task to learn a denoising model with limited supervised data.

Another challenge is how to handle unknown types of noise in seismic data. Different assumptions about the type of noise have been made when designing the denoising methods. The most common noise assumption is additive white Gaussian noise (AWGN) [2,5,6]. Real noise can be regarded as the addition of independent random variables with different distributions, so the normalized

© Springer Nature Switzerland AG 2021
S. Ramanna et al. (Eds.): IJCRS 2021, LNAI 12872, pp. 44–56, 2021.
https://doi.org/10.1007/978-3-030-87334-9_4

sum is close to the Gaussian distribution. However, Gaussian noise is only an approximation and simulation of real noise, so it cannot guarantee the noise reduction effect on real noise.

In this paper, we propose a new seismic denoising method with three techniques to handle them effectively. First, a large number of paired clean-noisy data is generated using GAN according to actual noise. One subtask is to generate high-quality noise samples. We use a fast smooth patch search algorithm [7] to extract noise blocks from noisy data. They are the learning goals of our GAN. The generated noise samples are treated as additive noise. The other subtask is to generate noisy data. We use the direct approach to superimpose the noise samples on synthesis clean data. In this way, a large number of data pairs can be constructed using only a small amount of data.

Second, a deep denoising autoencoder (DDAE) is pre-training using these data. This DDAE is a fully convolutional autoencoder using rectified linear unit (ReLU) as the activation function. The input is the noisy data, the output is the predicted clean signal, and the optimization objective is to minimize the residual. The training process terminates when the user-specified number of iterations is reached. To obtain the best parameters of the model, the iteration with the least loss is solicited.

Third, the DDAE is further trained using a transfer learning technique on a few field data. Unlike the supervised pre-training process, our transfer learning technique adopts the unsupervised scheme. Because the field data are not provided as clean-noisy pairs. Moreover, unlike the mean squared error (MSE) loss employed in pre-training, a customized loss function is employed. It is a revision of the loss function proposed in [8].

Synthetic and field data are used to evaluate the proposed method qualitatively and quantitatively. Results show that the proposed DDAE-GAN can suppress the random noise while preserving the useful seismic signals.

2   Related Work

We first briefly introduce several traditional methods. Then two denoising methods based on noise modeling and autoencoder are introduced respectively.

2.1   Seismic Noise Reduction Methods

Most seismic data are inevitably mixed with noise due to human or environmental factors. Noise may disrupt many seismic processes, such as geological interpretation and velocity analysis [9]. Many traditional denoising methods have been proposed. For instance, the f-x method [10–12] processes in the Fourier transform domain, MSSA [13] performs denoising in the frequency-space domain, and wavelet denoising [14] uses wavelet transform to suppress noise.

2.2   Noise Modeling Based Denoising Methods

So far, most denoising methods are based on noise modeling. BM3D [15] is known as the golden algorithm for image denoising. It is based on the Gaussian model and combined with non-local self-similar models and sparse models to achieve a good denoising effect. The multi-scale denoising algorithm based on Gaussian hypothesis [16] is on the basis of non-local Bayes [17]. Mixture of Gaussian (MoG) is used to approximate various continuous distributions to improve denoising performance [18]. However, most of these methods use a single fixed noise model, which may limit its generalization ability and further affect denoising performance. As a generative model, GAN can not only learn complex distributions, but also generate data with the same distribution. For example, the GCBD [7] algorithm applies GAN to real image noise modeling to generate a large number of data sets.

2.3   AutoEncoder Based Denoising Methods

The unsupervised nature of autoencoder (AE) makes it widely used in denoising. The sparse denoising autoencoder realizes the denoising of mixed noise by adding the sparse condition [19]. The performance of using autoencoder to achieve blind image denoising is better than the BM3D [15]. Moreover, DDAE based on transfer learning is used to suppress random seismic noise [8]. These examples illustrate the maturity of autoencoders in denoising.

3   DDAE-GAN Based Blind Denoiser

This section presents our work in detail. Figure 1 presents an overview of our method (DDAE-GAN).

3.1   Paried Data Constructing

Our first task is to construct paired training data sets. Similar to [7], we use GAN to generate noise instead of noisy data. Therefore, we need to extract the noise blocks from the noisy data as the learning goal of GAN.

Noise Block Extraction

This is a key step for GAN to simulate unknown noise. The quality of noise block extraction determines the quality of samples generated by GAN. We use a fast smooth patch search algorithm [7] to get the noise blocks.

Let $\mathbf{p} = (p_{ij})_{d \times d}$ be a global patch sampled from the seismic data, and $\mathbf{q} = (q_{ij})_{h \times h}$ be a local patch sampled from $\mathbf{p}$. A smooth block is often denoted by $\mathbf{s}$ and if and only if it satisfies

$$|\overline{\mathbf{q}} - \overline{\mathbf{p}}| \leq \mu \cdot |\overline{\mathbf{p}}|, \tag{1}$$

and

$$|\sigma^2(\mathbf{q}) - \sigma^2(\mathbf{p})| \leq \gamma \cdot \sigma^2(\mathbf{p}), \tag{2}$$

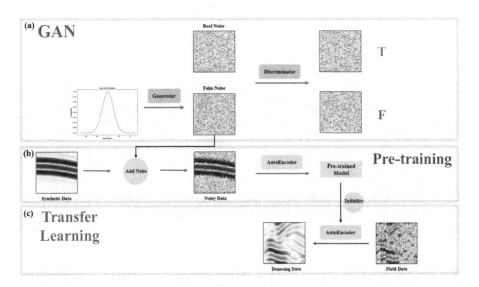

Fig. 1. Architecture of the proposed DDAE-GAN. (a) Generate noise data using GAN. (b) Use the generated supervised data for pre-training. (c) Use transfer learning to denoise field data.

where $\bar{\cdot}$ and $\sigma^2(\cdot)$ denote the mean and variance of the matrix elements, respectively, and $\mu, \gamma \in (0, 1)$ are parameters.

Given a smooth block $\mathbf{s}$, according to [7], we obtain a noise block as follows

$$\mathbf{v} = v(\mathbf{s}) = \mathbf{s} - \bar{\mathbf{s}}. \tag{3}$$

where $\mathbf{s}$ denotes a smooth block, see more details in [7].

Noise Modeling with GAN

The noise blocks obtained in the previous step is the learning goal of GAN. We adopt Wasserstein GAN (WGAN) [20] to learn the noise distribution. WGAN has the same structure as DCGAN [21], which is depicted in Fig. 2. The objective function for our task is

$$\mathcal{L}_{WGAN} = \max_{w \in \mathcal{W}} \mathbb{E}_{x \sim \mathbb{P}_r} \left[f_w(x) \right] - \mathbb{E}_{z \sim p(z)} \left[f_w \left(g_\theta(z) \right) \right], \tag{4}$$

where f_w is a discriminator network with parameter w and the last layer is not a nonlinear activation layer.

3.2 Pre-training

Network Settings

Our DDAE (Fig. 3) is composed of input, noise, encoder and decoder layers. We use the synthesized signal plus the noise generated by GAN as the pre-trained

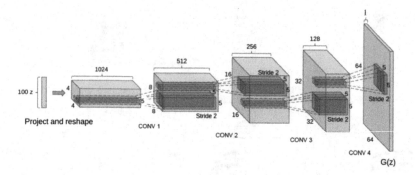

Fig. 2. [21] DCGAN generator used for real noise modeling. A 100-dimensional uniform distribution z is transformed into 64 × 64 noise data through four fractionally-strided convolutions.

data set of DDAE to obtain the best parameters. Let $\mathbf{s} = (s_{ij})_{d \times d}$ be the clean data and $\mathbf{n}' = (n'_{ij})_{d \times d}$ be the noise samples generated by GAN. In the noise layer, $\mathbf{s}$ and $\mathbf{n}'$ form noisy data $\mathbf{d}' = \mathbf{s} + \mathbf{n}'$. Our goal is to remove the noise part $\mathbf{n}'$ from the noisy data $\mathbf{d}'$ to obtain the clean signal $\mathbf{s}$. The input data $\mathbf{d}'$ is compressed into multiple abstraction levels through the coding layer to extract important features $\mathbf{z}$. The decoding layer attempts to use the extracted features $\mathbf{z}$ to reconstruct clean seismic data $\mathbf{s}'$.

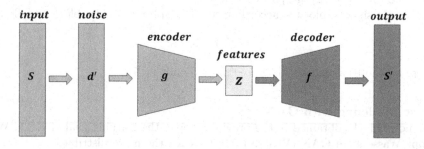

Fig. 3. Architecture of the proposed DDAE.

Generally, seismic data is two-dimensional or three-dimensional data with spatial transformation. We propose that DDAE uses convolutional layers instead of fully connected layers. On the one hand, the fully connected network ignores the location information of the data. On the other hand, the number of parameters will limit the depth of the network.

We use the controlled variable method to obtain the optimum DDAE topology. For non-linearity purposes, a non-linear layer is added after each convolutional layer. We have tested several activation functions such as the hyperbolic tangent (tanh), sigmoid and rectified linear unit (ReLU). We found that using the ReLU function can make DDAE reach optimal performance. The ReLU function is defined as $A(x) = \max(0, x)$.

According to the experimental results, we found that three encoding and decoding layers are sufficient to achieve a good denoising effect. The optimal number of output channels for each convolutional layer is as follows. In the encoder, the number of output channels of the first convolutional layer is 32, the second layer is 16, and the third layer is 8. In the decoder, the number of output channels of the first transposed convolutional layer is 16, the second layer is 32, and the third layer is the number of input data channels. The proposed DDAE's network topology is shown in Fig. 4.

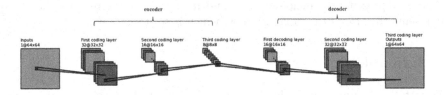

Fig. 4. Our DDAE's network topology.

Network Training
We use the training method of data block, that is, using a fixed-size window to slide the input data $\mathbf{d}'$. The window slides at a fixed step each time, and data samples are generated. The window size determines the size of the sample data. Once the window size is determined, the sliding step determines the number of sample data. We also use a rollover strategy for data enhancement. The block training is well adapted to the key information of seismic data which is only locally relevant. At the same time, it avoids the waste of computing resources during the training. In the entire network tuning process, a large input window can result in too few samples and slower training speed. We test four group small input window size: 32×32, 48×48, 64×64, and 96×96. Result shows that the 64×64 window size is the best.

Finally, we use mean square error (MSE) as the loss function of the pre-trained model. The loss function is defined as

$$\min \frac{1}{2} \sum_{N}^{N} \left(\mathbf{s}' - \mathbf{s}\right)^2, \tag{5}$$

where N represents input window size, which we set to 64×64.

In Adam optimizer, the optimum η, β_1, β_2, and ε parameters are 0.005, 0.9, 0.999, and 10^{-8}. During training, we use 500 epochs and a batch size of 256.

3.3   Transfer Learning

For DDAE that processes field data, we use a new loss function, which is based on [8]. Then the pre-trained model is used to initialize the model. The DDAE only needs to fine-tune the weight of each node in the network layer by layer to achieve a good denoising effect. The defined loss function is only related to the denoised data s' and the removed noise n'. The loss function defined for field data is defined as

$$\frac{[N \sum s'n' - \sum s' \sum n']^2}{N^2[\sum s'^2 - (\sum s')^2][\sum n'^2 - (\sum n')^2]}, \tag{6}$$

where s' represents the denoised data, and n' represents the removed noise. N represents the size of the input window, which is set to 64×64.

4   Examples

We test the proposed algorithm on synthetic data and field data respectively. The denoising performance of the proposed method compare the denoising performance with two denoising methods (MSSA, wavelet transform). Three parts of the experiment were carried out:

(1) Single Gaussian noisy data is used for evaluation.
(2) We also use mixture noise evaluation to show that the proposed method can handle more complex noise.
(3) Finally, we use field data to evaluate the applicability of the proposed method to real-world problems. Peak signal-to-noise ratio (PSNR) is used as a quantitative assessment of denoising performance, expressed as

$$PSNR = 10 \times \log_{10}(MAX_s^2/MSE), \tag{7}$$

where MAX_s denotes the maximum value of seismic data. MSE denotes the mean square error, that is, the square error of the clean data and the denoised data. The unit of PSNR is dB, and the larger the value, the less distortion.

4.1   Synthetic Examples

Gaussian Noise
We create synthetic data with a sampling rate of 2 ms, which includes 83 traces and 1000 time samples. The data contains strong amplitude signals and weak amplitude signals (Fig. 5(a)). We add single Gaussian noise to the synthesized data to form noisy data (Fig. 5(b)), and its PSNR is 16 dB.

Figures 6(a), 6(b) and 6(c) display the denoised results of MSSA, wavelet transform and the proposed method, respectively. MSSA and wavelet transform have a lot of residual noise, but the proposed method basically does not. In the corresponding, the differences between the results of three methods and the ground truth are shown in Figs. 6(d), 6(e) and 6(f). The denoised results of MSSA and wavelet transform are quite different from ground truth, while the proposed has only some subtle differences. In addition, the PSNR of the proposed method is 49.52 dB, which is higher than 27.53 dB and 22.81 dB by MSSA and Wavelet transform. Results show that the proposed method can suppress the noise while protecting the effective signal for Gaussian noisy data.

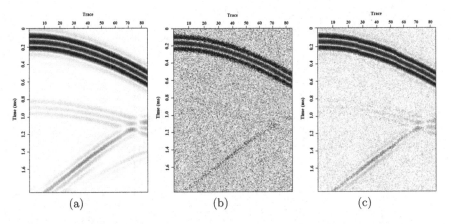

$$(a) \qquad\qquad (b) \qquad\qquad (c)$$

Fig. 5. Synthetic examples. (a) Clean dataset. (b) Single Gaussian noisy dataset (PSNR = 16 dB). (c) Mixture noisy dataset (PSNR = 25.19 dB).

Mixture Noise

Besides Gaussian noise, We also evaluate the denoising performance of several methods for complex noise. The mixture noise used in our experiments comes from [22]. It includes 70% Gaussian noise obeying N(0,1) distribution, 20% Gaussian noise obeying N(0,0.01) distribution and 10% uniform noise. Figure 5(c) shows an example of mixture noise with a PSNR of 25.19 dB.

Figures 7(a), 7(b) and 7(c) display the denoised results of the complex noisy data using MSSA, wavelet transform and the proposed method, respectively. We can observe that there is still a lot of residual noise in MSSA and wavelet transform. Although there is no significant residual noise in the proposed method, some details of the weak signal are missing. Figures 7(d), 7(e) and 7(f) show that the proposed method has less signal leakage than MSSA and wavelet transform. And the PSNRs of the denoised results are 11.66 dB, 29.45 dB, 37.67 dB. It can be seen from the denoised results and the PSNRs that the proposed method can also remove more complex noises.

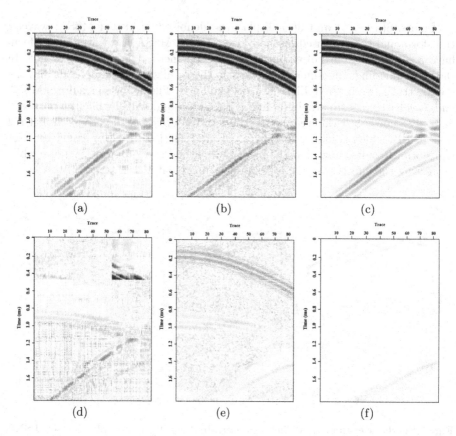

Fig. 6. Single Gaussian noisy data denoising results. Denoised result using (a) MSSA (PSNR = 27.53 dB), (b) wavelet transform (PSNR = 22.81 dB), (c) the proposed method (PSNR = 49.52 dB). Differences between the clean data and the denoising result using (d) MSSA, (e) wavelet transform, (f) the proposed method.

4.2   Field Examples

We further evaluated the denoising performance of the three methods on more complex field data. The field data used in the experiment consists of 1,200 traces and 800 time samples (Fig. 8). Although most of the signals in the data are relatively strong and continuous, they are contaminated by a large amount of incoherent noise, resulting in blurred signals. The denoised results corresponding to MSSA, wavelet transform and the proposed method are shown in Figs. 9(a), 9(b) and 9(c), respectively. The removed noise extracted by MSSA, wavelet transform and the proposed method are shown in Figs. 9(d), 9(e) and 9(f). We can observe that three methods can suppress random noise. But the denoising result of the proposed method is smoother and removes more noise than the other two. In addition, there is no obvious continuous signal in the noise part

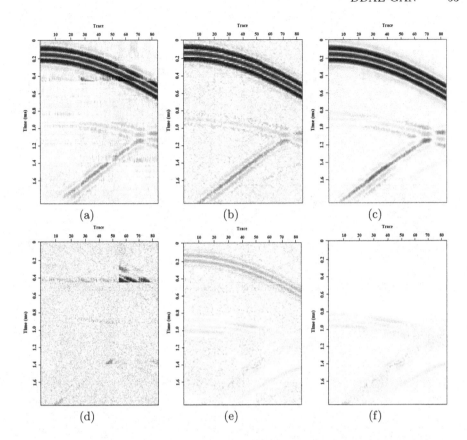

Fig. 7. Mixture noisy data denoising results. Denoised result using (a) MSSA (PSNR = 11.66 dB), (b) wavelet transform (PSNR = 29.45 dB), (c) the proposed method (PSNR = 37.67 dB). Differences between the clean data and the denoising result using (d) MSSA, (e) wavelet transform, (f) the proposed method.

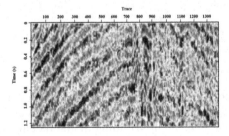

Fig. 8. Field data.

(Fig. 9(f)), indicating that the proposed method has a certain degree of fidelity. Therefore, the proposed method can suppress the noise while protecting the effective signal.

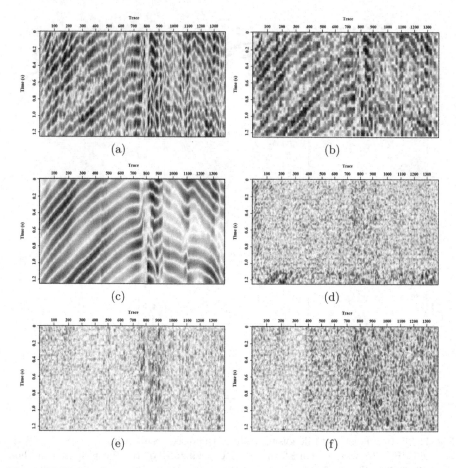

Fig. 9. Field data denoising results. Denoised result using (a) MSSA, (b) wavelet transform, (c) the proposed method. Removed noise using (d) MSSA, (e) wavelet transform, (f) the proposed method.

5   Conclusions

In this paper, we proposed a method of fusing three technologies. GAN was employed to learn the noise distribution in actual seismic data and constructed a large number of noisy-clean supervised data. Supervised data was used for DDAE for pre-training, and the pre-training model redefines the loss function to denoise the field data. We have assessed the proposed method based on synthetic noise and real noise examples. The results showed that the proposed method had better denoising performance than MSSA and wavelet transform. However, a small part of the effective signal was still lost in the synthesized noise part. Next, we will consider better protection of effective signals while suppressing noise.

References

1. Kai, Z., Zuo, W., Gu, S., Lei, Z.: Learning deep CNN denoiser prior for image restoration. In: 2017 IEEE Conference on Computer Vision and Pattern Recognition (CVPR) (2017)
2. Kai, Z., Zuo, W., Chen, Y., Meng, D., Lei, Z.: Beyond a gaussian denoiser: residual learning of deep CNN for image denoising. IEEE Trans. Image Process. **26**(7), 3142–3155 (2016)
3. EhRet, T., Davy, A., Morel, J.M., Facciolo, G., Arias, P.: Model-blind video denoising via frame-to-frame training. In: 2019 IEEE/CVF Conference on Computer Vision and Pattern Recognition (CVPR) (2019)
4. Lin, H., Wang, S., Li, Y.: A branch construction-based CNN denoiser for desert seismic data. IEEE Geosci. Remote Sens. Lett. **99**, 1–5 (2020)
5. Bae, W., Yoo, J., Ye, J.C.: Beyond deep residual learning for image restoration: persistent homology-guided manifold simplification. In: Computer Vision & Pattern Recognition Workshops (2017)
6. Burger, H.C., Schuler, C.J., Harmeling, S.: Image denoising: can plain neural networks compete with BM3D? In: 2012 IEEE Conference on Computer Vision and Pattern Recognition (CVPR) (2012)
7. Chen, J., Chen, J., Chao, H., Ming, Y.: Image blind denoising with generative adversarial network based noise modeling. In: 2018 IEEE/CVF Conference on Computer Vision and Pattern Recognition (CVPR) (2018)
8. Saad, O.M., Chen, Y.: Deep denoising autoencoder for seismic random noise attenuation. Geophysics **85**(4), V367–V376 (2020)
9. Chen, Y., Sergey, F.: Random noise attenuation using local signal-and-noise orthogonalization. Geophysics **80**(6), WD1-WD9 (2015)
10. Canales, Luis, L.: Random noise reduction. In: SEG Technical Program Expanded Abstracts, p. 329(1984)
11. Liu, G., Chen, X., Du, J., Wu, K.: Random noise attenuation using f-x regularized nonstationary autoregression. Geophysics **77**(2), 61 (2012)
12. Naghizadeh, M., Sacchi, M.: Multicomponent f-x seismic random noise attenuation via vector autoregressive operators. Geophysics **77**(2), 91 (2012)
13. Oropeza, V., Sacchi, M.: Simultaneous seismic data denoising and reconstruction via multichannel singular spectrum analysis. Geophysics **76**(3), V25–V32 (2011)
14. Donoho, D., L.: De-noising by soft-thresholding. IEEE Trans. Inf. Theory **41**(3), 613–627 (2002)
15. Yang, D., Sun, J.: Bm3d-net: a convolutional neural network for transform-domain collaborative filtering. IEEE Signal Process. Lett. **25**(1), 55–59 (2017)
16. Lebrun, M., Colom, M., Morel, J.M.: Multiscale image blind denoising. IEEE Trans. Image Process. **24**(10), 3149–3161 (2015)
17. Lebrun, M., Buades, A., Morel, J.M.: A nonlocal Bayesian image denoising algorithm. Siam J. Imaging Sci. **6**(3), 1665–1688 (2013)
18. Zhu, F., Chen, G., Heng, P.A.: From noise modeling to blind image denoising. In: 2016 IEEE Conference on Computer Vision and Pattern Recognition (CVPR) (2016)
19. Ye, X., Lin, W., Xing, H., Le, H.: Denoising hybrid noises in image with stacked autoencoder. In: IEEE International Conference on Information & Automation (2015)
20. Arjovsky, M., Chintala, S., Bottou, L.: Wasserstein gan (2017)

21. Radford, A., Metz, L., Chintala, S.: Unsupervised representation learning with deep convolutional generative adversarial networks (2016)
22. Zhao, Q., Meng, D., Xu, Z., Zuo, W., Zhang, L.: Robust principal component analysis with complex noise. In: Xing, E.P., Jebara, T., (eds.) Proceedings of the 31st International Conference on Machine Learning, Proceedings of Machine Learning Research, **32**, 55–63. PMLR, Bejing, China, 22–24 Jun 2014

Classification of Multi-class Imbalanced Data: Data Difficulty Factors and Selected Methods for Improving Classifiers

Jerzy Stefanowski[(✉)] [ID]

Institute of Computing Science, Poznan University of Technology, Poznan, Poland
jerzy.stefanowski@cs.put.poznan.pl

Abstract. The multiple class imbalanced problem is still less investigated than its binary counterpart. In particular, the sources of its difficulties have not been sufficiently studied so far. Therefore, in this paper we summarize the few literature works on the difficulty factors and present our own latest research results. The binary method for an identification of the types of minority examples is generalized for multiple imbalance classes. The second part of this paper presents three our recent methods for learning classifies from multi-class imbalanced data which exploit information on the aforementioned difficulty factors.

Keywords: Multi-class imbalanced data · Data difficulty factors · Re-sampling methods · Rule classifiers

1 Introduction

In imbalanced data at least one class, further called the minority class, contains a much smaller number of examples than other majority classes. Imbalanced classes pose serious difficulties for learning classifiers as the algorithms are biased towards the majority class examples and fail to recognize the instances from the minority class as accurate as possible [5,10].

Most of current research have been placed on constantly proposing new algorithms and less on studying why this class imbalanced problem is so difficult. However, some researchers have attempted to better understand the nature of the imbalance data and key properties of its underlying distribution. They noticed that the *class imbalance ratio* is not necessarily the only, or main, problem causing this performance decrease. Imbalanced data are often affected by other *difficulty factors*, which in turn cause the degradation of classification performance, sometimes even stronger than the global imbalance ratio [8,13,29,31]. The data difficulty factors are related to characteristics of class distribution, such as decomposition of the class into rare sub-concepts, overlapping between classes or presence of rare minority examples inside the majority class regions. With respect to data distribution characteristics Napierala et al. proposed in

© Springer Nature Switzerland AG 2021
S. Ramanna et al. (Eds.): IJCRS 2021, LNAI 12872, pp. 57–72, 2021.
https://doi.org/10.1007/978-3-030-87334-9_5

[27] to distinguish different *types of examples* – safe or unsafe to be learnt (e.g. borderline, rare or outliers) and present the methods for their identification.

Nevertheless this analysis and most of the methods concern binary imbalanced problems only. Despite this, in some applications it is necessary to deal with *multiple imbalanced classes* and to improve the recognition of more than one of the minority classes. Such multi-class imbalanced data occur, e.g., in medical diagnosis (where few important and rare diseases may occur), technical diagnostics with several degrees of the device failures, text categorization, etc.

The multi-class imbalanced problems are so far less investigated than their binary counterpart. The number of specialized approaches is definitely much smaller. In general, the multi-class learning problems are recognized as harder than two class ones, however the sources of these difficulties have not been sufficiently studied so far. The essential questions to be examined are as follows:

- Should the previously identified binary imbalanced data factors be adapted to multiple classes?
- Does the nature of the multi-class problems lead itself to rather new and different factors that cause deterioration of classifier's predictions?

So far, only a few hypotheses on such issues can be found in the literature. Therefore the first part of this paper is devoted to discussing the already identified data difficulty factors and presenting our own latest research results. Then, we discuss how the earlier binary method for an identification of the types of examples and their level of difficulty [27] can be generalized for multiple classes [22]. We describe the usage of the specialized grid clustering [21] to discover sub-concepts within minority classes and to find rare examples or outliers.

The other contribution includes a brief presentation of three recent methods, introduced by the author and his coauthors, for multi-class imbalanced data which exploit information on the aforementioned difficulty factors. These are SOUP resampling method [11], the rule induction multi-class BRACID algorithm [26] and a multi-class extension of Roughly Balanced Bagging ensemble [23]. The paper ends with a discussion of open problems and further perspectives for research on multi-class imbalanced problems.

2   Related Works on Classification of Multi-class Imbalanced Data

The current approaches to multi-class imbalances are usually divided into the following categories [5]:

- binary decomposition approaches that transform the multi-class problem into the set of binary ones and apply existing methods for improving binary problems,
- specialized approaches, which could be further split intro multi-class preprocessing, variants of cost-sensitive learning, algorithm modifications – including dedicated ensembles.

Following the authors of [5], the most popular are decomposition approaches which are based on the ensembles previously proposed to solve complex multi-class tasks[19]. The most often used frameworks are:

One-versus-all ensemble (OVA), which constructs binary classifiers to recognize a particular class against the remaining ones aggregated into one class [7]. During prediction, the test instance is classified by all base classifiers and is assigned to the class of the most confident base classifier.

One-versus-one ensemble (OVO), which constructs binary classifiers for all pairs of classes. The training set for a particular base classifier contains learning examples from the selected pair of classes only. The prediction for the new instance can be taken by majority voting of base classifiers' predictions or by weighted voting with confidence scores, or more complex aggregations [6, 19].

These frameworks can be easily used in combination with techniques for binary imbalanced data. Moreover, they are often used with oversampling or undersampling approaches [6].

The specialized *multi-class imbalanced re-sampling* methods, e.g., oversampling Static-SMOTE, Global-CS or MDO [1, 35], attempt to increase the cardinalities of minority class towards the size of the biggest class.

The selective hybrid re-sampling is done in SPIDER3 [33], where relations between classes are captured by pre-defined misclassification costs. *SMOTE and Clustered Undersampling Technique* (SCUT) [2] applies EM clustering for each majority class, and some examples are randomly removed from these clusters. The minority classes are oversampled with the standard SMOTE.

The final group of specialized methods aims at modifying neural networks or ensembles. The authors either try to integrate over-sampling in the network or propose different loss functions that direct the training the networks towards better recognition of minority classes. Boosting algorithms are also combined with specialized re-sampling, see e.g. [32].

3   Difficulty Factors in Imbalanced Data

3.1   Earlier Studies on Binary Imbalanced Classes

Imbalanced data are characterized with a *global imbalance ratio*. For binary classes it defined as a ratio of the majority class cardinality and minority one or a percentage of the minority class in all examples in the dataset. Besides this ratio researchers such as [8, 13, 31] noticed that other characteristics of examples distributions in the attribute space, called *data difficulty factors* also deteriorate classifier predictions. They mainly include:

– the fragmentation of the minority class into smaller, rare sub-concepts [15],
– the impact of too strong overlapping between classes,
– the presence of small, isolated groups of minority examples located deeply inside the majority class region.

The first factor comes from experimental observations that the minority class usually does not form a homogeneous region (single concept) in the attribute space, but is scattered into smaller sub-concepts spread over the space, often surrounded by examples from the majority class. Experimental studies, e.g. [15, 29] demonstrated its important impact.

The second factor corresponds to high *overlapping* between regions of minority and majority class examples in the attribute space. In particular it may occur in the complex boundary regions of both classes which are not clearly separated and contain mixed instances from minority and majority classes. Sanchez et al. also demonstrated that the local imbalanced ratio in the overlapping region is more influential than the global imbalanced ratio [8].

The third factor corresponds to *rare cases*, which are defined as isolated, very small, groups of minority examples (e.g., containing 1–3 examples) located more deeply inside regions of the other class [27]. They could be even single examples lying either inside this class or in empty regions of the attribute space. This is different from the first factor, which refers to the decomposition of the minority class into larger sub-clusters containing more examples than rare cases.

A related view on data difficulty factors leads to distinguishing different *types of minority examples*, usually called as *safe* or *unsafe*, based on the number of minority and majority class examples near them [18].

The special method for an identification of more detailed four types minority examples was proposed by Napierala and Stefanowski in [27]. It is based on analyzing class labels of examples in their local neighborhood defined either by k-nearest neighbours or by kernels. For instance, if $k = 5$ neighbourhood is considered, then the example is labelled as a safe example if all five or four its neighbors belong to its class. If three or two neighbors belong to the same class as the considered example, then it is labelled as borderline. If there are not the same class examples in k neighbourhood it is an outlier and a rare one for the remaining proportion.

Besides using labels of example types which depend on such proportions, the authors of [27] defined a coefficient expressing a *safe level* of the given example x – being a local estimator of conditional probability of its assignment to the target class as

$$p(C|x) = \frac{k_C}{k}, \tag{1}$$

where C is the class of example x, k is the number of neighbours and k_C is the number of neighbours which belongs to class C.

Experimental studies on the role of the aforementioned factors have shown that data complexities occur in imbalanced datasets, may play a key role in explaining the difference between the performance of various classifiers [27] and proposing new algorithms for improving classifiers.

3.2 Multi-class Difficulties

Researchers working on multi-class imbalanced data often argue that these data are more difficult than binary ones. However, this and some other hypotheses were already considered for standard, balanced, machine learning tasks. For instance, the claim that decision boundaries between multiple classes are more complex and non-linear than simpler boundaries for binary classes follows the older works, in particular in the context of specialized ensembles such as pairwise coupling [14] (which inspires binary decomposition OVO), see e.g. a chapter in [19]. However, there are other newer hypotheses or observations resulting from experiments with multi-class imbalanced datasets. We summarize them below:

- Wang and Yao [32] analysing their experimental results made an observation that different predictive accuracy may be related to considering various configuration of types (sizes) of classes. They distinguish two configuration *multi-majority* and *multi-minority* referring to the datasets with only one majority or only one minority class and all the other classes being of the same type. Following experiments they concluded that that multi-majority class configurations were more difficult than multi-minority ones.
- Buda et al. [4] paid attention to yet another configuration of multiple classes – *gradual imbalance*, which contains classes of linearly growing sizes.
- Krawczyk claimed that a given class can be a minority class with respect to some classes, and at the same time the majority one to another subset of classes [17]. It makes re-sampling approaches difficult[1].

The needs for transferring the idea of types of minority examples [27] into multi-class data was discussed in [17,22]. Independently, the authors of [30] also adopted the binary example types to the multi-class setting, however in the simplest one-vs-all manner. They studied the performance of classifiers trained on a dataset with oversampled minority examples of one type (using the brute force strategy for testing many variants of random oversampling examples of the particular type). Their results showed improvements of classifiers for almost all datasets, although the authors did not present any methods for tuning the degrees of oversampling nor selecting the variant of example type selection.

Lango has recently carried out a comprehensive experimental study with specially generated synthetic datasets [24]. His main conclusions are as follows:

- The class overlapping was the very influential factor when combined with the higher imbalances. Changing the imbalance ratio presented a limited impact on the recognition of datasets without class overlapping or its slight amount.
- The types of the class size configurations with multiple majority classes were more difficult than multi minority ones. In the second configuration, recognition of the smallest classes were worse than in the former one. The gradual class size configuration with the intermediate classes played a special role between them depending whether the these classes are closer to minority or majority classes.

---

[1] In our opinion this hypothesis may be particularly interesting for the gradual imbalance configuration, where some classes may be *intermediate* ones with respect to their sizes. Furthermore we share Krawczyk's view that it may lead to ambiguities in the decision on the degree of modifications of the examples in oversampling or undersampling. It will be even more difficult when such classes overlap, which the author did not take into account.

- The analysis of interrelations between different types of classes showed that the increase of overlapping between the minority and majority classes led to the stronger deterioration of classifier performance than between minority ones. The impact of the intermediate classes depends on the direction of overlapping with other classes. Its overlapping with the minority class caused faster deterioration of the recognition of minority class than itself, so it played a similar role to majority classes.
- An increasing the number of classes was the most influential for a smaller number of classes.

4   Identifying Types of Examples in Multi-class Imbalanced Data

The generalization of types of examples for multiple classes should take into account at least some of the difficulty factors.

Napierala et al. noticed in [22] that analyzing mutual relations between classes shows that some minority classes can be treated as more closely related to each other than to the majority class. As discussed in the previous section the degree of overlapping between various classes may be different. Thus the new multi-class type of examples may also strongly depend on their relations to other classes. For instance, a given example may be of a borderline type for certain classes and at the same time a safe example for the remaining classes. However using existing binary decomposition approaches to estimate data difficulty or the similar adaptation from [30] cannot properly handle these situations.

These motivations have led Napierala et al. to model relations between multiple imbalanced classes by means of additional information about *similarity between pairs of classes*. This information could be either acquired from users - experts or more automatically estimated from class distributions in the attribute space [22]. It means that one needs information which classes can be treated as more similar to each other than to the rest of the classes. Furthermore, this class similarity may correspond to the expert's interpretation of a mutual position of examples in the neighborhood of the example from a given class. An intuition behind this neighborhood is the following: if example x from a given class has some neighbors from other classes, then neighbors from the class with higher similarity are more preferred.

Let us come back to a medical diagnosis case considered in [22]. Two classes corresponding to similar types of the same asthma should be considered as closer to each other than similar to other types of non-asthma as they need completely different therapies.

Defining it more precisely, it is assumed that for each pair of classes C_i, C_j the degree of their similarity is defined as a real valued number $\mu_{ij} \in [0; 1]$. Similarity of a class to itself is defined as $\mu_{ii} = 1$. The degree of similarity does not have to be symmetric, i.e. for some classes C_i, C_j it may happen that $\mu_{ij} \neq \mu_{ji}$.

Although the values of μ_{ij} are defined individually for each dataset, the general recommendation of [22] is to have higher similarities ($\mu_{ig} \to 1$) for other minority classes C_g, while similarities to majority classes C_h should be rather low ($\mu_{ih} \to 0$). This recommendation follows the earlier discussed data difficulty factors, in particular on higher difficulty of the multi-majority case.

In the case of missing expert's preferences to defining these class similarities for the given dataset, the authors of [11] proposed to use the heuristics which follows class sizes as the basic symptom of class interrelations. This is defined as:

$$\mu_{ij} = \frac{\min(|C_i|, |C_j|)}{\max(|C_i|, |C_j|)} \tag{2}$$

where $|C_i|$ is the number of examples of C_i class.

These degrees of similarities are used to generalize the idea of an identification of types of examples. If one considers the k nearest neighborhood, then determining the number of examples from the majority class in the neighborhood of the example allows to assess how safe the example is, and then to establish its type. Let us start from defining the safe level for the multiple classes.

Considering a given example x belonging to the minority class C_i its safe level is defined with respect to l classes of examples in its neighborhood as:

$$safe(x_{C_i}) = \frac{\sum_{j=1}^{l} n_{C_j} \mu_{ij}}{k} \tag{3}$$

where μ_{ij} is a degree of similarity, n_{C_j} is a number of examples from class C_j inside the considered neighborhood of x and k is a total number of neighbors. The general interpretation of the safe level of the example is as follows: the lower the value, the more unsafe (difficult) is the example.

The safe levels could be exploited in two ways: either as the direct value, or by transforming the continuous levels into discrete intervals corresponding to types of example (as done in [27]). In Sect. 6 we will show how to use safe levels in SOUP preprocessing methods and how types of the examples are used inside BRACID rule induction algorithm.

In Table 1 we present some of experimental results from [22], which show that the recognition of minority classes is related to their average safe levels.

Table 1. Sensitivity of minority classes for three classifiers and average safe levels in these classes for new-thyroid, ecoli and cleveland datasets

	CART			NBayes			3NN			Average safe level		
	Min1	Min2	Min3	Min1	Min2	Min3	Min1	Min2	Min3	Min1	Min2	Min3
NT	0.94	0.83		0.94	0.86		0.71	0.89		0.77	0.78	
EC	0.60	0.85	0.78	0.68	0.30	0.90	0.48	0.75	0.84	0.57	0.91	0.82
CL	0.28	0.11	0.07	0.14	0.25	0.15	0.08	0.00	0.00	0.29	0.32	0.34

5   Discovering Split of Classes into Sub-concepts and Rare Examples

An identification of sub-concepts in the minority class is typically done by using clustering algorithms. Nearly all approaches exploit k-means algorithms which are run on examples of a single class, without analyzing their relation to remaining classes. Japkowicz et al. showed how the discovered clusters in both minority and majority classes could be used for random oversampling them [15]. The survey [31] covers other clustering approaches and presents their applications. The use of density algorithms such as DBCAN was considered much less frequently. However, the use of clustering algorithms for real-world datasets is still a non-trivial task, in particular tuning their parameters.

In [21], authors introduced a completely different *grid-based algorithm*, called *ImGrid*. The algorithm works in the following steps: 1) dividing the attribute space into grid cells, 2) joining similar adjacent cells taking into account their minority class distributions, 3) labeling examples according to difficulty factors, 4) forming minority sub-clusters.

The number of cells and the division of the attribute range into a number of intervals are estimated with a special heuristics [21]. The cells of the grid are joined based on example distributions, where each cell should contain enough examples, and by means of the statistical tests for the comparison of two discrete distributions. For binary classes it is done with *Barnard's test*. Clusters are formed after joining several cells. Each cluster is assigned one of four difficulty labels: safe, borderline, rare, or outlier, following rules developed by Napierala in [27]. To sum up, unlike other clustering ImGrid simultaneously does two things: detects clusters and categorizes them. More precisely it detects minority sub-clusters, outliers, rare cases, and class overlapping in binary imbalanced data. Furthermore, due to its small dependency on parameter tuning, ImGrid could be used to analyze real world datasets easier than previous algorithms.

Recently, it was generalized for multiple classes [16]. The main changes are the following. A special variant of Pearson χ^2 test (inspired by the solution from ChiMerge discretization) is used to evaluate similarities of class distributions in adjacent cells. Moreover, new heuristics for ordering cells are introduced in order to get larger clusters. Then, the new rules for labelling clustered cells were introduced as a multiple class generalization of intervals over the safe level, which were earlier considered in [27]. They are better suited for handling overlapping between several classes and identifying rare cases and outliers.

The multiple class ImGrid was validated on 12 synthetic datasets showing its ability to re-discover a structure of three classes.

6   Multi-class Hybrid Resampling Algorithm SOUP

Following a critical discussion of earlier resampling multi-class techniques, such as Global-CS or Static-SMOTE, the authors of [11] introduced a new method called called *Similarity Oversampling and Undersampling Preprocessing* (abbrev.

SOUP), which combines undersampling with oversampling and exploits the information about the difficulty of examples according to their safe levels.

The authors of SOUP decided that, all majority classes are undersampled and all minority classes are oversampled to the cardinality being the median of the sizes of the biggest minority and the smallest majority class. The resulting resampled dataset has a balanced class distribution, but also with a reasonable size, which is not present in other multi-class resampling methods.

The resampling is done following information on the safe levels of the examples presented in Sect. 4. The undersampling of the majority classes is performed by removing the most unsafe examples. It means that it removes the examples located near the boundaries with minority classes or inside their regions. On the other hand, the oversampling of minority classes is performed in the opposite direction, i.e. the safest examples are duplicated as firsts, enhancing the representation of clear minority concepts.

As the safe level of a particular example in the final distribution is changing while performing consecutive steps of resampling, the classes are ordered. Undersampling majority classes is done from the biggest to the smallest one while the minority classes are oversampled from the smallest to the biggest one. Moreover after each resampling step safe levels of all examples are recomputed.

The experiments [11] showed that SOUP outperformed baseline classifiers and Static-SMOTE and Global-CS – the two popular pre-processing methods for multi-class imbalances. Moreover SOUP is slightly better then OVO with re-sampling and competitive to MRBBag (discussed in Sect. 8). Selected results from [11] for using J.8 trees are presented in Table 2.

Table 2. Comparison of specialized multi-class methods vs. SOUP and multiple RBBagging – with respect to G-mean for selected real-world data sets

Dataset	Baseline tree	Global CS	Static SMOTE	OVA Oversam.	OVO Oversam.	SOUP	mRBBag
balancescale	0.0	0.340	0.080	0.302	0.526	0.614	0.683
car	0.847	0.940	0.897	0.184	0.939	0.938	0.917
cleveland	0.000	0.000	0.032	0.287	0.288	0.256	0.155
cmc	0.483	0.478	0.452	0.510	0.509	0.520	0.517
dermatology	0.945	0.952	.927	0.082	0.921	0.960	0.960
ecoli	0.728	0.719	0.738	0.000	0.805	0.0.721	0.768
flare	0.446	0.570	0.431	0.000	0.544	0.575	0.546
glass	0.625	0.715	0.699	0.000	0.698	0.667	0.405
led7digit	0.785	0.770	0.756	0.120	0.779	0.790	0.778
vehicle	0.912	0.912	0.915	0.164	0.923	0.909	0.943
winequality	0.421	0.464	0.356	0.456	0.492	0.448	0.525

7   Multi-class Variant of BRACID Algorithm

7.1   Rule Induction from Binary Imbalanced Data with BRACID

Although induction of rules from examples is one of the well studied tasks in machine learning, rule-based classifiers have not been studied in the context of imbalanced data as intensively as other algorithms. A fairly small number of rule classifiers specialized for imbalanced data has been introduced so far, for their review see e.g. [26]. BRACID (the acronym of Bottom-up induction of Rules And Cases for Imbalanced Data) is the most accurate of these algorithms,

To handle the data difficulty factors, the authors of BRACID [26] decided to use a *hybrid representation of rules and single instances*, where more general rules cover larger, homogeneous regions with more examples and instances should handle non-linear class borders and rare minority cases or outliers. The rules are induced in a special *Bottom-up rule sequential process*. It starts from the set of the most specific rule (single, seed learning examples) and in the next iteration it tries to generalize its condition in the direction of the nearest neighbour example from the same class, provided that it does not decrease the classification abilities of the whole rule set evaluated with measures specific for imbalanced data.

An exploitation of *types of difficulty* of learning examples estimated by analysis the k-nearest neighborhood of seed examples is one of the main features of BRACID. The difficult type [27] assigned to each seed example influences the rule generalization, as for the unsafe minority example it is possible to generate additional rules covering it. As a result, the number of minority class rules, as well as their support, are increased and they are more likely to win with the stronger majority rules while classifying new instances. For details see [26].

7.2   Generalizing BRACID for Multiple Imbalanced Classes

As BRACID was proposed for binary classes only. In a recent paper [25] two ways of its generalizations for multiple classes were studied: (1) exploitation of binary decomposition ensemble frameworks OVO and using the original binary BRACID within them; (2) generalization of BRACID with a new scheme for inducing a single set of rules from all multiple classes.

The second generalization partly follows a typical sequential schema for an iterative induction of rules from successive classes. In each iteration, for each class the temporary training dataset is constructed. It contains positive examples from the considered class and the negative examples from all other classes (similar to the OVA approach). BRACID algorithm is run on such data and only rules describing the considered class are added to the final set of rules, while the other class rules are discarded. At the end, the complete set contains rules from all classes. An important modification of rule generalization takes into types of classes, i.e. whether the positive class is a minority or majority one. More precisely when the majority class is considered then (1) the internal k-nearest neighbor generalization is done to a single nearest example for *safe* seed examples and (2) to one, best of rules induced by generalization to k nearest examples for

Table 3. Comparison of rule classifiers – PART, OVO Bracid and multiple BRACID – with respect to G-mean for selected real-world data sets.

Dataset	PART	OVO-B	m-BRACID
Balancescale	0.3136	0.4086	0.6789
Car	0.7925	0.9022	0.9004
Cleveland	0.0597	0.1750	0.2322
cmc	0.4431	0.4897	0.4691
Dermatology	0.8943	0.9204	0.9082
Ecoli	0.6373	0.7404	0.7976
Flare	0.1716	0.3904	0.4639
Glass	0.3174	0.1883	0.4256
Led7digit	0.7918	0.7736	0.7713
Vehicle	0.9147	0.9221	0.9323
Winequality	0.2917	0.4529	0.5338

unsafe example. For the minority class it is unchanged. This modification limits the number of produced rules for majority classes.

In [25] experiments on similar multi-class datasets as [11] were done. Their results show that this generalization of BRACID is better than the adaptive using of the binary BRACID within OVO ensemble, both with respect to higher predictive abilities and the number of rules. Some of these comparative results are shown in Table 3. In case of producing still too many rules they can be post-pruned with the special weighted coverage algorithm [28].

8   Multi-class Extension of Bagging Ensemble

Generalizations of bagging ensembles are quite effective for binary imbalanced data. Lango et al. studied in [23] *Roughly Balanced Bagging*, which is one of the most efficient under-sampling bagging for binary imbalanced classes and it often works better than generalizations of boosting. It exploits a random *under-sampling* before generating component classifiers, which reduces the presence of the majority class examples inside each bootstrap sample of the finally constructed bagging. The random number of majority examples to be sampled to the bootstrap is estimated according to the negative *binomial distribution*, while the number of sampled minority examples is equal to the size of the minority class inside the original training dataset. Finally, these numbers of examples are sampled from each class with replacement and predictions of the learned based classifiers are aggregated with the majority voting.

Lango and Stefanowski proposed in [23] its generalization to Multi-class Roughly Balanced Bagging (further abbreviated as MRBBag). The main modification concerns a construction of bootstrap samples, which is realized in the following way. The number of examples to be sampled from each class to the bootstrap is estimated from the multinomial distribution, which is defined by the following probability mass function:

$$p(n_1, n_2, ..., n_c) = \frac{n!}{n_1! n_2! \cdots n_c!} p_1^{n_1} p_2^{n_2} \cdots p_c^{n_c}$$

where $p_1, p_2, ..., p_c$ and $n = \sum_{i=1}^{c} n_i$ are the parameters of the distribution.

The authors handle the multi-class imbalance problem by obtaining roughly balanced bootstrap samples also with respect to class probabilities, so they fix values $p_1, p_2, ..., p_c$ to the same constant value equal to $\frac{1}{c}$, such that $\sum_{i=1}^{c} p_i = 1$. This parameterizes the upper formula. After learning component, base classifier the final decision of the classifier is constructed by the majority voting. For the pseudocode of this algorithm see [23].

In [23] MRBBag algorithm, constructed with J4.8 trees, was evaluated on several artificial and UCI real-life imbalanced datasets. It outperformed other tree and general ensemble classifiers with respect to G-mean and averaged F1-score (both adapted for the multi-class evaluation). Refer also to its performance in Table 2. Moreover MRBBag was further extended to deal with feature selection for highly dimensional data, see details in [23]. This variant was successfully applied to solve the task of categorization of twitter short text messages [20].

9   Software Implementations of Specialized Algorithms for Multi-class Imbalanced Data

The methods for dealing with binary imbalanced data are already implemented in various software libraries. The representatives are: `imbalanced-learn` with scikit-learn in Python, KEEL, WEKA and its extensions in Java or several R libraries such as IRIC or ClimbR. In case of methods for multi-class imbalanced data there are nearly no public available software implementations. In the past year two open source software kits were proposed: `multi-imbalance` Python library [9], Matlab toolkit Multiple-imbalance [34].

The first library was developed by the author's co-operators and it implements state-of-the-art approaches for multi-class imbalanced problems, which are divided into three general categories: (1) binary decomposition approaches (OVO, OVA and ECOC), (2) specialized pre-processing (Global-CS, Mahalanobis Distance Oversampling (MDO), Static-SMOTE, SPIDER3 and SOUP), and (3) other ensembles (MRBagging and SOUP-bagging). So it covers methods discussed in this paper.

On the other hand, the Matlab toolkit contains 18 methods, mainly variants of Adaboost or ECOC and specialized tree classifiers.

10   Future Research Directions and Conclusions

Looking at the current literature, we could expect that many new methods will still be proposed to improve the classification of imbalanced data, including multi-class variants. It is hoped, however, that these expected proposals will go beyond simple adaptations of known approaches or exploitation of the binary decomposition frameworks, and in particular they will take advantage of the aforementioned data difficulty factors. Below some personal opinions are expressed as to the research directions.

Cost sensitive learning for multiple classes should estimate misclassification costs for each example also with their difficulty levels. The current proposals are too much oriented to global imbalance ratios.

In terms of further research on data difficulties, it is necessary to more carefully explore the differences in the impact of overlapping between different class types and in the context of different class size configurations. In particular, this applies to a more detailed analysis of the intermediate classes in the so-called gradual configurations that appear to be more difficult than configurations with sharp changes of class size between minority and majority ones.

In particular the role of rare examples for many classes, which previously had a large impact on deteriorating the classification of imbalanced binary data, has not been sufficiently studied for multiple classes yet.

New preprocessing methods should be developed for better dealing with overlapping between various classes as they are more critical than in the binary problems. It should also be assessed to what extent changes (e.g. by resampling) in the size of overlapping classes will affect the recognition of other classes.

Other approaches for discovery sub-concepts in multiple classes could be still studied, in particular with exploiting density based clustering.

Deeper research on specialized artificial neural networks should be undertaken. The current few studies are too focused on including random re-sampling or relatively simple modifications of the optimized loss function. This is desirable given the current strong interest in image recognition or natural language processing using deep neural networks.

An open question concerns multi-class and highly dimensional datasets. Feature random sampling does not take into account internal relations between classes. Furthermore, more research is needed on the specialized construction of new features, projections of the original ones into new representation space, like in embedded representations in deep networks or similarity learning.

Nearly all current research were done on static multi-class imbalanced data. On the other hand data streams with concept drifts occur in many modern applications of Big Data [12]. They are naturally imbalanced and the global imbalance ratio may vary over time. However, the data factors such as class split into factors, overlapping or presence of rare case may also change (similarly to typical drifts) and their drifts are definitely local as it was recently shown for binary imbalanced streams [3]. Their experiments demonstrated that these drifts deteriorate predictive performance of popular stream classifiers and posed needs for the developments of new specialized online algorithms. However such studies should be done with more complex multiple classifiers. Furthermore new online clustering algorithms for detection of the class split, their changes over time and appearance of new classes in the streams are necessary.

References

1. Abdi, L., Hashemi, S.: To combat multi-class imbalanced problems by means of over-sampling techniques. IEEE Trans. Knowl. Data Eng. **28**(1), 238–251 (2016)
2. Agrawal, A., Herna, L.V., Paquet, E.: SCUT: multi-class imbalanced data classification using SMOTE and cluster-based undersampling. In: International Joint Conference on Knowledge Discovery, Knowledge Engineering and Knowledge Management (IC3K), vol. 01, pp. 226–234 (2015)
3. Brzezinski, D., Minku, L.L., Pewinski, T., Stefanowski, J., Szumaczuk, A.: The impact of data difficulty factors on classification of imbalanced and concept drifting data streams. Knowl. Inf. Syst. **63**(6), 1429–1469 (2021)
4. Buda, M., Maki, A., Mazurowski, M.A.: A systematic study of the class imbalance problem in convolutional neural networks. CoRR abs/1710.05381 (2017)
5. Fernández, A., García, S., Galar, M., Prati, R.C., Krawczyk, B., Herrera, F.: Learning from Imbalanced Data Sets. Springer, Heidelberg (2018)
6. Fernández, A., López, V., Galar, M., Del Jesus, M.J., Herrera, F.: Analysing the classification of imbalanced data-sets with multiple classes: binarization techniques and ad-hoc approaches. Knowl. Based Syst. **42**, 97–110 (2013)
7. Galar, M., Fernández, A., Barrenechea, E., Sola, H., Herrera, F.: An overview of ensemble methods for binary classifiers in multi-class problems: experimental study on one-vs-one and one-vs-all schemes. Pattern Recogn. **44**, 1761–1776 (2011)
8. Garcia, V., Sanchez, J., Mollineda, R.: An empirical study of the behaviour of classifiers on imbalanced and overlapped data sets. In: Rueda, L., Mery, D., Kittler, J. (eds) Progress in Pattern Recognition, Image Analysis and Applications. CIARP 2007. Lecture Notes in Computer Science, **4756**, 397–406. Springer, Berlin, Heidelberg (2007). https://doi.org/10.1007/978-3-540-76725-1_42
9. Grycza, J., Horna, D., Klimczak, H., Lango, M., Plucinski, K., Stefanowski, J.: multi-imbalance: open source python toolbox for multi-class imbalanced classification. In: Dong, Y., Ifrim, G., Mladenić, D., Saunders C., Van Hoecke S. (eds) Machine Learning and Knowledge Discovery in Databases. Applied Data Science and Demo Track - European Conference, ECML PKDD, Proceedings, Part V. Lecture Notes in Computer Science, **12461**, 546–549. Springer, Cham (2020). https://doi.org/10.1007/978-3-030-67670-4_36
10. He, H., Ma, Y.: Imbalanced Learning: Foundations, Algorithms, and Applications. Wiley-IEEE Press, Hoboken (2013)
11. Janicka, M., Lango, M., Stefanowski, J.: Using information on class interrelations to improve classification of multiclass imbalanced data: a new resampling algorithm. Int. J. Appl. Math. Comput. Sci. **29**, 769–781 (2019)
12. Japkowicz, N., Stefanowski, J.: A machine learning perspective on big data analysis. In: Japkowicz, N., Stefanowski, J. (eds) Big Data Analysis: New Algorithms for a New Society. Studies in Big Data, **16**, 1–31. Springer, Cham (2016). https://doi.org/10.1007/978-3-319-26989-4_1
13. Japkowicz, N., Stephen, S.: The class imbalance problem: a systematic study. Intell. Data Anal. **6**(5), 429–449 (2002)
14. Jelonek, J., Stefanowski, J.: Experiments on solving multiclass learning problems by n2-classifier. In: Nédellec, C., Rouveirol, C. (eds) Machine Learning: ECML-1998. ECML 1998. Lecture Notes in Computer Science (Lecture Notes in Artificial Intelligence), **1398**. LNCS(LNAI), 172–177. Springer, Berlin, Heidelberg (1998). https://doi.org/10.1007/BFb0026687

15. Jo, T., Japkowicz, N.: Class imbalances versus small disjuncts. SIGKDD Explor. **6**(1), 40–49 (2004)
16. Kocur, Z.: Clustering algorithm for multi-class imbalanced data to improve classification quality. Ph.D. thesis, Poznan University of Technology (2020)
17. Krawczyk, B.: Learning from imbalanced data: open challenges and future directions. Prog. Artif. Intell. **5**(4), 221–232 (2016)
18. Kubat, M., Matwin, S.: Addressing the curse of imbalanced training sets: one-side selection. In: Proceedings of the 14th International Conference on Machine Learning ICML-1997, pp. 179–186 (1997)
19. Kuncheva, L.: Combining Pattern Classifiers. Methods and Algorithms, 2nd edn. Wiley, Hoboken (2014)
20. Lango, M.: Tackling the problem of class imbalance in multi-class sentiment classification: an experimental study. Found. Comput. Decis. Sci. **44**, 151–178 (2019)
21. Lango, M., Brzezinski, D., Firlik, S., Stefanowski, J.: Discovering minority sub-clusters and local difficulty factors from imbalanced data. In: Discovery Science - 20th International Conference, DS 2017, Proceedings, pp. 324–339 (2017)
22. Lango, M., Napierała, K., Stefanowski, J.: Evaluating difficulty of multi-class imbalanced data. In: Proceedings of 23rd International Symposium on Methodologies for Intelligent Systems, pp. 312–322 (2017)
23. Lango, M., Stefanowski, J.: Multi-class and feature selection extensions of roughly balanced bagging for imbalanced data. J. Intell. Inf. Syst. **50**(1), 97–127 (2018)
24. Lango, M., Stefanowski, J.: What makes multi-class imbalanced problems difficult? (2021). (manuscript under review)
25. Naklicka, M., Stefanowski, J.: Two ways of extending Bracid rule-based classifiers for multi-class imbalanced data. In: Nuno, M., Paula, B., Luis, T., Nathalie, J., Michal, W., Shuo, W. (eds) Proceedings of the Third International Workshop on Learning with Imbalanced Domains: Theory and Applications, co-located with ECML–PKDD 2012, Proceedings of Machine Learning Research (2021)
26. Napierala, K., Stefanowski, J.: BRACID: a comprehensive approach to learning rules from imbalanced data. J. Intell. Inf. Syst. **39**(2), 335–373 (2012)
27. Napierała, K., Stefanowski, J.: Types of minority class examples and their influence on learning classifiers from imbalanced data. J. Intell. Inf. Syst. **46**, 563–597 (2016)
28. Napierala, K., Stefanowski, J., Szczech, I.: Increasing the interpretability of rules induced from imbalanced data by using Bayesian confirmation measures. In: Appice, A., Ceci, M., Loglisci, C., Masciari, E., Raś, Z. (eds) New Frontiers in Mining Complex Patterns. NFMCP 2016. Lecture Notes in Computer Science, **1031**84–98. Springer, Cham (2016). https://doi.org/10.1007/978-3-319-61461-8_6
29. Prati, R., Batista, G., Monard, M.: Class imbalance versus class overlapping: an analysis of a learning system behavior. In: Proceedings of 3rd Mexican International Conference on Artificial Intelligence, pp. 312–321 (2004)
30. Seaz, J., Krawczyk, B., Wozniak, M.: Analyzing the oversampling of different classes and types in multi-class imbalanced data. Pattern Recogn. **57**, 164–178 (2016)
31. Stefanowski, J.: Dealing with data difficulty factors while learning from imbalanced data. In: Mielniczuk, J., Matwin, S. (eds.) Challenges in Computational Statistics and Data Mining, **605**, 333–363. Springer, Cham (2016). https://doi.org/10.1007/978-3-319-18781-5_17
32. Wang, S., Yao, X.: Mutliclass imbalance problems: analysis and and potential solutions. IEEE Trans System Man Cybern. Part B. **42**(4), 1119–1130 (2012)

33. Wojciechowski, S., Wilk, S., Stefanowski, J.: An algorithm for selective preprocessing of multi-class imbalanced data. In: Proceedings of the 10th International Conference on Computer Recognition Systems CORES 2017, Polanica Zdroj. Advances in Intelligent Systems and Computing, vol. 578, pp. 238–247 (2017)
34. Zhang, C., Bi, J., Xu, S., Ramentol, E., Fan, G., Qiao, B., Fujita, H.: Multi-imbalance: an open-source software for multi-class imbalance learning. Knowl. Based Syst. **174**, 137–143 (2019)
35. Zhou, Z.H., Liu, X.Y.: On multi-class cost sensitive learning. Comput. Intell. **26**(3), 232–257 (2010)

Core Rough Set Models and Methods

General Rough Modeling of Cluster Analysis

A. Mani$^{(\boxtimes)}$

Indian Statistical Institute, Kolkata 203, B. T. Road, Kolkata 700108, India
amani.rough@isical.ac.in
https://www.logicamani.in

Abstract. In this research a general theoretical framework for clustering is proposed over specific partial algebraic systems by the present author. Her theory helps in isolating minimal assumptions necessary for different concepts of clustering information in any form to be realized in a situation (and therefore in a semantics). *It is well-known that of the limited number of proofs in the theory of hard and soft clustering that are known to exist, most involve statistical assumptions.* Many methods seem to work because they seem to work in specific empirical practice. A new general rough method of analyzing clusterings is invented, and this opens the subject to clearer conceptions and contamination-free theoretical proofs. Numeric ideas of validation are also proposed to be replaced by those based on general rough approximation. The essential approach is explained in brief and supported by an example.

Keywords: Cluster validation · Clustering frameworks · General rough sets · Mereology · Contamination problem · Ontology · Axiomatic granular computing

1 Introduction

Hard and soft clustering processes are based on ideas of optimization that contribute to uncertainty, vagueness and indeterminacy in associated proofs and measures. Often convergence of algorithms or cluster validity are not proven, and the ones proved proceed from questionable statistical and topological assumptions [1,2] about the context associated with a dataset. To see this consider the problem of clustering six objects into two clusters. Among other things, certain possible clusters may not be reasonable. In this situation, what does it mean to consider the collection of all possible clusters relative to a purely combinatorial perspective? This is just one of the ways in which statistical proofs may lose universality and relevance. Aggregation operations are also known to become paradoxical in the context of statistical tests and decision theory [3]. The level of context dependency in the use of clustering techniques in the AI (and ML) literature is severe – an important heuristic is to try every technique that is known to work in related application contexts (or use cases). This scenario suggests that it can be useful to build a minimalist framework

© Springer Nature Switzerland AG 2021
S. Ramanna et al. (Eds.): IJCRS 2021, LNAI 12872, pp. 75–82, 2021.
https://doi.org/10.1007/978-3-030-87334-9_6

for exploring proofs, ontology, and associated methodology (and also because application contexts are loaded with excess baggage).

The basic problem is of developing a reasonable framework with its assumptions and not of a language of expression (though the former requires some of the latter). Metric logics and variants (see for example [4]) in particular, cannot add much to the ideas of validation for distance-based clustering because they merely intend to express the same facts and methods in a restricted language. Higher order rough frameworks for analyzing soft and hard clustering are proposed in this research over axiomatic granular rough sets as they are far more capable of handling knowledge evolution. New concepts of clustering (including a definition) are also proposed.

The essence of the introduced frameworks revolve around the following ideas:

A A language (or model) that can express clustering related information should include at least one ternary predicate δ (with δabc interpreted as a is closer to b than to c in some sense).

B Since all kinds of clustering involve approximations of some kind (that may possibly be ontologically justified) in their definition, computation or validation, it is necessary to permit generalized approximations that are associated with the intrinsic structure of the data.

C It is important to avoid making wild external numeric approximations about the data, and contaminating it [5,6]. This is also about the data being able to speak by itself.

D Granular approximations (in the axiomatic sense) are better suited to handle meaning and evolution of knowledge (but a number of rough approximations may get excluded by restricting to higher granular operator spaces/partial algebras).

E Concepts of aggregation, and commonality can be partial (as they are in real life). It does not always make sense to combine objects or concepts for example. Further they may be unrelated to fundamental part-of relations in the context.

F While numeric valuations are best avoided, in some contexts they can be meaningful. Frameworks should be able to handle this possibility.

Points **D**, **E**, **F** are already considered in previous work of the present author [5,6] in the context of higher granular operator spaces/partial algebras and variants. These will be used with minimal explanation.

1.1 Background

The reader is expected to be familiar with the literature on abstract granular approaches to general rough sets [5,6], and some mereology [5,7,8]. Further concepts such as those of higher granular operator spaces/partial algebras, contamination and admissible granulation will be assumed.

The process or concept of *cluster validation* generally refers to exploring the quality of one or more clustering methods and possibly comparing them. In almost all cases, true class information is not available (that is if one avoids looking at anything apart from the dataset) and validation methods are inherently not rigorous even in comparison to statistical ones used in supervised learning. Further they are subjective, highly contextual, are not generalizable, and *assume some heuristics that are not well understood and in some cases even the values produced may not be clear (see* [9] *for example).*

Rough clustering refers to clustering methods that are enhanced with rough set theoretical ideas of what a rough cluster ought to be. These may sometimes be seen as a two layered process in which the last layer is about interpreting membership from a rough perspective. A rough cluster is seen as a pair consisting of a lower and upper approximation of an object or as some other representation of a rough object [10–13]. The present research though related, is about building a general rough framework for all clustering contexts, and differs in purpose and methods.

2 New Rough Semantic Approaches

Let $S = \langle \underline{S}, \Sigma^{\underline{S}} \rangle$ be a partial algebraic system over a signature Σ (the superscript indicates interpretation on $\underline{S}$). For any subset Σ_o of Σ, $S_o = \langle \underline{S}, \Sigma_o^{\underline{S}} \rangle$ will be referred to as a *reduct* of S. For two terms s, t, $s \overset{\omega}{=} t$ shall mean, if both sides are defined then the two terms are equal (the quantification is implicit), while $s \overset{\omega^*}{=}$ t shall mean if either side is defined, then the other is and the two sides are equal (the quantification is implicit). *As points can be regarded as singletons, the point-subset distinction that is explicitly assumed in clustering theory can be discarded.* That is the partial algebraic system can include all. Apart from parthood, ternary predicates of the form δ that satisfy some of the inner (i-coh) and near (n-coh) coherence conditions below are of interest:

$$(\forall a, b)\, \delta bba \qquad \text{(i-coh)}$$
$$(\forall a, b, c)\, (\delta abc \longrightarrow \delta bac) \qquad \text{(n-coh)}$$
$$(\forall a, b)\, \neg\delta abb \qquad \text{(i-coh-2)}$$
$$(\forall a, b, c)\, (\delta abc \longrightarrow \neg\delta acb) \qquad \text{(strict n-coh)}$$
$$(\forall a, b, c)\, (\delta abc\ \&\ \delta aeb \longrightarrow \neg\delta aec) \qquad \text{(trans-1)}$$

Intended meanings of δabc are a *is closer to* b *than* c *in some sense,* a *is more similar to* b *than* c *in some sense* and variants thereof. This predicate covers the intent of using metrics, similarities, dissimilarities, proximities, descriptive proximities, kernels and other functions for the purpose.

For an arbitrary subset K of a set H to qualify as a cluster (under a large number of additional constraints), many conditions that depend on δ (essentially) are typically required to be satisfied. These can be written with restricted quantification as in $(\forall a, b \in K)(\forall c \in K^c)\,\delta abc$. But such an expression is neither elegant nor general enough or useful from a logical perspective. A better strategy is to identify clusters with a unary predicate κ in H and express such conditions with an additional binary *part of* predicate **P**.

In describing clustering contexts or processes, it may not always be reasonable to combine objects (or groups formed at some step) into a new object, or their combination may be regarded as a plural object (like some collection of subsets of a union [5]) and it can also happen that parthood does not define the mereological sum always [7]. Such a partial sum operation $\oplus$ can be expected to satisfy all of the following properties:

$$a \oplus b \stackrel{\omega^*}{=} b \oplus a;\; a \oplus a \stackrel{\omega}{=} a;\; a \oplus (b \oplus c) \stackrel{\omega}{=} (a \oplus b) \oplus c \quad (\omega^* - \text{com}; \omega - \text{id}; \omega - \text{asso})$$

$$\delta abc \longrightarrow \delta(a \oplus a)bc;\; \delta abc \longrightarrow \delta a(b \oplus b)c;\; \delta abc \longrightarrow \delta ab(c \oplus c) \quad (\delta - \text{sum1}; \delta - \text{sum2}; \delta - \text{sum3})$$

In general clustering contexts, it is often the case that an additional external ordered algebraic system is used to measure (or evaluate) ideas of nearness or proximities (in a descriptive, spatial or generalized metric sense [14]). These may be partially ordered sets or even the semi-ring of positive reals. A minimalist structure that is always present (as an algebraic reduct) is a parthood space [15] in which it is possible to express some idea of comparison (in a perspective of containment) that may not necessarily be transitive.

The definition of δ can be made explicit with the help of additional maps $f : S^2 \longmapsto H$ (this may not be required as shown in the example below 2) that satisfies one or more of

$$\delta abc \longrightarrow \mathbf{P}f(a, b)f(a, c);\; \mathbf{P}f(a, b)f(a, c) \longrightarrow \delta abc;\; \delta abc \longleftrightarrow \mathbf{P}f(a, b)f(a, c) \quad (\text{def1}; \text{def2}; \text{def0})$$

In some cases as in clustering contexts that depend on a metric, the stronger version def0 actually holds.

Given the above concepts, it is possible to specify a number of formula using higher order constructs that correspond to definitions of clusters. A specific general form is (with A^* being a subobject of A and A° being a generalized complement)

$$(\forall a \in A)(\forall b \in A^*)(\forall c \in A^\circ)\,\delta abc \qquad \text{(clue)}$$

It is not the most general form because the dependence on specific clusters like A takes other forms. Under additional conditions, the following form has the potential to specify clusters.

$$(\forall a \in A)(\forall b \in B)(\forall c \in E)\,\delta abc \qquad \text{(gclue)}$$

B in particular can depend on A, E and possible ideas of being an A. The dependence between B, E and A might also be of a higher order nature in that the concept of a cluster is in relation to the collection of all clusters. Further additional properties may be satisfied by the collection of all clusters. The former can all be expressed with the help of an additional unary operation $\kappa : H \longmapsto H$ that enables identification of cluster members.

Based on the above considerations, the following partial algebraic systems appears to be optimal for theoretical studies on clustering in the mentioned perspective.

Definition 1. *A partial algebraic system of the form* $S = \langle \underline{S}, \Sigma^{\underline{S}} \rangle$ *with* $\Sigma = \{P, \delta, \oplus, \kappa, \leqslant, \vee, \wedge, l, u, \top, \bot\}$ *of type* $(2, 3, 2, 1, 2, 2, 2, 2, 1, 1, 0, 0)$ l, u *being operators* $: \underline{S} \longmapsto \underline{S}$ *satisfying the following* ($\underline{S}$ *is replaced with* S *if clear from the context.* $\vee$ *and* $\wedge$ *are idempotent partial operations,* κ *a unary predicate for identifying clusters, and* **P** *is a binary parthood predicate) will be referred to as a* minimal soft clustering system *(MSS) whenever the conditions* i-coh, n-coh, i-coh-2, trans1, clos1 *and* PT1, PT2, G1, G2, G3, G4, G5, UL1, UL2, UL3, TB *of the definition of high granular operator space (GGS)* [5] *hold:*

$$(\forall x)Pxx; \ (\forall x, b)(Pxb \ \& \ Pbx \longrightarrow x = b) \qquad \text{(PT1; PT2)}$$

$$a \vee b \stackrel{\omega}{=} b \vee a; \ a \wedge b \stackrel{\omega}{=} b \wedge a; \ (a \vee b) \wedge a \stackrel{\omega}{=} a; \ (a \wedge b) \vee a \stackrel{\omega}{=} a \qquad \text{(G1; G2)}$$

$$(a \wedge b) \vee c \stackrel{\omega}{=} (a \vee c) \wedge (b \vee c); \ (a \vee b) \wedge c \stackrel{\omega}{=} (a \wedge c) \vee (b \wedge c) \qquad \text{(G3; G4)}$$

$$(a \leqslant b \leftrightarrow a \vee b = b \leftrightarrow a \wedge b = a) \qquad \text{(G5)}$$

$$(\forall a \in \mathbb{S})Pa^l a \ \& \ a^{ll} = a^l \ \& \ Pa^u a^{uu}; \ (\forall a, b \in \mathbb{S})(Pab \longrightarrow Pa^l b^l \ \& \ Pa^u b^u) \qquad \text{(UL1; UL2)}$$

$$\bot^l = \bot \ \& \ \bot^u = \bot \ \& \ PT^l\top \ \& \ PT^u\top; \ (\forall a \in \mathbb{S})P\bot a \ \& \ Pa\top \qquad \text{(UL3; TB)}$$

In the context of the above definition, if the condition strict n-coh (lclu) is also satisfied, then the MSS will be referred to as *strict* (*rough*).

$$(\forall a, b, c)(\delta abc \longrightarrow \neg \delta acb); \ (\forall a)(\kappa a \longrightarrow \kappa a^l) \qquad \text{(strict n-coh; lclu)}$$

If the signature in Definition 1 is $\Sigma^* = \{P, \gamma, \delta, \oplus, \kappa, \leqslant, \vee, \wedge, l, u, \top, \bot\}$ instead and the additional conditions on the granulation of a GGS [5] are also satisfied, then the resulting system will be referred to as a *granular MSS* (GMSS). Granularity in this sense is essential for construction of knowledge as in [5].

The conditions defining admissible granulations mean that every approximation is representable by granules in an algebraic way, that every granule coincides with its lower approximation (granules are lower definite), and that all pairs of distinct granules are part of definite objects (those that coincide with their own lower and upper approximations).

Internal measures that characterize the quality of clusters in terms of deviance from associated approximations (and therefore lack of coherence) in a GMSS $\mathbb{H}$ in which $\vee$, and $\wedge$ are set union and intersection respectively and $\setminus$ is a partial set difference operation are defined below:

Definition 2. *In the GMSS $\mathbb{H}$ mentioned, let $\mathcal{C}$ be a clustering on $\top$. then the* lower deficit $(C^\flat)$ *of a cluster* $C \in \mathcal{C}$ *will be the set* $(C \setminus C^l)^u$ *(if defined), and its* upper deficit $(C^\eth)$ *will be the set* $(C^u \setminus C)^u$ *(if defined). Further C will be* lu-valid *iff* $c^l = c^u = C$, l-pre-valid *if and only if* $(\exists V \in \mathbb{S})V^l = C$, *and* l-traceable *if and only if* $(\exists V \in \mathbb{S})V = C^l$. *Analogous concepts of u-pre validity can be defined. In addition, if all clusters in $\mathcal{C}$ are* l-pre-valid *(resp.* lu-valid, u-pre-valid, l-traceable, u-traceable) *then $\mathcal{C}$ will itself be said to be* l-pre-valid *(resp.* lu-valid, u-pre-valid, l-traceable, u-traceable).

Proposition 1. *In the context of Definition 2, if the l-deficit (resp. u-deficit) of a cluster C is computable, then it must necessarily be l-traceable (resp. u-traceable).*

The central idea of lu-validity (and weakenings thereof) is that of representability in terms of granules and approximations. These do not test the key predicate δ for validation, and the aspect is left to the process of construction of rough approximations. By contrast, the $*$-deficits are an internal measure of what is lacking or what is in excess.

Delta Methodology

The framework of **MSS** *introduced permits evaluation of clusterings relative to parthood and approximation operations. Because the predicate (or equivalent partial operation) and approximations are constructed by the user, the system is not inherently constraining in any way.* The proposed *delta methodology* consists of the following steps:

Step-1 Define most of the MSS for the context (except for κ and δ possibly). In addition associate mereo-ontologies with the system.
Step-2 Do feature selection if required and form a reduct of the original MSS
Step-3 Compute clusters as per desired algorithm
Step-4 Either define the new clustering in the MSS or form a new MSS with the clustering
Step-5 Investigate through minimal additional assumptions and possible definitions of δ.

Delta Methodology: Example

For this example, the reader needs to refer to Sect. 6.3.3 of [6] by the present author. Let $H = \{x_1, x_2, x_3, x_4\}$ and T a tolerance on it generated by $\{(x_1, x_2), (x_2, x_3)\}$. Denoting the statement that the granule generated by x_1 is (x_1, x_2) by $(x_1 : x_2)$, let the granules be the set of predecessor neighborhoods: $\mathcal{G} = \{(x_1 : x_2), (x_2 : x_1, x_3), (x_3 : x_2), (x_4 :)\}$. The different approximations (lower (l), upper (u) and bited upper (u_b)) are then as in [6] (the symbols are changed here, and $\sim$ is an equivalence on $\wp(H)$ defined by $A \sim B$ if and only if $A^l = B^l$ & $A^{u_b} = B^{u_b}$. Now let, $\underline{S} = \wp(H)$, $\mathbf{P} = \subseteq = \leqslant$, and consider the following possible definitions of δ:

$$\delta abc \text{ if and only if } \mathbf{P}(a \cup b)(a \cup c) \ \& \ \neg\mathbf{P}(a \cup c)(a \cup b) \tag{E1}$$

$$\delta abc \text{ if and only if } \mathbf{P}(a \cap c)^l(a \cap b)^l \ \& \ \neg\mathbf{P}(a \cap b)^l(a \cap c)^l \tag{E2}$$

$$\delta abc \text{ if and only if } \mathbf{P}(a \cup b)^u(a \cup c)^u \tag{uE1}$$

$$\delta abc \text{ if and only if } \mathbf{P}(a \cup b)(a \cup c) \tag{E0}$$

Suppose some clustering technique produces the clustering $\mathcal{C}$ defined by: $\mathcal{C} = \{\{x_1, x_3\}, \{x_2, x_3\}, \{x_2, x_4\}\}$. *Then one can see that it is compatible with definitions E1 and E0. Further the lower and upper deficits of the* $\{x_2, x_4\}$ *are* $\{x_1, x_2, x_3\}$ *and*

$\{x_1, x_2, x_3\}$ respectively. Note that the granulation $\mathcal{G}$ can also be seen as a clustering relative to uE1. In the context, the MSS is $S = \langle \underline{S}, \subseteq, \delta, \oplus, \kappa, \cup, \cap, l, u, H, \emptyset \rangle$, with κ being specified by $\mathcal{C}$. It is also easy to extend it to a GMSS.

Future Work: A much extended version of this research for a logic-oriented audience is under revision. This includes a detailed critique of the issues with existing cluster validation techniques. Newer methods in the context of this research based on prototypes, rationality and other criteria and novel rough clustering methods are part of forthcoming papers of the present author. A joint paper on related empirical studies in education research is in progress. Researchers and practitioners take a soft view of validation in both soft and hard clustering. The proposed framework affords a better way of formalizing the soft aspect. More work on this is obviously motivated.

Acknowledgment. This research is supported by a Women Scientist grant of the Department of Science and Technology.

References

1. Bouveyron, C., Celeux, G., Murphy, B., Raftery, A.: Model-Based Clustering and Classification for Data Science: With Applications in R. Cambridge University Press, Cambridge (2019)
2. Hennig, C., Meila, M., Murtagh, F., Rocci, R.: Handbook of Cluster Analysis, 1st edn. CRC Press, Boca Raton (2016). Edited Volume, Chapman and Hall
3. Haunsperger, D.: Aggregated statistical rankings are arbitrary. Soc. Choice Welf. **20**, 261–272 (2003)
4. Djordjevic, R., Ikodinovic, N., Stojanovic, N.: A propositional metric logic with fixed finite ranges. Fundamenta Informaticae **174**, 185–199 (2020)
5. Mani, A.: Comparative approaches to granularity in general rough sets. In: Bello, R., Miao, D., Falcon, R., Nakata, M., Rosete, A., Ciucci, D. (eds.) IJCRS 2020. LNCS (LNAI), vol. 12179, pp. 500–518. Springer, Cham (2020). https://doi.org/10.1007/978-3-030-52705-1_37
6. Mani, A.: Algebraic methods for granular rough sets. In: Mani, A., Cattaneo, G., Düntsch, I. (eds.) Algebraic Methods in General Rough Sets. Trends in Mathematics, pp. 157–335. Springer, Cham (2018). https://doi.org/10.1007/978-3-030-01162-8_3
7. Burkhardt, H., Seibt, J., Imaguire, G., Gerogiorgakis, S., eds.: Handbook of Mereology. Philosophia Verlag, Munich (2017)
8. Gruszczyński, R., Varzi, A.: Mereology then and now. Log. Log. Philos. **24**, 409–427 (2015)
9. Kim, M., Ramakrishna, R.S.: New indices for cluster validity assessment pattern. Pattern Recognit. Lett. **26**, 2353–2363 (2005)
10. Düntsch, I., Gediga, G.: Rough set clustering. In: Hennig, C., Meila, M., Murtagh, F. (eds.) Handbook of Cluster Analysis. CRC Press, Boca Raton, pp. 575–594 (2016)
11. Peters, G.: Rough clustering utilizing the principle of indifference. Inf. Sci. **277**, 358–374 (2014)
12. Mitra, S.: An evolutionary rough partitive clustering. Pattern Recognit. Lett. **25**(12), 1439–1449 (2004)
13. Zhou, J., Pedrycz, W., Miao, D.: Shadowed Sets in the characterization of rough-fuzzy clustering. Pattern Recognit. **44**(8), 1738–1749 (2011)

14. Concilio, A.D., Guadagni, C., Peters, J., Ramanna, S.: Descriptive proximities. properties and interplay between classical proximities and overlap. Math. Comput. Sci. **12**(1), 91–106 (2018)
15. Mani, A.: Functional extensions of knowledge representation in general rough sets. In: Bello, R., Miao, D., Falcon, R., Nakata, M., Rosete, A., Ciucci, D. (eds.) IJCRS 2020. LNCS (LNAI), vol. 12179, pp. 19–34. Springer, Cham (2020)

Possible Coverings in Incomplete Information Tables with Similarity of Values

Michinori Nakata[1(⊠)], Norio Saito[1], Hiroshi Sakai[2], and Takeshi Fujiwara[3]

[1] Faculty of Management and Information Science, Josai International University,
1 Gumyo, Togane, Chiba 283-8555, Japan
nakatam@ieee.org, saitoh_norio@jiu.ac.jp

[2] Department of Mathematics and Computer Aided Sciences, Faculty of Engineering,
Kyushu Institute of Technology, Tobata, Kitakyushu 804-8550, Japan
sakai@mns.kyutech.ac.jp

[3] Faculty of Informatics, Tokyo University of Information Sciences, 4-1 Onaridai,
Wakaba-ku, Chiba 265-8501, Japan
fujiwara@rsch.tuis.ac.jp

Abstract. Rough sets are described by an approach using possible coverings in an incomplete information table with similarity of values. Lots of possible coverings are derived in an incomplete information table. This seems to cause difficulty due to computational complexity, but it is not, because the family of possible coverings has a lattice structure. Four approximations that make up a rough set are derived by using only two coverings: the minimum and maximum possible ones which are derived from the minimum and the maximum possible indiscernibility relations that are equal to the intersection and the union of those from possible tables. The approximations are equal to those derived using the minimum and the maximum possibly indiscernible classes.

Keywords: Rough sets · Incomplete information · Possible coverings · Possible indiscernibility relations

1 Introduction

Rough sets by Pawlak [1] are based on equality of values characterizing objects. The rough sets are used as an effective method for deriving significant rules from a variety of data. The rough sets are usually used to complete information tables with no similarity of objects. However, similar objects often appear in the real world. Furthermore, data values with incomplete information appear everywhere. So, it is not sufficient for information processing in the real world unless we deal with similarity of objects and values with incomplete information. This requires an extension of rough sets.

An approach most frequently used to deal with incomplete information comes from the way that Kryszkiewicz proposed [2]. The approach takes into account

© Springer Nature Switzerland AG 2021
S. Ramanna et al. (Eds.): IJCRS 2021, LNAI 12872, pp. 83–89, 2021.
https://doi.org/10.1007/978-3-030-87334-9_7

only one possibility that a value with incomplete information has. Information loss occurs. As a result, the approach causes poor results [3,4].

Therefore, considering the possibilities that a value has, we develop rough sets based on an approach using possible indiscernibility relations under the similarity of values. A possible indiscernibility relation is a possible world in possible world semantics, although Lipski used a possible table as a possible world.

2 Rough Sets from Coverings in Complete Information Tables

A complete information table consists of U, a non-empty finite set of objects, $D(a_i)$, the domain of attribute a_i, and AT, a non-empty finite set of attributes where $a_i \in AT : U \rightarrow D(a_i)$. Indiscernibility relation $R_{a_i}^{\delta}$[1] meaning indistinguishability of objects on attribute $a_i \in AT$ under threshold δ_{a_i}[2] is:

$$R_{a_i}^{\delta} = \{(o, o') \in U \times U \mid SIM_{a_i}(o, o') \geq \delta_{a_i}\}, \tag{1}$$

where $SIM_{a_i}(o, o')$ expresses what degree objects o and o' are similar for attribute a_i and δ_{a_i} is the similarity threshold for a_i.

$$SIM_{a_i}(o, o') = sim(a_i(o), a_i(o')), \tag{2}$$

where $sim(a_i(o), a_i(o'))$ is the similarity degree of $a_i(o)$ and $a_i(o')$, which is given by experts such that it is reflexive, symmetric, and not transitive[3].

Using $R_{a_i}^{\delta}$, indiscernible class $C(o)_{a_i}^{\delta}$ of o on a_i is expressed in:

$$C(o)_{a_i}^{\delta} = \{o' \mid (o, o') \in R_{a_i}^{\delta}\}. \tag{3}$$

$C(o)_{a_i}^{\delta}$[4] is a tolerance class.

Family $\mathcal{C}_{a_i}$ of indiscernible classes on a_i is:

$$\mathcal{C}_{a_i} = \{C \mid o \in U \wedge C = C(o)_{a_i}\}. \tag{4}$$

Clearly, $\cup_{C \in \mathcal{C}_{a_i}} C = U$. $\mathcal{C}_{a_i}$ is a covering, which is unique for a_i. Using covering $\mathcal{C}_{a_i}$, lower approximation $\underline{apr}_{a_i}(\mathcal{O})$ and upper approximation $\overline{apr}_{a_i}(\mathcal{O})$ of set $\mathcal{O}$ for a_i are:

$$\underline{apr}_{a_i}(\mathcal{O}) = \{o \in U \mid C(o) \in \mathcal{C}_{a_i} \wedge C(o) \subseteq \mathcal{O}\}, \tag{5}$$

$$\overline{apr}_{a_i}(\mathcal{O}) = \{o \in U \mid C(o) \in \mathcal{C}_{a_i} \wedge C(o) \cap \mathcal{O} \neq \emptyset\}. \tag{6}$$

---

[1] Unless confusion may arise, R_{a_i} is used.
[2] δ is used in place of δ_{a_i} if no confusion arises.
[3] Therefore, $R_{a_i}^{\delta}$ becomes a tolerance relation.
[4] $C(o)$ or $C(o)_{a_i}$ is used in place of $C(o)_{a_i}^{\delta}$ if no confusion arises.

3 Rough Sets from Possible Coverings in Incomplete Information Tables

A value with incomplete information is expressed in a disjunctive set of possible value. So, v in $a_i(o)$ may be the actual one.

There are lots of possible coverings derived from an incomplete information table [5,6], although some authors deal with only one covering [7,8]. A possible indiscernibility relation creates a possible covering. FPR_{a_i}, the family of possible indiscernibility relations, is:

$$FPR_{a_i} = \{PR \mid PR = SR_{a_i} \cup e \wedge e \in \mathcal{P}(MPPR_{a_i})\}, \tag{7}$$

where $\mathcal{P}(MPPR_{a_i})$ denotes the power set of $MPPR_{a_i}$ and $MPPR_{a_i}$ is expressed as:

$$MPPR_{a_i} = \{\{(o',o),(o,o')\} \mid (o',o) \in MPR_{a_i}\}, \tag{8}$$

$$MPR_{a_i} =$$
$$\{(o,o') \in U \times U \mid \exists u \in a_i(o) \exists v \in a_i(o') sim(u,v) \geq \delta_{a_i}\} \backslash SR_{a_i}, \tag{9}$$

$$SR_{a_i} =$$
$$\{(o,o') \in U \times U \mid (o = o') \vee (\forall u \in a_i(o) \forall v \in a_i(o') sim(u,v) \geq \delta_{a_i})\}, \tag{10}$$

where a pair in SR_{a_i} is called a certain one whereas a pair in MPR_{a_i} a possible one. FPR_{a_i} constitutes a lattice based on set inclusion. SR_{a_i} and $SR_{a_i} \cup MPR_{a_i}$ are the minimum and the maximum possible indiscernibility relations in FPR_{a_i}, respectively. A possible indiscernibility relation does not always correspond to a possible table, but the following proposition hold.

Proposition 1. $PR_{a_i,min} = \cap PTR_{a_i}$, $PR_{a_i,max} = \cup PTR_{a_i}$, where $PR_{a_i,min}$ and $PR_{a_i,max}$ are the minimum and the maximum possible indiscernibility relations, and PTR_{a_i} is the indiscernibility relation derived from a possible table using formula (1).

Example 1. Let complete information table $IT0$ be as follows:

U	ITO a_1	a_2	U	PT1 a_1	a_2	U	PT2 a_1	a_2
o_1	$< a >$	$< x >$	o_1	$< a >$	$< x >$	o_1	$< a >$	$< x >$
o_2	$< b,e >$	$< y >$	o_2	$< b >$	$< y >$	o_2	$< e >$	$< y >$
o_3	$< c >$	$< x,y >$	o_3	$< c >$	$< x,y >$	o_3	$< c >$	$< x,y >$
o_4	$< d >$	$< y >$	o_4	$< d >$	$< y >$	o_4	$< d >$	$< y >$
o_5	$< e >$	$< z >$	o_5	$< e >$	$< z >$	o_5	$< e >$	$< z >$

$PT1$ and $PT2$ are possible tables of $IT0$ for a_1. Let similarity degree $sim(u, v)$ on domain $D(a_1) = \{a, b, c, d, e\}$ be as follows:

$$sim(u, v) = \begin{pmatrix} 1 & 0.2 & 0.9 & 0.6 & 0.1 \\ 0.2 & 1 & 0.8 & 0.8 & 0.2 \\ 0.9 & 0.8 & 1 & 0.2 & 0.3 \\ 0.6 & 0.8 & 0.2 & 1 & 0.9 \\ 0.1 & 0.2 & 0.3 & 0.9 & 1 \end{pmatrix}.$$

In $IT0$, let δ_{a_1} be 0.75. $<b, e>$ is the disjunctive set that means b or e. SR_{a_1} is obtained as $\{(o_1, o_1), (o_1, o_3), (o_2, o_2), (o_2, o_4), (o_3, o_3), (o_3, o_1), (o_4, o_2), (o_4, o_4), (o_4, o_5), (o_5, o_5), (o_5, o_4)\}$. MPR_{a_1} is $\{(o_2, o_3), (o_3, o_2), (o_2, o_5), (o_5, o_2)\}$. Using formulae (7)–(10), family FPR_{a_1} of possible indiscernibility relations is $\{PR_1, \cdots, PR_4\}$, and 4 possible indiscernibility relations are:

$$PR_1 = \{(o_1, o_1), (o_1, o_3), (o_2, o_2), (o_2, o_4), (o_3, o_3), (o_3, o_1), (o_4, o_2), (o_4, o_4),$$
$$(o_4, o_5), (o_5, o_5), (o_5, o_4)\},$$

$$PR_2 = \{(o_1, o_1), (o_1, o_3), (o_2, o_2), (o_2, o_4), (o_3, o_3), (o_3, o_1), (o_4, o_2), (o_4, o_4),$$
$$(o_4, o_5), (o_5, o_5), (o_5, o_4), (o_2, o_3), (o_3, o_2)\},$$

$$PR_3 = \{(o_1, o_1), (o_1, o_3), (o_2, o_2), (o_2, o_4), (o_3, o_3), (o_3, o_1), (o_4, o_2), (o_4, o_4),$$
$$(o_4, o_5), (o_5, o_5), (o_5, o_4), (o_2, o_5), (o_5, o_2)\},$$

$$PR_4 = \{(o_1, o_1), (o_1, o_3), (o_2, o_2), (o_2, o_4), (o_3, o_3), (o_3, o_1), (o_4, o_2), (o_4, o_4),$$
$$(o_4, o_5), (o_5, o_5), (o_5, o_4), (o_2, o_3), (o_3, o_2), (o_2, o_5), (o_5, o_2)\}.$$

PR_2 and PR_3 are also equal to indiscernibility relations $PT1R$ and $PT2R$ from possible tables $PT1$ and $PT2$, respectively. PR_1 and PR_4 are the minimum and the maximum possible indiscernibility relations, respectively. And $PR_1 = PT1R \cap PT2R$ and $PR_4 = PT1R \cup PT2R$ hold.

From $PR_j \in FPR_{a_i}$, possibly indiscernible class $C(o)_{a_i,j}$ is derived:

$$C(o)_{a_i,j} = \{o' \mid (o, o') \in PR_j \wedge PR_j \in FPR_{a_i}\}. \tag{11}$$

Proposition 2. If $PR_{a_i,k} \subseteq PR_{a_i,l}$, then $C(o)_{a_i,k} \subseteq C(o)_{a_i,l}$.

This means that the family of possibly indiscernible classes for an object has a lattice structure for set inclusion.

$PC_{a_i,j}$, the possible covering from possible indiscernibility relation $PR_{a_i,j}$, is:

$$PC_{a_i,j} = \{e \mid e = C(o)_{a_i,j} \wedge o \in U\}. \tag{12}$$

From Proposition 2 family FPC_{a_i} of possible coverings is a lattice for $\sqsubseteq$[5].

---

[5] $\sqsubseteq$ is defined as $\mathcal{E} \sqsubseteq \mathcal{E}'$ if $\forall E \in \mathcal{E} \exists E' \in \mathcal{E}' \wedge E \subseteq E'$.

Example 2. Possibly indiscernible classes in each possible indiscernibility relation $PR_{a_1,j}$ with $j = 1,\ldots,4$ are obtained. For example, in $PR_{a_1,1}$ $C(o_1)_{a_1,1}$, $C(o_2)_{a_1,1}$, $C(o_3)_{a_1,1}$, $C(o_4)_{a_1,1}$, and $C(o_5)_{a_1,1}$ are $\{o_1,o_3\}$, $\{o_2,o_4\}$, $\{o_1,o_3\}$, $\{o_2,o_4,o_5\}$, and $\{o_4,o_5\}$, respectively. In $PR_{a_1,4}$ $C(o_1)_{a_1,4}$, $C(o_2)_{a_1,4}$, $C(o_3)_{a_1,4}$, $C(o_4)_{a_1,4}$, and $C(o_5)_{a_1,4}$ are $\{o_1,o_3\}$, $\{o_2,o_3,o_4,o_5\}$, $\{o_1,o_2,o_3\}$, $\{o_2,o_4,o_5\}$, and $\{o_2,o_4,o_5\}$, respectively. Using these possibly indiscernible classes, possible coverings are obtained as follows:

$$PC_{a_1,1} = \{\{o_1,o_3\},\{o_2,o_4\},\{o_2,o_4,o_5\},\{o_4,o_5\}\},$$
$$PC_{a_1,2} = \{\{o_1,o_3\},\{o_1,o_2,o_3\},\{o_2,o_3,o_4\},\{o_2,o_4,o_5\},\{o_4,o_5\}\},$$
$$PC_{a_1,3} = \{\{o_1,o_3\},\{o_2,o_4,o_5\}\},$$
$$PC_{a_1,4} = \{\{o_1,o_3\},\{o_1,o_2,o_3\},\{o_2,o_3,o_4,o_5\},\{o_2,o_4,o_5\}\}.$$

$PC_{a_1,1}$ and $PC_{a_1,4}$ are the minimum and the maximum coverings, respectively.

For the relationship between the minimum and maximum possible coverings and coverings from possible tables, we have the following proposition.

Proposition 3. $PC_{a_i,min} = \cap PTC_{a_i}$ and $PC_{a_i,max} = \cup PTC_{a_i}$ where $\cap PTC_{a_i} = \{\cap_{pt} C(o)_{a_i} \mid o \in U \wedge C(o)_{a_i} \in C_{a_i,pt}\}$, $\cup PTC_{a_i} = \{\cup_{pt} C(o)_{a_i} \mid o \in U \wedge C(o)_{a_i} \in C_{a_i,pt}\}$, and $C_{a_i,pt}$ is the covering derived from possible table pt.

By using possible covering PC_j, two approximations of $\mathcal{O}$ are:

$$\underline{apr}_{a_i,j}(\mathcal{O}) = \{o \in U \mid C(o) \subseteq \mathcal{O} \wedge C(o) \in PC_j\}, \tag{13}$$
$$\overline{apr}_{a_i,j}(\mathcal{O}) = \{o \in U \mid C(o) \cap \mathcal{O} \neq \emptyset \wedge C(o) \in PC_j\}. \tag{14}$$

Proposition 4. If $PC_k \sqsubseteq PC_l$ for possible indiscernibility coverings $PC_k, PC_l \in FPC_{a_i}$, then $\underline{apr}_{a_i,k}(\mathcal{O}) \supseteq \underline{apr}_{a_i,l}(\mathcal{O})$, and $\overline{apr}_{a_i,k}(\mathcal{O}) \subseteq \overline{apr}_{a_i,l}(\mathcal{O})$.

This shows that the families of approximations are also lattices.

Aggregating approximations in each possible covering, certain lower approximation $\underline{Sapr}_{a_i}(\mathcal{O})$ of $\mathcal{O}$, possible lower approximation $\underline{Papr}_{a_i}(\mathcal{O})$, certain upper approximations $\overline{Sapr}_{a_i}(\mathcal{O})$, and possible upper approximation $\overline{Papr}_{a_i}(\mathcal{O})$ are:

$$\underline{Sapr}_{a_i}(\mathcal{O}) = \{o \in U \mid \forall PC_j \in FPC_{a_i}\ o \in \underline{apr}_{a_i,j}(\mathcal{O})\}, \tag{15}$$
$$\underline{Papr}_{a_i}(\mathcal{O}) = \{o \in U \mid \exists PC_j \in FPC_{a_i}\ o \in \underline{apr}_{a_i,j}(\mathcal{O})\}, \tag{16}$$
$$\overline{Sapr}_{a_i}(\mathcal{O}) = \{o \in U \mid \forall PC_j \in FPC_{a_i}\ o \in \overline{apr}_{a_i,j}(\mathcal{O})\}, \tag{17}$$
$$\overline{Papr}_{a_i}(\mathcal{O}) = \{o \in U \mid \exists PC_j \in FPC_{a_i}\ o \in \overline{apr}_{a_i,j}(\mathcal{O})\}. \tag{18}$$

Using Proposition 4, these approximations are:

$$\underline{Sapr}_{a_i}(\mathcal{O}) = \underline{apr}_{a_i,max}(\mathcal{O}),\quad \underline{Papr}_{a_i}(\mathcal{O}) = \underline{apr}_{a_i,min}(\mathcal{O}), \tag{19}$$
$$\overline{Sapr}_{a_i}(\mathcal{O}) = \overline{apr}_{a_i,min}(\mathcal{O}),\quad \overline{Papr}_{a_i}(\mathcal{O}) = \overline{apr}_{a_i,max}(\mathcal{O}), \tag{20}$$

where $\underline{apr}_{a_i,min}(\mathcal{O})$ and $\overline{apr}_{a_i,min}(\mathcal{O})$ are the approximations in the minimum possible covering, and $\underline{apr}_{a_i,max}(\mathcal{O})$ and $\overline{apr}_{a_i,max}(\mathcal{O})$ are the approximations in the maximum possible covering.

Example 3. We go back to Example 2. Using (13) and (14) under $\mathcal{O} = \{o_2, o_4\}$, in $PC_{a_1,1}$ $\underline{apr}_{a_1,1}(\mathcal{O}) = \{o_2\}$, $\overline{apr}_{a_1,1}(\mathcal{O}) = \{o_2, o_4, o_5\}$, and in $PC_{a_1,4}$ $\underline{apr}_{a_1,4}(\mathcal{O}) = \emptyset$, $\overline{apr}_{a_1,4}(\mathcal{O}) = \{o_2, o_3, o_4, o_5\}$. Using (19) and (20), $\underline{Sapr}_{a_1}(\mathcal{O}) = \emptyset$, $\underline{Papr}_{a_1}(\mathcal{O}) = \{o_2\}$, $\overline{Sapr}_{a_1}(\mathcal{O}) = \{o_2, o_4, o_5\}$, and $\overline{Papr}_{a_1}(\mathcal{O}) = \{o_2, o_3, o_4, o_5\}$.

We obtain the following proposition from (19) and (20).

Proposition 5. $\underline{Sapr}_{a_i}(\mathcal{O}) = \{o \mid C(o)_{a_i,max} \subseteq \mathcal{O}\}$, $\underline{Papr}_{a_i}(\mathcal{O}) = \{o \mid C(o)_{a_i,min} \subseteq \mathcal{O}\}$, $\overline{Sapr}_{a_i}(\mathcal{O}) = \{o \mid C(o)_{a_i,min} \cap \mathcal{O} \neq \emptyset\}$, and $\overline{Papr}_{a_i}(\mathcal{O}) = \{o \mid C(o)_{a_i,max} \cap \mathcal{O} \neq \emptyset\}$, where minimum possibly indiscernible class $C(o)_{a_i,min}$ and maximum possibly indiscernible class $C(o)_{a_i,max}$ are derived using formula (11) in $PR_{a_i,min}$ and $PR_{a_i,max}$, respectively.

This shows that approximations obtained in this approach are equal to those derived using the minimum and the maximum possibly indiscernible classes $C(o)_{a_i,min}$ and $C(o)_{a_i,max}$. There exist possible tables from which these classes can be derived. Here, $C(o)_{a_i,min}$ and $C(o)_{a_i,max}$ are expressed in $\{o' \in U \mid (o = o') \vee \forall u \in a_i(o) \forall v \in a_i(o') sim(u,v) \geq \delta_{a_i}\}$, and $\{o' \in U \mid \exists u \in a_i(o) \exists v \in a_i(o') sim(u,v) \geq \delta_{a_i}\}$, respectively.

4   Conclusions

We have dealt with rough sets based on possible coverings under possible world semantics in incomplete information tables with similarity of values. Lots of coverings are derived in an incomplete information table, but this does not cause difficulty due to computational complexity, because the family of possible coverings is a lattice with the minimum and maximum elements. Four approximations are derived using only the minimum and maximum possible indiscernibility relations that are equal to the intersection and the union of indiscernibility relations from possible tables, respectively. And these approximations are equal to those derived using the minimum and the maximum possibly indiscernible classes that are obtained from possible tables. This justifies the approach.

References

1. Pawlak, Z.: Rough Sets: Theoretical Aspects of Reasoning about Data. Kluwer Academic Publishers, Dordrecht (1991). https://doi.org/10.1007/978-94-011-3534-4
2. Kryszkiewicz, M.: Rules in incomplete information systems. Inf. Sci. **113**, 271–292 (1999)

3. Nakata, M., Sakai, H.: Applying rough sets to information tables containing missing values. In: Proceedings of 39th International Symposium on Multiple-Valued Logic, pp. 286–291. IEEE Press (2009). https://doi.org/10.1109/ISMVL.2009.1
4. Yang, T., Li, Q., Zhou, B.: Related family: a new method for attribute reduction of covering information systems. Inf. Sci. **228**, 175–191 (2013)
5. Couso, I., Dubois, D.: Rough sets, coverings and incomplete information. Fundamenta Informaticae **108**(3–4), 223–347 (2011)
6. Lin, G., Liang, J., Qian, Y.: Multigranulation rough sets: from partition to covering. Inf. Sci. **241**, 101–118 (2013). https://doi.org/10.1016/j.ins.2013.03.046
7. Chen, D., Wang, C., Hu, Q.: A new approach to attribute reduction of consistent and inconsistent covering decision systems with covering rough sets, Inf. Sci. **177**, 3500–3518 (2007). https://doi.org/10.1016/j.ins.2007.02.041
8. Zhang, X., Mei, C. L., Chen, D. G., Li, J.: Multi-confidence rule acquisition oriented attribute reduction of covering decision systems via combinatorial optimization. Knowl.-Based Syst. **50**, 187–197 (2013). https://doi.org/10.1016/j.knosys.2013.06.012

Attribute Reduction Using Functional Dependency Relations in Rough Set Theory

Mauricio Restrepo[1]([⊠]) and Chris Cornelis[2]

[1] Universidad Militar Nueva Granada, Bogotá, Colombia
mauricio.restrepo@unimilitar.edu.co
[2] Ghent University, Ghent, Belgium
chris.cornelis@ugent.be
https://www.umng.edu.co, https://www.ugent.be

Abstract. This paper presents some functional dependency relations defined on the attribute set of an information system. We establish some basic relationships between functional dependency relations, attribute reduction, and closure operators. We use the partial order for dependencies to show that reducts of an information system can be obtained from the maximal elements of a functional dependency relation.

Keywords: Functional dependency relations · Attribute reduction · Closure operators · Rough sets

1 Introduction

The concept of functional dependency (FD) was introduced by Armstrong in 1974, [1]. FD relations have been used in database theory for different purposes. Dependency relations were extensively studied in the rough set community already during 1980's and 1990's [10–12]. Also, a one-to-one and onto correspondence between FD relations and closure operators was established in [6]. A notion of a generalized dependency relation between subsets of an arbitrary set was introduced by Chiaselotti and Infusino, connecting with formal context analysis in [3]. Some algorithms based on indiscernibility relations were introduced by Qu and Fu in [9] and Zhang et al. in [16] to discover dependencies in datasets. All these works are closely related to the attribute reduction problem.

Rough set theory, proposed by Z. Pawlak, is based on an indiscernibility relation between objects of a non-empty set U, called *Universe* [7]. Attribute reduction in information systems is a fundamental aspect of rough sets and it has been successfully applied in many fields, such as machine learning and data mining.

The reduction of attributes is a fundamental problem in data analysis, since it allows to build measurements that are simpler and easier to interpret. Concerning to the theory

This work was supported by Universidad Militar Nueva Granada's VICEIN Special Research Fund, under project CIAS 3144-2020, and by the Odysseus program of the Research Foundation-Flanders.

© Springer Nature Switzerland AG 2021
S. Ramanna et al. (Eds.): IJCRS 2021, LNAI 12872, pp. 90–96, 2021.
https://doi.org/10.1007/978-3-030-87334-9_8

of rough sets, there is a great variety of works on attribute reduction [15,16], to mention only a few.

In this paper, we use a partial order on dependencies to obtain the maximal elements. We show that the reducts of an information system can be obtained from the maximal elements of this functional dependency relation. Since each functional dependency relation defines a closure operator, the properties of this type of operator are studied, according to [6]. The study of attribute reduction using a dependency relation contributes to understanding this important problem.

The content of this paper is organized as follows. Section 2 presents some preliminary concepts regarding rough set theory, functional dependency relations, and closure operators. Section 3 presents and studies, from a different perspective, the properties of a functional dependency relation on an attribute set of an information system. Finally, Sect. 4 presents the main conclusions of the paper and describes future work.

2   Preliminaries

2.1   Rough Sets

The rough set approach was proposed by Pawlak in 1982 as a tool for dealing with imperfect knowledge and incomplete information [7]. The main concept of this theory is the indiscernibility relation defined on a finite set U. The elements of this set U are called objects, which are described by a finite set of attributes A. The pair (U, A) is called an information system.

Each subset of attributes $P \subseteq A$ defines an equivalence relation:

$$IND(P) = \{(x,y) \in U \times U : f_a(x) = f_a(y), \forall a \in P\}, \tag{1}$$

where $f_a(x)$ is the value of an object $x \in U$ for an attribute $a \in A$.

Table 1 shows an information system where $U = \{1,2,3,4,5,6\}$ is the set of objects, organized in rows, and $A = \{a,b,c,d\}$ is the set of attributes, organized in columns. According to the information above, we have that $f_a(2) = B$, while $f_b(1) = A$.

If $X \subseteq U$, the operators:

$$\underline{apr}(X) = \{x \in U : [x]_P \subseteq X\}, \quad \overline{apr}(X) = \{x \in U : [x]_P \cap X \neq \emptyset\} \tag{2}$$

are called the lower and upper approximations of X.

Example 1. For the information system shown in Table 1, it is easy to see that the sets of attributes $P = \{a,b\}$, $Q = \{a,c\}$, and $R = \{a,b,c\}$ define the same partition of U: $\mathbb{P} = \{\{1\},\{2\},\{3\},\{4,5\},\{6\}\}$.

2.2   Reducts for Information Systems

Attribute reduction involves searching for particular subsets of attributes. Generally, it implies removing attributes that have no significance in determining indiscernible elements. We say that a is a superfluous attribute of P if $[x]_P = [x]_{P-\{a\}}$ for all $x \in U$; otherwise, a is called indispensable in P [4].

In this paper, we want to establish a difference between reductions and reducts.

Table 1. An information system.

	Attributes			
Object	a	b	c	d
1	A	A	B	A
2	B	B	B	C
3	C	A	A	A
4	C	B	B	C
5	C	B	B	A
6	A	B	A	B

Definition 1 [4]. *The set P is independent if all of its attributes are indispensable. The subset Q of P is a reduction of P if Q is independent and $[x]_Q = [x]_P$ for all $x \in U$. A reduction of A is called a reduct.*

2.3 Functional Dependency Relations

An axiomatic description of a functional dependency relation was introduced in [1], using precise concepts and terminology of relational models of data. A dependency, denoted as $X \to Y$, means that the values of attributes Y are determined by the values of attributes X, i.e. two objects with the same values of X will necessarily have the same values for Y. We present a definition given by Matús in [6].

Definition 2 [6]. *Let A be a finite set. $\mathcal{N} \subseteq \mathscr{P}(A) \times \mathscr{P}(A)$ is a functional dependency (FD) relation if for all $I, J, K \subseteq A$ the following properties are fulfilled:*

1. *If $I \supseteq J$ then $(I, J) \in \mathcal{N}$.*
2. *If $(I, J) \in \mathcal{N}$ and $(J, K) \in \mathcal{N}$, then $(I, K) \in \mathcal{N}$.*
3. *If $(I, J) \in \mathcal{N}$ and $(I, K) \in \mathcal{N}$, then $(I, J \cup K) \in \mathcal{N}$.*

Order Relation. A partial order can be defined on $\mathcal{N}$. If $(A, B), (A', B') \in \mathcal{N}$, then $(A, B) \geq (A', B')$ if and only if $A \subseteq A'$ and $B \supseteq B'$. It is easy to show that this relation satisfies the following:

1. $(A, B) \geq (A, B)$, (reflexive).
2. If $(A, B) \geq (A', B')$ and $(A', B') \geq (A, B)$, then $A = A'$ and $B = B'$, (anti-symmetric).
3. If $(A, B) \geq (A', B')$ and $(A', B') \geq (A'', B'')$, then $(A, B) \geq (A'', B'')$, (transitive).

Maximal Elements. An element $(A, B) \in \mathcal{N}$ is maximal if and only if for all $(A', B') \in \mathcal{N}$ such that $(A', B') \geq (A, B)$, we have that $A = A'$ and $B = B'$.

2.4   Closure Operators

In ordered sets, the functions that preserve order relations are very important. Closure operators are a special class of order-preserving functions. We present some concepts about ordered structures, according to Blyth and Järvinen [2,5].

Definition 3. *An order-preserving map* $c : \mathscr{P}(A) \to \mathscr{P}(A)$ *is a **closure operator** if*

1. $P \subseteq c(P)$, *(extensive).*
2. $c(P) = c[c(P)]$, *(idempotent).*

Some equivalent definitions of a closure operator on the attribute set were defined in [13] and we present one of them, as follows:

Definition 4 [13]. *For each* $P \subseteq A$ *subset of attributes, a closure operator can be defined as:*

$$c(P) = \{a \in A : [x]_P = [x]_{P \cup \{a\}} \text{ for all } x \in U\} \tag{3}$$

In [6], Matús also established a Galois connection between FD relations and closure systems, since that each functional dependency relation defines a closure operator, and each closure operator defines a functional dependency relation.

If $\mathscr{N}$ is a functional dependency relation, the function $c_{\mathscr{N}} : \mathscr{P}(A) \to \mathscr{P}(A)$ defined as:

$$c_{\mathscr{N}}(I) = \bigcup \{J \subseteq A : (I,J) \in \mathscr{N}\} \tag{4}$$

is a closure operator.

If c is a closure operator, the relation

$$\mathscr{N}_c = \{(I,J) : J \subseteq c(I)\} \tag{5}$$

is a functional dependency relation.

Equations (4) and (5) were introduced in [6] to show the Galois connection between functional dependencies and closure operators.

2.5   Relationships on Attribute Sets

The following results are well-known in rough set theory and are useful relationships between equivalence classes and approximation operators for different sets of attributes [13].

Proposition 1 [13]. *If* $P,Q \subseteq A$, $a \in A$ *and* $P \subseteq Q$, *then*

1. $[x]_P \supseteq [x]_Q$, *for all* $x \in U$.
2. $\underline{apr}_P(X) \subseteq \underline{apr}_Q(X)$, *for all* $X \subseteq U$.
3. $[x]_{P \cup \{a\}} = [x]_P \cap [x]_a$, *for all* $x \in U$.

Example 2. The sets of attributes $P_1 = \{a,b,d\}$, $P_2 = \{a,c,d\}$, and $A = \{a,b,c,d\}$ have the same partition $\mathbb{P} = \{\{1\},\{2\},\{3\},\{4\},\{5\},\{6\}\}\}$.

3   Functional Dependency Relations

A first functional dependency (FD) relation for rough set theory was introduced by Pawlak in [8] and later Ślezak in [14], using the indiscernibility relation, shown in Eq. (1). If P and Q are sets of attributes, $P \to Q$ if and only if $IND(P) = IND(P \cup Q)$.

Let (U,A) be an information system, where U is a finite set and A is a finite set of attributes. It is possible to define a functional dependency relation on A, as follows.

Definition 5. *If* $P,Q \subseteq A$ *are subsets of attributes, we define:*

$$(P,Q) \in \mathcal{N}_1 \text{ if and only if } [x]_P = [y]_P \Rightarrow [x]_Q = [y]_Q, \ \forall x, y \in U \tag{6}$$

It is easy to see that $P \supseteq Q$ implies that $(P,Q) \in \mathcal{N}_1$. This dependency is called a trivial dependency. If $(P,Q) \in \mathcal{N}$, it will be represented by means of $P \to Q$ and $(P,Q) \notin \mathcal{N}_1$ will be represented with $P \nrightarrow Q$. Also, it is easy to prove the following proposition.

Proposition 2. *The relation* $\mathcal{N}_1$ *is a functional dependency on A.*

This order relation is stable considering the partial order on $\mathcal{N}_1$ for minor elements, as follows:

Proposition 3. *If* $(P,Q) \in \mathcal{N}_1$ *and* $(P,Q) \geq (P',Q')$*, then* $(P',Q') \in \mathcal{N}_1$.

Proof. If $(P,Q) \in \mathcal{N}_1$, then $[x]_P = [y]_P \Rightarrow [x]_Q = [y]_Q$. If $[x]_{P'} = [y]_{P'}$, we have that $[x]_P = [y]_P$ since $P \subseteq P'$, so $[x]_Q = [y]_Q$. In particular, $[x]_{Q'} = [y]_{Q'}$ since $Q \supseteq Q'$.

The maximal elements for this order relation on dependencies are: $\{a,b\} \to \{a,b,c\}$, $\{a,b,d\} \to A$, $\{a,c,d\} \to A$, $\{b,c,d\} \to A$, as well as all the trivial dependencies $P \to P$, with $P \subseteq A$.

Let us consider the closure operator $c_{\mathcal{N}_1}$ defined using $\mathcal{N}_1$, according to Equation (4). The following proposition establishes the equivalence between closure operators c and $c_{\mathcal{N}_1}$.

Proposition 4. $c = c_{\mathcal{N}_1}$.

Proof. We will see that $c(P) \subseteq c_{\mathcal{N}_1}(P)$ and $c_{\mathcal{N}_1}(P) \subseteq c(P)$, for all $P \subseteq A$.

1. $c \leq c_{\mathcal{N}}$. If $a \in c(P)$, then $[x]_P = [x]_{P \cup \{a\}}$ for all $x \in U$. In particular, $[x]_P \subseteq [x]_{P \cup \{a\}}$ for all $x \in U$, $\mathbb{P}_P \leq \mathbb{P}_{P \cup a}$, and $(P, P \cup a) \in \mathcal{N}_1$, so $a \in \cup \{Q : (P,Q) \in \mathcal{N}_1\}$.
2. $c_{\mathcal{N}} \leq c$. If $a \in c_{\mathcal{N}}(P)$, $a \in Q_0$ with $(P,Q_0) \in \mathcal{N}_1$. $\mathbb{P}_P \leq \mathbb{P}_{Q_0}$, then $[x]_P \subseteq [x]_{Q_0} = [x]_{Q_0 - a} \cap [x]_a$. So, $[x]_P \subseteq [x]_a$ and $[x]_P \cap [x]_a = [x]_P$, then $a \in c(P)$.

According to Proposition 9 in [13], the fixed points of the closure operator c, i.e. the sets $P \subseteq A$ such that $c(P) = P$, are important to describe the reducts of an information system. We use the order relation defined on dependencies to characterize the reducts.

Proposition 5. (P,A) *is maximal in* $\mathcal{N}_1$ *if and only if P is a reduct of A.*

Proof. Let us suppose that $(P,A) \in \mathcal{N}_1$ is maximal. If P is not a reduct, there exists a reduct P' with $P' \subset P$. Therefore, $(P',A) > (P,A)$ and (P,A) is not maximal.

If P is a reduct, then $(P,A) \in \mathcal{N}_1$. Now, if $(P',Q') \in \mathcal{N}_1$ and $(P',Q') \geq (P,A)$ then $P' \subseteq P$ and $Q' \supseteq A$. So $Q' = A$ and $(P',A) \in \mathcal{N}_1$. If $P' \subset P$, P contains dispensable attributes and it is not a reduct. Therefore, $P' = P$ and (P,A) is maximal.

4   Conclusions

The reducts of an information system can be obtained from the maximal elements of a partial order defined on pairs of subsets of attributes that define dependencies.

The importance of this proposal is to initiate a systematic study of the attribute reduction problem and its relationships with different structures that are defined on the set of attributes, as opposed to other structures defined on the set of objects. Future research aims at applying the definition of functional dependency to other generalizations of rough set theory and looking for connections with matroid theory.

References

1. Armstrong, W.: Dependency structures of database relationships. In: Information Processing, pp. 580–583. North Holland Publishing (1974)
2. Blyth, T.S.: Lattices and Ordered Algebraic Structures. Universitext. Springer, London (2005). https://doi.org/10.1007/b139095
3. Chiaselotti, G., Infusino, F., Notions from rough sets theory in a generalized dependency relation context. Int. J. Approx. Reason. **98**, 25–61 (2018)
4. Greco, S., Matarazzo, B., Slowinski, R.: Rough sets theory for multicriteria decision analysis. Eur. J. Oper. Res. **129**, 1–47 (2001)
5. Järvinen, J.: Lattice theory for rough sets. In: Peters, J.F., Skowron, A., Düntsch, I., Grzymała-Busse, J., Orłowska, E., Polkowski, L. (eds.) Transactions on Rough Sets VI. LNCS, vol. 4374, pp. 400–498. Springer, Heidelberg (2007). https://doi.org/10.1007/978-3-540-71200-8_22
6. Matús, F.: Abstract functional dependency structures. Theor. Comput. Sci. **81**, 117–126 (1991)
7. Pawlak Z.: Rough sets. Int. J. Comput. Inf. Sci. **11**(5), 341–356 (1982)
8. Pawlak Z.: Information systems theoretical foundations. Inf. Syst. 6(3), 205–218 (1981)
9. Qu, Y., Fu, X.: Rough set based algorithm of discovering functional dependencies for relation database. In: 4th International Conference on Wireless Communications, Networking and Mobile Computing, Dalian, 2008, pp. 1–4 (2008). https://doi.org/10.1109/WiCom.2008.2526
10. Novotný, M., Pawlak, Z.: Independence of attributes. Bull. Polish Acad. Sci. Math. **36**, 459–465 (1988)
11. Novotný, M.: Dependence spaces of information systems. In: Orłowska, E. (ed.) Incomplete Information: Rough Set Analysis. Studies in Fuzziness and Soft Computing, vol. 13, pp. 193-246. Physica, Heidelberg (1998). https://doi.org/10.1007/978-3-7908-1888-8_7
12. Rauszer, C.M.: An equivalence between indiscernibility relations in information systems and a fragment of intuitionistic logic. In: Skowron, A. (ed.) SCT 1984. LNCS, vol. 208, pp. 298–317. Springer, Heidelberg (1985). https://doi.org/10.1007/3-540-16066-3_25
13. Restrepo M., Cornelis C., Attribute reduction from closure operators and matroids in Rough Set Theory. In: Bello, R., Miao, D., Falcon, R., Nakata, M., Rosete, A., Ciucci, D. (eds.) IJCRS 2020. LNAI, vol. 11499, pp. 183–192. Springer, Cham (2020). https://doi.org/10.1007/978-3-030-52705-1_13
14. Ślezak, D.: Rough sets and functional dependencies in data: foundations of association reducts. In: Gavrilova, M.L., Tan, C.J.K., Wang, Y., Chan, K.C.C. (eds.) Transactions on Computational Science V. LNCS, vol. 5540, pp. 182–205. Springer, Heidelberg (2009). https://doi.org/10.1007/978-3-642-02097-1_10

15. Yang T., Li Q.: Reduction about approximation spaces of covering generalized rough sets. Int. J. Approx. Reason. **51**, 335–345 (2010)
16. Zhang, B., Qian, S., Wei, C.: Research of reduct algorithm based on functional dependency. In: International Conference on Computer and Communication Technologies in Agriculture Engineering, Chengdu, 2010, pp. 312–315 (2010). https://doi.org/10.1109/CCTAE.2010.5543312

The RSDS: A Current State and Future Plans

Zbigniew Suraj$^{(\boxtimes)}$ ⓘ and Piotr Grochowalski ⓘ

Institute of Computer Science, Rzeszów University, Rzeszów, Poland
{zbigniew.suraj,piotrg}@ur.edu.pl

Abstract. This paper provides a brief overview of the Rough Set Database System (the RSDS for short) for creating bibliographies on rough sets and related fields, as well as sharing and analysis. The current version of the RSDS includes a number of modifications, extensions and functional improvements compared to the previous versions of this system. The system was made in the client-server technology. Currently, the RSDS contains over 38 540 entries from nearly 42 860 authors. This system works on any computer connected to the Internet and is available at http://rsds.ur.edu.pl.

Keywords: Rough set · Soft computing · Ontology · Data mining · Knowledge discovery · Pattern recognition · Machine learning · Database system

1 Generally About the RSDS

The concept of the rough set has its origin in Pawlak's seminal article from 1982 [5]. The rough set theory [6] is a formal theory derived from the fundamental research into the logical properties of information systems [4]. This theory is a simple and effective methodology for database mining or knowledge discovery in relational databases. In its abstract form, it is a new area of soft mathematics [2], related closely to the fuzzy set theory initiated by Zadeh [26]. Rough and fuzzy sets are complementary generalizations of classical sets. The rapid development of these two approaches formed the basis of the "soft computing" [27] which includes, in addition to rough sets, at least fuzzy logic, neural networks, probabilistic reasoning, belief networks, machine learning, evolutionary computing and chaos theory [1].

For some time we have seen a systematic, global increase of interest in the rough set theory and its applications [3,7,8,24,25]. However, on the other hand, there is a lack of publicly available bibliographic databases that facilitates access to literature and other tools needed by users of rough sets.

The purpose of this paper is to present the latest issue of the RSDS as compactly as possible, which to some extent appears to fill this gap. The system offers a wide range of functional possibilities, including bibliography creation, modifying, downloading, analyzing, visualizing and more capabilities. The bibliography

© Springer Nature Switzerland AG 2021
S. Ramanna et al. (Eds.): IJCRS 2021, LNAI 12872, pp. 97–102, 2021.
https://doi.org/10.1007/978-3-030-87334-9_9

included in the RSDS is formatted according to the BibTeX specification [28]. It consists of the following publication types: articles, inproceedings, incollections, books, techreports, proceedings, inbooks, phdtheses, mastertheses, manuals and unpublished. In addition to the bibliography, the system also includes: (1) information about software related to rough sets, (2) bibliographies of famous people working in this field, (3) personal data of the authors of publications available in the system. The access to the system depends on whether the user is logged in or not. If the user is logged in, he/she can enter data into the system, edit and classify the data he/she has entered. However, all users, logged in or not, can download the bibliography from the system and save it to an RTF or BibTeX file. Thanks to this possibility, the user can quickly and easily create a literature list for his own needs without having to delve into the structure of available file formats.

An important feature of this system is that it can successfully act as experimental environment for researches related to, inter alia, broadly understood information processing based on methods and techniques in the field of ontology and rough sets as well as advanced data analysis using the methods and techniques of statistics and graph theory. The current version of the RSDS was made in the client-server technology, which is classified as a modern database management technology.

Original and useful functionalities of the RSDS, as far as we know, rather unheard of in other database systems with a similar purpose as our system are:

– Searching according to predefined classifiers [12], as well as using the ontological search method [16,21], thanks to which the search for the desired information becomes more accurate and effective.
– Possibility of searching for information in the system with the use of an interactive world map, illustrating who and where in the world is working on the development of the rough set theory and its applications. To date, we have identified 72 research groups worldwide with 2 416 active members.
– Easy and convenient access to a number of statistics on the data contained in the RSDS and their graphical representation, such as: the number of authors, the number of and types of publications, number and years of publications, information on the percentage of all authors who wrote a certain number of publications at specified intervals, a list of indicators characterizing publishing collaboration between authors [15,17]. The results obtained from such an analysis may be useful in determining the structure of research groups in relation to rough sets, research interests of members of these groups, mutual cooperation between groups and research identification trends. It is also worth adding that the data are analyzed using different statistical methods and graph theory methods [23].
– Determining both two Pawlak numbers and the numbers of individual authors indicating the strength of the publication relationship between the authors represented in the RSDS database [18–20].
– Searching for information about software supporting research and experiments based on the rough set methodology and biographies of prominent people dealing with the rough set theory and its applications.

The main page of the RSDS looks like Fig. 1.

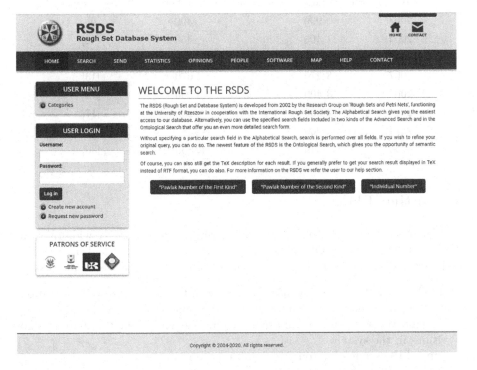

Fig. 1. The main page

Significant functional and visualization changes in the current version of the system include:

- implementation of a new ontological search method based on fuzzy logic,
- extension of the module for calculating Pawlak numbers and individual authors numbers,
- update of the statistical and graphical data analysis module,
- rebuilding the interactive world map and adding the *Help* section,
- updating the system engine and rebuilding the user interface,
- modernizing the existing system functionalities,
- increasing the role of system administrators and amenities for registered users,
- entering data status and modification,
- extension of system-user communication.

Historically speaking, the first version of the RSDS was released almost two decades ago, i.e. in 2003 [9], subsequent editions appeared in 2005 [11], 2008 [14] and 2013 [22].

In the Table 1 we present the quantitative changes in the system's bibliographic data over the years.

Table 1. The quantitative changes in the RSDS

Year of issue	Number of publications	Number of authors	Source
2003	900	400	[9]
2004	1 400	450	[10]
2005	1 900	815	[11]
2007	3 270	1 670	[13]
2008	3 400	1 900	[14]
2010	3 800	2 250	[17]
2014	4 000	2 380	[22]
2020	38 549	42 859	[30]

2   Further Plans

As mentioned above, the system provides users with a relatively wide range of different functionalities. Of course, they can still be expanded and improved in various ways. In the nearest future, we plan to implement the following tasks:

- develop a method of automatic verification of the correctness of the relationship between the concepts appearing in the general ontology [17],
- increase the efficiency of information retrieval and semantic analysis algorithms in the system,
- expand the scope of automatic data processing in the system,
- add new functionalities regarding automatic user profile detection and data search on the Internet,
- carry out experiments to verify the practical usefulness of our methods of detecting behaviour patterns implemented in the system.

3   Final Remarks

We briefly introduced the RSDS, which brings together different types of computer tools for the rough set community. These tools in a natural and effective way support the user in editing, analysing and downloading bibliographies. Additionally, the system provides a wide range of methods of visualizing the results of data analysis contained in the system. A unique feature of the system is the possibility of using it as an experimental platform in the scientific research. An equally important, desirable and practical feature of the RSDS is its expandability: you can easily connect other tools to the system. It is worth to underline that the RSDS was designed and made in accordance with modern design and programming techniques.

Using this system makes it possible to exchange information between academics and practitioners who are interested in the foundations and applications of rough sets. The developers of the RSDS sincerely hope that it will result in a marked increase of interest in rough set theory and related approaches. The research results published in the works represented in this system will stimulate the further development of the foundations, methods and real-life applications of these approaches in intelligent systems [8].

It is not possible to present all aspects of the possibilities and use of the RSDS in one paper with a significant limitation of its length. Therefore, we are planning to prepare a separate paper to fully present the benefits of using the RSDS.

The RSDS is being developed by the Rough Set and Petri Net Research Group [31], operating at Rzeszów University in cooperation with the International Rough Set Society [29], among others.

Acknowledgment. The authors are very grateful to Professor Andrzej Skowron for his constructive comments and encouragement to work on the development of the RSDS. In addition, the authors also appreciate anonymous reviewers for helpful comments.

References

1. Chakraverty, S., Sahoo, D.M., Mahato, N.R.: Concepts of Soft Computing: Fuzzy and ANN with Programming. Springer, Singapore (2019). https://doi.org/10.1007/978-981-13-7430-2
2. Devlin, K.: Goodbye Descartes: The End of Logic and the Search for a New Cosmology of the Mind. Wiley, Hoboken (1997)
3. Pal, S.K., Polkowski, L., Skowron, A. (eds.): Rough-Neural Computing. Techniques for Computing with Words. Springer, Heidelberg (2004). https://doi.org/10.1007/978-3-642-18859-6
4. Pawlak, Z.: Information systems - theoretical foundations. Inf. Syst. **6**, 205–218 (1981)
5. Pawlak, Z.: Rough sets. Int. J. Comput. Inf. Sci. **11**(1982), 205–218 (1982)
6. Pawlak, Z.: Rough Sets - Theoretical Aspects of Reasoning About Data. Kluwer, Dordrecht (1991)
7. Pedrycz, W., Skowron, A., Kreinovich, V. (eds.): Handbook of Granular Computing. Wiley, Hoboken (2008)
8. Skowron, A., Suraj, Z. (eds.): Rough Sets and Intelligent Systems - Professor Zdzisław Pawlak in Memoriam. Intelligent Systems Reference Library, vols. 42, 43, Springer, Heidelberg (2013). https://doi.org/10.1007/978-3-642-30344-9
9. Suraj, Z., Grochowalski, P.: The rough sets database system: an overview. Bull. Int. Rough Set Soc. **7**(1/2), 75–81 (2003). (Ed. by, Tsumoto, S., Miyamoto, S.) Shimane, Japan
10. Suraj, Z., Grochowalski, P.: The rough sets database system: an overview. In: Tsumoto, S., et al. (eds.) RSCTC 2004. LNAI, vol. 3066, pp. 841–849. Springer, Heidelberg (2004). https://doi.org/10.1007/978-3-540-25929-9_107
11. Suraj, Z., Grochowalski, P.: The rough set database system: an overview. In: Peters, J.F., Skowron, A. (eds.) Transactions on Rough Sets III. LNCS, vol. 3400, pp. 190–201. Springer, Heidelberg (2005). https://doi.org/10.1007/11427834_9
12. Suraj, Z., Grochowalski, P.: Functional extension of RSDS system. In: Greco, S., et al. (eds.) RSCTC 2006. LNCS, vol. 4259, pp. 786–795. Springer, Heidelberg (2006). https://doi.org/10.1007/11908029_81

13. Suraj, Z., Grochowalski, P.: On basic possibilities of RSDS system. In: Wakulicz-Deja, A. (ed.) Decision Support Systems, pp. 179–185. University of Silesia, Katowice (2007)
14. Suraj, Z., Grochowalski, P.: The rough sets database system. In: Peters, J.F., Skowron, A. (eds.) Transactions on Rough Sets VIII. LNCS, vol. 5084, pp. 307–331. Springer, Heidelberg (2008). https://doi.org/10.1007/978-3-540-85064-9_14
15. Suraj, Z., Grochowalski, P.: Patterns of collaborations in rough set research. In: Bello, R., Falcón, R., Pedrycz, W., Kacprzyk, J. (eds.) Granular Computing: At the Junction of Rough Sets and Fuzzy Sets. SFSC, vol. 224, pp. 79–92. Springer, Heidelberg (2008). https://doi.org/10.1007/978-3-540-76973-6_5
16. Suraj, Z., Grochowalski, P.: Toward intelligent searching the rough set database system (RSDS): an ontological approach. Fundam. Inform. **101**, 115–123 (2010)
17. Suraj, Z., Grochowalski, P.: Some comparative analyses of data in the RSDS system. In: Yu, J., Greco, S., Lingras, P., Wang, G., Skowron, A. (eds.) RSKT 2010. LNCS, vol. 6401, pp. 8–15. Springer, Heidelberg (2010). https://doi.org/10.1007/978-3-642-16248-0_7
18. Suraj, Z., Grochowalski, P., Lew, Ł.: Discovering patterns of collaboration in rough set research: statistical and graph-theoretical approach. In: Yao, J., Ramanna, S., Wang, G., Suraj, Z. (eds.) RSKT 2011. LNCS, vol. 6954, pp. 238–247. Springer, Heidelberg (2011). https://doi.org/10.1007/978-3-642-24425-4_33
19. Suraj, Z., Grochowalski, P., Lew, Ł.: Pawlak collaboration graph and its properties. In: Kuznetsov, S.O., Ślęzak, D., Hepting, D.H., Mirkin, B.G. (eds.) RSFDGrC 2011. LNCS, vol. 6743, pp. 365–368. Springer, Heidelberg (2011). https://doi.org/10.1007/978-3-642-21881-1_56
20. Suraj, Z., Grochowalski, P., Lew, Ł.: Pawlak collaboration graph of the second kind and its properties. In: Proceeding of the Workshop on CS&P 2011, Pułtusk, Poland, pp. 512–522, September 2011, Białystok University of Technology (2011)
21. Suraj, Z., Grochowalski, P., Pancerz, K.: On knowledge representation and automated methods of searching information in bibliographical data bases: a rough set approach. In: [8], vol. 43
22. Suraj, Z., Grochowalski, P.: About new version of the RSDS system. Fundam. Inform. **135**, 503–519 (2014)
23. Wilson, R.J.: Introduction to Graph Theory. Longman (1985)
24. Yao, J.T., Onasanya, A.: Recent development of rough computing: a scientometrics view. Stud. Comput. Intell. **708**, 21–45 (2017)
25. Yao, J.T.: The impact of rough set conferences. In: Mihálydeák, T. et al. (eds.) IJCRS 2019. LNCS, vol. 11499, pp. 383–394. Springer, Cham (2019). https://doi.org/10.1007/978-3-030-22815-6_30
26. Zadeh, L.A.: Fuzzy sets. Inf. Control **8**, 338–353 (1965)
27. Zadeh, L.A.: Fuzzy logic, neural networks and soft computing. Commun. ACM, 77–84 (1994)
28. BibTeX. http://www.bibtex.org/
29. International Rough Set Society. https://www.roughsets.org/
30. Rough Set Database System. http://rsds.ur.edu.pl/
31. Rough Set and Petri Net Research Group. http://rspn.univ.rzeszow.pl/

Many-Valued Dynamic Object-Oriented Inheritance and Approximations

Andrzej Szałas[1,2](✉)

[1] Institute of Informatics, University of Warsaw, Banacha 2, 02-097 Warsaw, Poland
andrzej.szalas@mimuw.edu.pl
[2] Department of Computer and Information Science, Linköping University,
581 83 Linköping, Sweden
andrzej.szalas@liu.se

Abstract. The majority of contemporary software systems are developed using object-oriented tools and methodologies, where constructs like classes, inheritance and objects are first-class citizens. In the current paper we provide a novel formal framework for many-valued object-oriented inheritance in rule-based query languages. We also relate the framework to rough set-like approximate reasoning. Rough sets and their generalizations have intensively been studied and applied. However, the mainstream of the area mainly focuses on the context of information and decision tables. Therefore, approximations defined in the much richer object-oriented contexts generalize known approaches.

1 Introduction

Rough sets and their generalizations have intensively been studied and applied, there is a vast literature on related techniques and problems (see, e.g., [4,16–19,24,29,30] and references there). The mainstream of the area mainly focuses on the context of information and decision tables. This approach has proved very useful in applications where information structures can be derived from or modeled by relational approaches. On the other hand, the majority of contemporary software systems are developed using object-oriented tools, where constructs like classes, inheritance and objects are first-class citizens. In particular, such constructs allow one to define complex taxonomies and offer means to specialize concepts' definitions when moving from general to more specific ones. In the current paper we provide a novel definition of many-valued object-oriented inheritance in the context of rule-based query languages and show its relationship to rough set-like approximations. That is, we generalize approximations in two ways: from classical to many-valued and from those based on information/decision tables to more general, based on more complex object-oriented structures.

As the technical engines we use a many-valued framework o^nQL developed in [25] and many-valued nested structures defined in [8]. More precisely, we first extend o^nQL with object-oriented inheritance. Next, we indicate how nested

---

Supported by the Polish National Science Centre grant 2017/27/B/ST6/02018.

S. Ramanna et al. (Eds.): IJCRS 2021, LNAI 12872, pp. 103–119, 2021.
https://doi.org/10.1007/978-3-030-87334-9_10

structures represent $O^n QL$ with inheritance. Since $O^n QL$ encompasses a class of rule-based query languages, including ASP [2,9,10,13,23] and 4QL [14,15], nested structures serve as an abstract representation of rule-based object-oriented languages and are further used to define approximations.

The considered structures reflect (object-oriented) databases or belief bases so we deal with finite domains. To simplify notation and considerations we restrict the language to its propositional version. In this context, rules with first-order variables may be seen as concisely expressed schemata which can later be translated to propositional logic using grounding.[1]

The original contribution of the paper includes:

- a novel definition of object-oriented inheritance in the context of many-valued query languages;
- a representation of objects in the context of many-valued nested structures;
- generalizations of rough set-like approximations of the used semantical structures;
- tractability results for computing queries and approximations.

The rest of the paper is structured as follows. First, in Sect. 2, we introduce the formal framework used in the paper by reminding rule-based query languages, many-valued logics and nested structures. Next, in Sect. 3, we discuss and define many-valued inheritance. In Sect. 4 we show how approximations may be interpreted in the introduced framework. Finally, Sect. 5 concludes the paper.

2 Preliminaries

2.1 Many-Valued Logics

Let us start with the syntax of many-valued logics considered in this paper. The connectives we will use are $\neg, \wedge, \vee, \rightarrow, \dot{\wedge}, \dot{\vee}$, where $\neg, \wedge, \vee$ and $\rightarrow$ are traditional connectives expressing *negation, conjunction, disjunction* and *implication*, and the connectives $\dot{\wedge}, \dot{\vee}$ are respectively *doxastic conjunction* and *doxastic disjunction* [8].

Definition 1 (Formulas). *The following BNF grammar defines the syntax of formulas, where $\mathcal{F}$ denotes the set of formulas and $\mathcal{P}$ denotes the set of propositional variables:*
$$\langle \mathcal{F} \rangle ::= \mathcal{P} \mid \neg \langle \mathcal{F} \rangle \mid \langle \mathcal{F} \rangle \wedge \langle \mathcal{F} \rangle \mid \langle \mathcal{F} \rangle \vee \langle \mathcal{F} \rangle \mid \langle \mathcal{F} \rangle \rightarrow \langle \mathcal{F} \rangle \mid$$
$$\langle \mathcal{F} \rangle \dot{\wedge} \langle \mathcal{F} \rangle \mid \langle \mathcal{F} \rangle \dot{\vee} \langle \mathcal{F} \rangle$$
When $p \in \mathcal{P}$, p is also called a positive literal *and $\neg p$ is a* negative literal. □

Before defining the semantics of the considered logics let us indicate the rule-based context in which they are employed. In order to evaluate rules expressed in many-valued logics, one typically needs two orderings:

- *truth ordering* for evaluating truth value of rule's premises;

---

[1] This technique is actually used, e.g., in SAT solver-based implementations of ASP.

– *information ordering* for fusing knowledge from different rules when they contribute to a conclusion involving the same propositional variable.

Truth ordering is then used to define the semantics of traditional connectives while information ordering is used for doxastic connectives. For example, orderings for ASP are shown in Fig. 1.

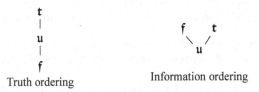

Truth ordering   Information ordering

Fig. 1. Orderings for Kleene logic used in ASP.

Another example of truth and information ordering is shown in Fig. 2. These orderings are used in the 4QL rule language, based on a four-valued logic with non-classical truth values u and i denoting "unknown" and "inconsistent".[2]

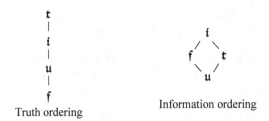

Truth ordering   Information ordering

Fig. 2. Orderings used in 4QL.

Conjunction and disjunction are interpreted as the minimum and the maximum wrt respective ordering. To illustrate the idea consider the following two rules:

$$p :\text{-} q, r. \tag{1}$$
$$\neg p :\text{-} s. \tag{2}$$

According to rule (1), p obtains the truth value of the conjunction of q and r. The rule (2) assigns the truth value of s to $\neg p$. The rules' premises are evaluated using truth ordering so, assuming that q, r, s are true, the rule (1) results in p being true and the rule (2) assigns t to $\neg p$ (i.e., f to p). Therefore, inconsistent information

---

[2] Observe that removing the value i from these orderings result in orderings shown in Fig. 1.

about p is derived. In 4QL we use doxastic disjunction (wrt information ordering) to fuse the results what assigns the value i to p.[3] Note that using truth ordering one would obtain the value t giving, without a good reason, a higher priority to the first rule.

As widely accepted [22,26], in order to define the semantics of an n-valued logic, $\mathcal{L}_n$, one first has to fix:

- a set of *truth values*, $\{\tau_1, \ldots, \tau_n\}$, where $n \geq 2$;
- a non-empty set $\mathcal{D}$ of *designated* truth values, acting as true.

We additionally fix a set $\mathcal{U}$ of *ignorance representing* truth values acting as unknown, proposed in [8]. In the current paper it is needed to define the semantics of inheritance and approximations.

Rule languages, like ASP and 4QL, employ the Open World Assumption so are basically three-valued [5,7,21] with the third truth value representing ignorance. The paraconsistent rule language 4QL is four-valued [14,15] with the fourth truth value representing inconsistency and may be used to implement paraconsistent rough sets [28]. Therefore we will further require that we deal with at least three truth values with the third truth value, u, representing ignorance. Semantically, all logic we consider in the paper are extensions of Kleene three-valued logic K_3 [12].

While the semantics of traditional connectives is based on *truth ordering*, doxastic connectives reflect *information ordering*. Let us now formally define the semantics of many-valued logics we deal with.

Definition 2 (Truth Values). *By a set of truth values we shall mean* $\mathcal{T} = \{\tau_1, \ldots, \tau_n\}$ *such that* $\{f, u, t\} \subseteq \mathcal{T}$. *We distinguish three disjoint subsets of* $\mathcal{T}$, *closed under propositional connectives* $\vee, \wedge, \dot{\vee}, \dot{\wedge}$:

- *a set* $\mathcal{D} \subsetneq \mathcal{T}$, *called the set of* designated *truth values, such that* $t \in \mathcal{D}$, $f, u \notin \mathcal{D}$;
- *a set* $\mathcal{U} \subsetneq \mathcal{T}$, *called the set of* ignorance representing *truth values, such that* $u \in \mathcal{U}$, $f, t \notin \mathcal{U}$.   □

To define the semantics we need a mapping w, assigning truth values to propositions, $w : \mathcal{P} \longrightarrow \{\tau_1, \ldots, \tau_n\}$, and then its extension to all formulas of $\mathcal{L}$, reflecting the semantics of connectives,

$$w : \mathcal{F} \longrightarrow \{\tau_1, \ldots, \tau_n\}. \tag{3}$$

Definition 3 (Semantics of Connectives wrt Truth Values). *Let* $\mathcal{T}$ *be the set of truth values and* $\leq_t$, $\leq_i$ *be orderings (truth and information ordering, respectively).*

---

[3] Each rule is treated as a separate information source. In 4QL ignorance and inconsistencies may be resolved using heuristic/non-monotonic rules or additional information sources. For simplicity we skip these aspects here. For details see [14,15].

– *The semantics of negation* ¬ *is given by a function* $\nu : T \longrightarrow T$, *such that on truth values* f, u, t *we have:*[4]

$$\nu(f) \stackrel{\text{def}}{=} t, \quad \nu(u) \stackrel{\text{def}}{=} u, \quad \nu(t) \stackrel{\text{def}}{=} f. \qquad (4)$$

– *For* $\tau, \tau' \in T$,

$$\tau \wedge \tau' \stackrel{\text{def}}{=} \text{glb}_{\leq_t}\{\tau, \tau'\}; \quad \tau \vee \tau' \stackrel{\text{def}}{=} \text{lub}_{\leq_t}\{\tau, \tau'\}; \qquad (5)$$

$$\tau \dot{\wedge} \tau' \stackrel{\text{def}}{=} \text{glb}_{\leq_i}\{\tau, \tau'\}; \quad \tau \dot{\vee} \tau' \stackrel{\text{def}}{=} \text{lub}_{\leq_i}\{\tau, \tau'\}, \qquad (6)$$

where glb *and* lub *denote respectively the greatest lower and the least upper bound wrt the specified ordering, where* $\leq_t$ *and* $\leq_i$ *stand for truth and information ordering, respectively..*   □

Remark 1 (On Undefined Values of lub *and* glb*).* Note that lub, glb do not always have to be defined. For example, glb$\{f, t\}$ wrt information ordering for Kleene logic shown in Fig. 1 is undefined. Whenever it is referred to, the computation fails and the computed candidate for a model is rejected. This happens, for example, in ASP when, for some $p \in P$ a rule concludes that p is t and another rule concludes that $\neg p$ is t (i.e., p is f). Here inconsistency is encountered and the current candidate for a model is rejected since ASP models are required to be consistent.   □

Example 1 (The Semantics of Negation). The semantics of negation specified in (4) is abstract and concrete definitions are to be provided for particular logics. E.g., in the case of three- and four-valued logics with truth values included in $\{f, u, i, t\}$ one typically defines:

$$\nu(f) \stackrel{\text{def}}{=} t, \quad \nu(u) \stackrel{\text{def}}{=} u, \quad \nu(i) \stackrel{\text{def}}{=} i, \quad \nu(t) \stackrel{\text{def}}{=} f. \qquad (7)$$

Indeed, (7) with truth values $\{f, u, t\}$ is assumed in Kleene logic K_3 [12] and in many other three-valued logics (even though u may have various meanings, like "neutral", "meaningless", "nonsense", etc.). The definition for i is used, e.g., in Priest's three-valued logic [20] and the four-valued logic used in [14, 15, 28].   □

The definition can be extended to cover all formulas in the standard manner, assuming additionally that:

$$\tau \rightarrow \tau' \stackrel{\text{def}}{=} \nu(\tau) \vee \tau'. \qquad (8)$$

Of course, many-valued implication can be defined in many ways. Equation (8) can be substituted by any other definition, e.g.,

$$\tau \rightarrow_t \tau' \stackrel{\text{def}}{=} \tau \leq_t \tau'. \qquad (9)$$

The intention is that implication $\rightarrow$ determines inclusion in the standard sense:

$$A \subseteq B \text{ iff for every object } x \text{ in the domain, } x \in A \rightarrow x \in B. \qquad (10)$$

---

[4] ν is a function specific to a logic in question. The requirements (4) and (5) make the considered many-valued logics compatible with K_3 on connectives $\neg, \wedge, \vee$.

Remark 2 (The Status of Truth and Information Ordering). For the sake of clarity of presentation, as its semantical foundation, Definition 3 uses orderings $\leq_t$, $\leq_i$. Of course, one could employ logical matrices or other formalisms. The inheritance framework presented in the paper is independent on the form the semantics is presented.   □

2.2   Nested Structures

Let us now recall nested structures of [8]. They will be used to model any set of objects defined in the o^nQL framework.

Definition 4 (Worlds, Nested Structures). *Let $P \subseteq \mathcal{P}$ be a finite set of propositional variables. By a* world *over P we mean any function $w : P \longrightarrow T$.*

Let S be the smallest set containing worlds, such that whenever $\sigma_1, \ldots, \sigma_n \in S$ $(n \geq 0)$ then also $\{\sigma_1, \ldots, \sigma_n\} \in S$. Members of S are called nested structures *or* structures, *for simplicity.*   □

In the rest of the paper, for $p \in \mathcal{P}$, $\tau \in \{\tau_1, \ldots, \tau_n\}$ and a structure σ, by $\langle p, \tau \rangle$ we denote that $\sigma(p) = \tau$. In such cases, σ will always be known from the context.

While worlds assign truth values to given propositional variables, nested structures may be sets of arbitrary order with worlds serving as the "zero-order" elements.

2.3   Rule-Based Object-Oriented Query Languages

In o^nQL, like in other object-oriented languages, objects are created using classes as patterns. Classes, among others, contain rules which allow for a concise and uniform representation of the contents of objects. It is assumed that actual parameters used in object creation uniquely determine the object. Using parameters and possibly other data, rules are instantiated and can be used to compute the object's contents (the truth values of facts and perhaps other objects).

For example, consider a class specification in Program 1 (for details see [25]):[5] Assume that an object 'person(jim)' with 'mother(amy)' and 'father(john)' is created. Then the rule in Lines 7–10 is instantiated to:

```
sibling(Z) :- mother(amy), father(john),     % M=amy, F=john
              person(Z).mother(amy),
              person(Z).father(john),
              Z=/=jim.                        % N=jim
```

Therefore, the object consists of 'sibling(Z)' with 'Z' instantiated to domain objects (names) satisfying the rule's premises, i.e., having 'amy', 'john' as Jim's mother and father, and not being 'jim' himself.

---

[5] According to a convention used in rule languages, comma in rules' premises (bodies) is interpreted as a conjunction.

Program 1: Example of class definition.

```
1  class person(N: name){ % 'name' is the type of N
2  |   class mother(name) % mother of N
3  |   class father(name)  % father of N
4  |   prologue:           % set mother and father as N's parents
5  |   |  ...              % e.g., using external sources
6  |   rules:
7  |   |   sibling(Z) :- mother(M), father(F),
8  |   |               person(Z).mother(M), % Z has the same mother
9  |   |               person(Z).father(F),  % Z has the same father
10 |   |               Z≠N.                  % Z is not N
11 |  }
```

Given that Tom and Eve are Jim's siblings, the object 'person(jim)' both in Asp and 4QL can be modeled by the nested structure where, for simplicity, it is assumed that the values of all literals not listed in respective sets obtain the value u in these sets:

$$\text{person(jim)} = \big\{ \ \{\langle mother(amy), \mathbf{t}\rangle\}, \ \{\langle father(john), \mathbf{t}\rangle\}, \\ \langle sibling(tom), \mathbf{t}\rangle, \ \langle sibling(eve), \mathbf{t}\rangle \ \big\}. \tag{11}$$

In considering the relationship among rule-based and object-oriented languages the following additional remarks are in order.

Remark 3 (First-Order vs Propositional Literals). In the paper we consider propositional version of many-valued logics. Accordingly, nested structures are defined to include propositional information only. On the other hand, the nested structure (11) contains expressions like 'mother(amy)'. Though looking like first-order literals, they only contain constants, so can be treated as propositional variables with brackets added for a better readability.   □

Remark 4 (Methods). In object-oriented languages one specifies *methods* interfacing objects with their environment and specifying their behavior. Methods allow other components of object-oriented systems to access object data. In o^nQL methods are specified by propositional variables.   □

Remark 5 (Nested Structures vs Objects). Worlds in the sense of Definition 4 reflect the contents of *flat* objects (created on the basis of classes not containing other classes). Sets of worlds are containers of flat objects. The induction step in Definition 4 allows one to model objects of arbitrary finite order. That is, unrestricted finite nesting of classes is allowed. Therefore arbitrary objects of o^nQL can be modeled by nested structures.

Note also that programs in many languages, including Asp, may have many models. Given that $M_1, \ldots, M_r$ are all models of a program, the corresponding nested structure can be the set $\{M_1, \ldots, M_r\}$ consisting of these models.   □

Definition 5 (Object Expressions). *By a simple object expression we under-stand any expression of the form 'c(ā)', where 'c' is a class name and 'ā' is a tuple of constants of the domain, assuming type compatibility among formal parameters of 'c' and actual parameters in 'ā'.*[6]

By the set of object expressions, $\mathcal{O}$, we understand the smallest set of expres-sions including simple object expressions and such that whenever each expression in ā is in $\mathcal{O}$ then $c(ā) \in \mathcal{O}$, assuming type compatibility among formal parameters of 'c' and actual parameters in 'ā'. □

Expressions in $\mathcal{O}$ denote objects syntactically. Since the o^nQL framework is inde-pendent of the underlying rule language, we assume that a mapping, Σ reflecting the semantics of the rule language is given by the following two mappings:

$$\Sigma : \mathcal{O} \longrightarrow \mathcal{S}; \tag{12}$$
$$\Phi : \mathcal{S} \times \mathcal{P} \longrightarrow \{\tau_1, \ldots, \tau_n\}, \tag{13}$$

where $\mathcal{O}$ is the set of object expressions (see Definition 5), $\mathcal{S}$ is the set of nested structures (see Definition 4) and $\mathcal{P}$ is the set of propositional variables (see Def-inition 1).[7] The mapping Σ assigns to each object a nested structure and is dependent on the semantics of the underlying rule language. The mapping Φ reflects the semantics of methods in nested structures, where fusing informa-tion from different components of a nested structure is needed. For example, when nested structures serve as belief bases, doxastic disjunction can be used to combine results obtained from sub-components [8].

Example 2 (Illustrating Φ). Of course, Φ may be defined in various ways. The one introduced in [8] uses $\dot{\vee}$ to fuse information from different components of a nested structure, $\Phi(\mathcal{S}, p) \stackrel{\text{def}}{=} \bigvee_{w \in \mathcal{S}} \dot{} w(p)$. To illustrate the idea, consider the following nested structure consisting of two worlds, w_1 and w_2, e.g., representing the results of classifiers, each linked to a camera located at a different place:

$$\mathcal{S} = \{ \underbrace{\{\langle tall(jim), \mathsf{t}\rangle, \langle tall(amy), \mathsf{u}\rangle, \langle tall(john), \mathsf{t}\rangle\}}_{w_1},$$
$$\underbrace{\{\langle tall(jim), \mathsf{t}\rangle, \langle tall(amy), \mathsf{f}\rangle, \langle tall(john), \mathsf{f}\rangle\}}_{w_2} \}. \tag{14}$$

When $\dot{\vee}$ is defined over information ordering shown in Fig. 2, we have:

$$\Phi(\mathcal{S}, tall(jim)) = w_1(tall(jim)) \dot{\vee} w_2(tall(jim)) = \mathsf{t} \dot{\vee} \mathsf{t} = \mathsf{t};$$
$$\Phi(\mathcal{S}, tall(amy)) = w_1(tall(amy)) \dot{\vee} w_2(tall(amy)) = \mathsf{u} \dot{\vee} \mathsf{f} = \mathsf{f};$$
$$\Phi(\mathcal{S}, tall(john)) = w_1(tall(john)) \dot{\vee} w_2(tall(john)) = \mathsf{t} \dot{\vee} \mathsf{f} = \mathsf{i}.$$

□

---

[6] We assume here type compatibility in the sense of o^nQL [25].

[7] Recall that we assume that objects are created using classes as patterns. Therefore we require that actual parameters determine a unique object represented by a nested structure.

3 Many-Valued Dynamic Object Inheritance

In most object-oriented programming languages the approach to inheritance is rather rigid. This creates many problems especially when multiple inheritance is allowed. On the other hand, Common Lisp offers much more flexible approach called *method combination*. When several methods can be selected, one can decide how to combine their results, making dynamic inheritance user-programmable [1,27].

The approach we propose can be located somewhere between the rigid approaches and the Lisp-like method combination. In our approach, methods always return truth values as their results. We will therefore allow combining methods in a more restricted but, in our opinion, better controlled manner via *inheritance expressions* which can be analyzed and verified using the background logical formalisms.

Due to the heterogeneity and dynamics of contemporary AI/ML-based systems, in the paper we focus on inheritance at the object level. Of course, the method can be made static by shifting the approach to the class level. However, parameter passing and other syntactic conventions should also be addressed in this case.

The main concept for inheritance which we introduce is that of inheritance expressions defined as follows, where $\gamma \dashrightarrow \delta$ denotes that γ inherits from δ.

Definition 6 (Inheritance Expressions). *The following BNF grammar defines the syntax of* inheritance expressions, *where $\mathcal{I}$ denotes the set if inheritance expressions and $\mathcal{O}$ is the set of object expressions:*

$$\langle \mathcal{I} \rangle ::= \langle \mathcal{O} \rangle \mid \langle \mathcal{I} \rangle \dashrightarrow \langle \mathcal{I} \rangle \mid \neg \langle \mathcal{I} \rangle \mid \langle \mathcal{I} \rangle \wedge \langle \mathcal{I} \rangle \mid \langle \mathcal{I} \rangle \vee \langle \mathcal{I} \rangle \mid \langle \mathcal{I} \rangle \rightarrow \langle \mathcal{I} \rangle \mid$$
$$\langle \mathcal{I} \rangle \dot{\wedge} \langle \mathcal{I} \rangle \mid \langle \mathcal{I} \rangle \dot{\vee} \langle \mathcal{I} \rangle \qquad \Box$$

Observe that inheritance expressions are like formulas except that rather than propositional variables, object expressions serve as terminal symbols, and the inheritance symbol '$\dashrightarrow$' is also a connective.

Definition 7 (Semantics of Inheritance Expressions). *Semantics of inheritance expressions is defined inductively by a mapping $\Psi : \mathcal{I} \times \mathcal{P} \longrightarrow \{\tau_1, \ldots, \tau_n\}$ assigning truth values to propositional variables, where mappings Σ and Φ are specified by Eqs. (12)–(13) and $\mathcal{U}$ is the set of ignorance representing truth values (see Definition 2):*

- $\Psi(o, p) \stackrel{\text{def}}{=} \Phi(\Sigma(o), p)$, *where $o \in \mathcal{O}$;*
- $\Psi(\gamma \dashrightarrow \delta, p) \stackrel{\text{def}}{=} \begin{cases} \Psi(\gamma, p) \text{ when } \Psi(\gamma, p) \notin \mathcal{U}; \\ \Psi(\delta, p) \text{ otherwise}; \end{cases}$
- $\Psi(\neg\gamma, p) \stackrel{\text{def}}{=} \neg\Psi(\gamma, p)$;
- $\Psi(\gamma \circ \delta, p) \stackrel{\text{def}}{=} \Psi(\gamma, p) \circ \Psi(\delta, p)$, *where $\circ \in \{\wedge, \vee, \rightarrow, \dot{\wedge}, \dot{\vee}\}$.* $\qquad \Box$

According to Definition 7, in the case of inheritance operator $\gamma \dashrightarrow \delta$ one first looks for the most specific method defining a given method p (provided by γ). If it appears unknown, one looks for the value of p in the expression δ.

Given the values of propositional variables wrt inheritance expressions one can define the semantics of arbitrary queries as follows.

Definition 8 (Semantics of Queries). *The mapping Ψ is extended to arbitrary formulas as follows, where $\gamma \in \mathcal{I}$ and $\alpha, \beta \in \mathcal{F}$:*

- $\Psi(\gamma, \neg\alpha) \stackrel{\text{def}}{=} \neg\Psi(\gamma, \alpha);$
- $\Psi(\gamma, \alpha \circ \beta) \stackrel{\text{def}}{=} \Psi(\gamma, \alpha) \circ \Psi(\gamma, \beta)$, *where* $\circ \in \{\wedge, \vee, \rightarrow, \dot\wedge, \dot\vee\}.$ □

The following example shows a non-conflicting multiple inheritance.

Example 3 (Inheritance). Consider a smartphone which inherits properties of a standard phone and a camera. Figure 3 shows a suitable inheritance.

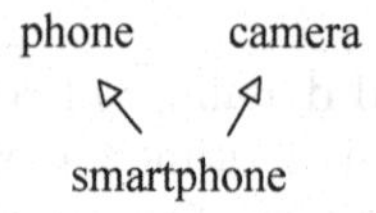

Fig. 3. Inheritance considered in Example 3.

Assume we have three objects with the following contents:[8]

$$\text{phone} = \{\, \langle \text{can\_call}, \mathbf{t} \rangle \,\}, \langle \text{can\_make\_video}, \mathbf{f} \rangle \,\}, \tag{15}$$

$$\text{camera} = \{\, \langle \text{can\_call}, \mathbf{f} \rangle \,\}, \langle \text{can\_make\_video}, \mathbf{t} \rangle \,\}, \tag{16}$$

$$\text{smartphone} = \{\, \langle \text{can\_call}, \mathbf{u} \rangle \,, \langle \text{can\_make\_video}, \mathbf{u} \rangle \,\}. \tag{17}$$

The following inheritance expressions reflect the inheritance shown in Fig. 3. For illustration purposes, the expressions specify two methods for information fusion:

$$\text{smartphone} \rightarrow (\text{phone} \vee \text{camera}); \tag{18}$$

$$\text{smartphone} \rightarrow (\text{phone} \dot\vee \text{camera}). \tag{19}$$

Using orderings shown in Fig. 2 and Definition 7, one obtains that:

$$\Psi((18), \text{can\_call}) = \big(\mathbf{u} \rightarrow (\mathbf{t} \vee \mathbf{f})\big) = \mathbf{t}; \tag{20}$$

$$\Psi((18), \text{can\_make\_video}) = \big(\mathbf{u} \rightarrow (\mathbf{f} \vee \mathbf{t})\big) = \mathbf{t}; \tag{21}$$

$$\Psi((19), \text{can\_call}) = \big(\mathbf{u} \rightarrow (\mathbf{t} \dot\vee \mathbf{f})\big) = \mathbf{i}; \tag{22}$$

$$\Psi((19), \text{can\_make\_video}) = \big(\mathbf{u} \rightarrow (\mathbf{f} \dot\vee \mathbf{t})\big) = \mathbf{i}. \tag{23}$$

As shown by the results (20)–(21), the choice of $\vee$ as the information fusion operator works well since properties inherited by 'smartphone' from 'phone' and 'camera' are not in conflict with each other. The results in (22)–(23) would be more suitable for conflicting information whose fusion is inconsistent (see, e.g., Example 4 below).   □

---

[8] For clarity, the unknown facts in 'smartphone' are listed explicitly.

As the second example let us consider "Nixon diamond", a classical scenario used in default reasoning to illustrate the problem of inconsistent (conflicting) conclusions.

Example 4 (Diamond Problem). The scenario is formulated as follows:

- usually, Quakers are pacifists;
- usually, Republicans are not pacifists;
- Nixon is both a Quaker and a Republican.

The question is whether Nixon is a pacifist or not?

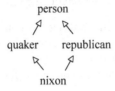

Fig. 4. Inheritance considered in the "Nixon diamond" example.

The corresponding inheritance is shown in Fig. 4, where:

- 'nixon' inherits from 'quaker' and 'republican' simultaneously;
- both 'quaker' and 'republican' inherit from 'person'.

Using inheritance expressions one has to decide how two conflicting conclusions should be combined. At least the following options can be considered:

$$\text{nixon} \dashrightarrow (\text{quaker} \wedge \text{republican}) \dashrightarrow \text{person}; \tag{24}$$

$$\text{nixon} \dashrightarrow (\text{quaker} \vee \text{republican}) \dashrightarrow \text{person}; \tag{25}$$

$$\text{nixon} \dashrightarrow (\text{quaker} \overset{.}{\wedge} \text{republican}) \dashrightarrow \text{person}; \tag{26}$$

$$\text{nixon} \dashrightarrow (\text{quaker} \overset{.}{\vee} \text{republican}) \dashrightarrow \text{person}. \tag{27}$$

To simplify the discussion, consider four objects with the following contents:

$$\text{nixon} = \text{person} = \{ \, \langle \text{pacifist}\, \mathfrak{u} \rangle \, \}, \tag{28}$$

$$\text{quaker} = \{ \, \langle \text{pacifist}\, \mathfrak{t} \rangle \, \}, \quad \text{republican} = \{ \, \langle \text{pacifist}\, \mathfrak{f} \rangle \, \}. \tag{29}$$

Using orderings shown in Fig. 2, Definition 7 results in:

$$\Psi((24), \text{pacifist}) = \left(\mathfrak{u} \dashrightarrow (\mathfrak{t} \wedge \mathfrak{f}) \dashrightarrow \mathfrak{u} \right) = \mathfrak{f}; \tag{30}$$

$$\Psi((25), \text{pacifist}) = \left(\mathfrak{u} \dashrightarrow (\mathfrak{t} \vee \mathfrak{f}) \dashrightarrow \mathfrak{u} \right) = \mathfrak{t}; \tag{31}$$

$$\Psi\big((26), \text{pacifist}\big) = \big(\mathsf{u} \dashrightarrow (\mathsf{t} \wedge \mathsf{f}) \dashrightarrow \mathsf{u}\big) = \mathsf{u}; \tag{32}$$

$$\Psi\big((27), \text{pacifist}\big) = \big(\mathsf{u} \dashrightarrow (\mathsf{t} \mathbin{\dot\vee} \mathsf{f}) \dashrightarrow \mathsf{u}\big) = \mathsf{i}. \tag{33}$$

In the case of (30) 'republican' defeats 'quaker'. The case of (31) is opposite: 'quaker' defeats 'republican'. The expressions in (32) and (33) do not priori- tize 'quaker' nor 'republican'. When one considers objects as independent non- prioritized information sources, the choice of information fusion using $\dot\vee$ is more appropriate.[9]   □

Remark 6 (Differences Between Inheritance and Shadowing). Notice the differ- ence between inheritance and belief shadowing as defined in [8]:

- inheritance focuses on methods; queries are computed after methods are eval- uated. Shadowing concentrates on queries without inheriting methods;
- inheritance prioritizes methods from the most specific to the most general ones. In shadowing the order is inverted: query evaluation starts from the outermost layer;
- while inheritance is defined for the context of object-oriented rule-based query languages, shadowing concentrates on belief bases.   □

Complexitywise, we have the following theorem, where we address data com- plexity. That is, we assume that the size of the query (α) is constant.

Theorem 1 (Data Complexity of Queries). *Assume that computing truth values of connectives and the values of Φ is tractable. Then for every $\gamma \in \mathcal{I}$ and $\alpha \in \mathcal{F}$, computing the truth value $\Psi(\gamma, \alpha)$ is tractable wrt the size of $\Sigma(o)$, where Σ and Φ are specified by (12)–(13) and o is an object involved in γ with the maximal size of $\Sigma(o)$.*   □

4   Approximations

Let us now discuss approximations in the context of many-valued logics we deal with (for an alternative approach see, e.g., [11]). Three-valued interpretations of approximations are very natural since the boundary region gives rise to the third truth value 'unknown' (see [4,5,7] and references there). Four-valued paracon- sistent rough sets have been discussed, e.g., in [28]. When more truth values are present, or non-classical truth values are interpreted differently, new phenomena can be observed. Let us first define standard approximations, where E is the elementary relation modeling indiscernibility, similarity, proximity, etc.

Definition 9 (Approximations, Elementary Relations, Approximate Sets). *Let $\mathbb{D}$ be a domain of elements (objects), $c \subseteq \mathbb{D}$ and let σ be a binary*

---

[9] In fact, it is compatible with belief fusion in 4QL and belief bases of [8].

relation on $\mathbb{D}$. *Then the* lower approximation c_E^+ *and the* upper approximation $c_E^\oplus$ of c wrt E *are:*

$$c_E^+ \overset{\text{def}}{=} \{x \mid \forall y\big(E(x,y) \to y \in c\big)\}; \tag{34}$$

$$c_E^\oplus \overset{\text{def}}{=} \{x \mid \exists y\big(E(x,y) \land y \in c\big)\}. \tag{35}$$

The set $c_E^\pm \overset{\text{def}}{=} c_E^\oplus \setminus c_E^+$ *is called the* boundary region *of* c *wrt* E. *The relation* E *is called the* elementary relation *for approximations* c_E^+, $c_\sigma^\oplus$. *The pair* $\langle c_\sigma^+, c_\sigma^\oplus \rangle$ *is called an* approximate set. □

Definition 9 can be adapted to the many-valued case assuming that c and E are many-valued rather than classical. The next definition serves the purpose, where quantifiers and the membership relation are understood as follows:[10]

– we treat $\forall$ and $\exists$ as the conjunction and disjunction over domain elements:

$$\forall y\big(\alpha(y)\big) \overset{\text{def}}{=} \bigwedge_{d \in \mathbb{D}} \alpha(d); \qquad \exists y\big(\alpha(y)\big) \overset{\text{def}}{=} \bigvee_{d \in \mathbb{D}} \alpha(d); \tag{36}$$

– we denote $y \in c$ by $c(y)$ and, for each $d \in \mathbb{D}$, treat each $c(d)$ as a propositional variable. In such a case, worlds provide information about truth values of each $c(d)$, thus for $d \in c$. Similarly, for $d, e \in \mathbb{D}$, we encode $E(d, e)$ as a propositional variable.

In classical rough sets there is no need to define the negative region (being the lower approximation of negative information) since it is the complement of the upper approximation of a given set. In the many-valued case approximations of the negative region appear useful and sometimes needed. Therefore the following definition covers also these cases.

Definition 10 (Many-Valued Approximations). *Let* $\mathcal{L}$ *be a many-valued logic with the set* $\mathcal{D}$ *of designated truth values. Let* $\mathbb{D}$ *be a domain of elements (objects),* $c \subseteq \mathbb{D}$ *and let* σ *be a binary relation on* $\mathbb{D}$. *Given that* c *and* E *are given by an inheritance expression* $\gamma \in \mathcal{I}$, *the* lower approximation c_E^+, *upper approximation* $c_E^\oplus$, negative lower approximation c_E^- *and the* negative upper approximation $c_E^\ominus$ *of* c *wrt* E *are:*[11]

$$c_E^+ \overset{\text{def}}{=} \{x \mid \Psi\big(\gamma, \forall y\big(E(x,y) \to y \in c\big)\big) \in \mathcal{D}\}; \tag{37}$$

$$c_E^\oplus \overset{\text{def}}{=} \{x \mid \Psi\big(\gamma, \exists y\big(E(x,y) \land y \in c\big)\big) \in \mathcal{D}\}; \tag{38}$$

$$c_E^- \overset{\text{def}}{=} \{x \mid \Psi\big(\gamma, \forall y\big(E(x,y) \to y \notin c\big)\big) \in \mathcal{D}\}; \tag{39}$$

$$c_E^\ominus \overset{\text{def}}{=} \{x \mid \Psi\big(\gamma, \exists y\big(E(x,y) \land y \notin c\big)\big) \in \mathcal{D}\};. \tag{40}$$

□

---

[10] Recall that the domains we deal with are finite.
[11] Observe that approximations are classical two-valued sets.

Example 5 (Many-Valued Approximations). Let, for some object o, $\Sigma(o) = \mathcal{S}$, where $\mathcal{S}$ is the nested structure defined by (14). $\mathcal{S}$ specifies the (four-valued) set $tall \subseteq \mathbb{D}$, where $\mathbb{D} = \{jim, amy, john\}$. Let E be the (reflexive and symmetric closure) of the relation consisting of $\langle jim, john \rangle$.

To illustrate the approach, consider two variants of the four-valued logic with truth values $\{f, u, i, t\}$ and orderings shown in Fig. 2 serving as the semantical basis for connectives. The variants differ in the choice of designated truth values:[12]

$$\text{(i)} \quad \mathcal{D} = \{t\}, \quad \text{(ii)} \quad \mathcal{D} = \{t, i\}.$$

Let us first calculate queries $low(x), up(x), nlow(x)$ and $nup(x)$ for each $x \in \mathbb{D}$, useful for calculating approximations:

$$low(x) \stackrel{\text{def}}{=} \Psi\big(o, \forall y\big(E(x,y) \rightarrow tall(y)\big)\big) \quad \text{– for lower approximation;}$$
$$up(x) \stackrel{\text{def}}{=} \Psi\big(o, \exists y\big(E(x,y) \wedge tall(y)\big)\big) \quad \text{– for upper approximation;}$$
$$nlow(x) \stackrel{\text{def}}{=} \Psi\big(o, \forall y\big(E(x,y) \rightarrow \neg tall(y)\big)\big) \quad \text{– for negative lower approximation;}$$
$$nup(x) \stackrel{\text{def}}{=} \Psi\big(o, \exists y\big(E(x,y) \wedge \neg tall(y)\big)\big) \quad \text{– for negative upper approximation.}$$

We have:

$$low(jim) = i, \quad up(jim) = t, \quad nlow(jim) = f, \quad nup(jim) = i,$$
$$low(amy) = f, \quad up(amy) = f, \quad nlow(amy) = t, \quad nup(amy) = t,$$
$$low(john) = i, \quad up(john) = t, \quad nlow(john) = f, \quad nup(john) = i.$$

Therefore, according to Definition 10, the approximations of *tall* wrt E are:[13]

$$variant(i) : \quad tall_E^+ = \emptyset, \qquad\qquad tall_E^\oplus = \{jim, john\},$$
$$tall_E^- = \{amy\}, \qquad tall_E^\ominus = \{amy\};$$
$$variant(ii) : \quad tall_E^+ = \{jim, john\}, \quad tall_E^\oplus = \{jim, john\},$$
$$tall_E^- = \{amy\}, \qquad tall_E^\ominus = \{jim, amy, john\}.$$

□

Since approximations can be expressed as queries, we have the following corollary from Theorem 1.

Corollary 1 (Data complexity of approximations). *Assume that computing truth values of connectives and the values of Φ is tractable. Let the domain $\mathbb{D}$ be fixed. Then for every $\gamma \in \mathcal{I}$, $c \subseteq \mathbb{D}$ and $E \subseteq \mathbb{D} \times \mathbb{D}$, computing the approximations c_E^+, $c_E^\oplus$ and $c_E^\pm$ is tractable wrt the size of $\Sigma(o)$, where Σ and Φ are specified by (12)–(13) and o is an object involved in γ with the maximal size of $\Sigma(o)$.* □

---

[12] The assumption that i is a designated truth value is not artificial – see, e.g., the Priest logic [20].

[13] Since approximations are two-valued sets, we list elements belonging to a set.

5   Conclusions

In the paper we have developed a tractable framework for object-oriented inheritance in many-valued query languages. The use of nested structures of [8] allowed us to define inheritance abstracting from a particular query language. In particular, it is compatible with $o^n QL$ [25] and can be adapted to languages like Asp or 4QL as well as their many-valued generalizations.

It is worth noticing that the class of propositional many-valued logics with connectives definable by computable functions,[14] including the logics behind Asp or 4QL, can be shown decidable. In fact, signed tableaux provide us with a useful verification technique (for an example of tableaux for Kleene logic and Asp see, e.g., [6]). Though the worst case complexity of tableaux is exponential, the technique is widely used.[15]

The research initiated in this paper can be continued in many directions. First, in traditional rough sets one frequently considers approximations obtained by removing part of information, e.g., selected attributes from information/decision tables. Using the framework provided in the current paper, such an extension can also be achieved. However, it deserves a separate study. Next, in many application areas an elementary relation and/or a crisp set are not explicitly given. Computational Asp-based methods for dealing with such cases in the framework of standard rough sets are provided in [7]. Along these lines one can abstract from elementary relations and consider nested structures as approximate sets containing literals. This direction could also be seen as a generalization of orthopairs [3]. Finally, algebraic properties of many-valued approximations in the context of inheritance expressions are worth further investigations.

References

1. Bobrow, D., Kahn, K., Kiczales, G., Masinter, L., Stefik, M., Zdybel, F.: Common-Loops: merging lisp and object-oriented programming. In: Cardenas, A., McLeod, D. (eds.) Research Foundations in OO and Semantic Database System, pp. 70–90. Prentice-Hall, Upper Saddle River (1990)
2. Brewka, G., Eiter, T., Truszczyński, M.: Answer set programming at a glance. Commun. ACM **54**(12), 92–103 (2011)
3. Ciucci, D.: Orthopairs: a simple and widely used way to model uncertainty. Fundam. Inform. **108**(3–4), 287–304 (2011)
4. Doherty, P., Łukaszewicz, W., Skowron, A., Szałas, A.: Knowledge Engineering. A Rough Set Approach. Springer, Heidelberg (2006)
5. Doherty, P., Szałas, A.: Stability, supportedness, minimality and Kleene answer set programs. In: Eiter, T., Strass, H., Truszczynski, M., Woltran, S. (eds.) Advances in Knowledge Representation, Logic Programming, and Abstract Argumentation. LNCS, vol. 9060, pp. 125–140. Springer, Cham (2014). https://doi.org/10.1007/978-3-319-14726-0_9

---

[14] The complexity of computing truth values returned by connectives is typically $\mathcal{O}(1)$.

[15] In fact, tableaux became a dominant verification technique in Semantic Web.

6. Doherty, P., Szałas, A.: Signed dual tableaux for Kleene answer set programs. In: Golińska-Pilarek, J., Zawidzki, M. (eds.) Ewa Orłowska on Relational Methods in Logic and Computer Science. Outstanding Contributions to Logic, vol. 17, pp. 233–252. Springer, Cham (2018). https://doi.org/10.1007/978-3-319-97879-6_9

7. Doherty, P., Szałas, A.: Rough set reasoning using answer set programs. Int. J. Approx. Reason. **130**, 126–149 (2021)

8. Dunin-Kęplicz, B., Szałas, A.: Shadowing in many-valued nested structures. In: 50th IEEE International Symposium on Multiple-Valued Logic, ISMVL, pp. 230–236. IEEE (2020)

9. Gebser, M., Kaminski, R., Kaufmann, B., Schaub, T.: Answer set solving in practice. In: Synthesis Lectures on AI and Machine Learning. Morgan and Claypool Pub., San Rafael (2012)

10. Gelfond, M., Kahl, Y.: Knowledge Representation, Reasoning, and the Design of Intelligent Agents - The Answer-Set Programming Approach. Cambridge University Press, Cambridge (2014)

11. Greco, S., Matarazzo, B., Słowiński, R.: Algebra and topology for dominance-based rough set approach. In: Ras, Z., Tsay, L.S. (eds.) Advances in Intelligent Information Systems, SCI, vol. 265, pp. 43–78. Springer, Heidelberg (2010). https://doi.org/10.1007/978-3-642-05183-8_3

12. Kleene, S.: On a notation for ordinal numbers. Symbol. Logic **3**, 150–155 (1938)

13. Leone, N., et al.: The DLV system for knowledge representation and reasoning. ACM TCL **7**(3), 499–562 (2006)

14. Małuszyński, J., Szałas, A.: Living with inconsistency and taming nonmonotonicity. In: de Moor, O., Gottlob, G., Furche, T., Sellers, A. (eds.) Datalog 2.0 2010. LNCS, vol. 6702, pp. 334–398. Springer, Heidelberg (2011). https://doi.org/10.1007/978-3-642-24206-9_22

15. Małuszyński, J., Szałas, A.: Partiality and inconsistency in agents' belief bases. In: Barbucha, D., Le, M., Howlett, R., Jain, L. (eds.) KES-AMSTA. Frontiers in AI and Applications, vol. 252, pp. 3–17. IOS Press, Amsterdam (2013)

16. Pal, S.K., Polkowski, L., Skowron, A. (eds.): Rough-Neuro Computing: Techniques for Computing with Words. Springer, Heidelberg (2003). https://doi.org/10.1007/978-3-642-18859-6

17. Pawlak, Z.: Rough sets. Int. J. Comput. Inf. Sci. **11**, 341–356 (1982)

18. Pawlak, Z.: Rough Sets. Theoretical Aspects of Reasoning About Data. Kluwer Academic Publishers, Dordrecht (1991)

19. Pawlak, Z., Skowron, A.: Rudiments of rough sets. Inf. Sci. **117**(1), 3–27 (2007)

20. Priest, G.: The logic of paradox. J. Philos. Logic **8**, 219–241 (1979)

21. Przymusiński, T.: Stable semantics for disjunctive programs. New Gener. Comput. **9**(3/4), 401–424 (1991)

22. Rescher, N.: Many-Valued Logic. McGraw Hill, New York (1969)

23. Simons, P., Niemelä, I., Soininen, T.: Extending and implementing the stable model semantics. Artif. Intell. **138**(1–2), 181–234 (2002)

24. Ślęzak, D., Synak, P., Toppin, G., Wróblewski, J., Borkowski, J.: Rough SQL - semantics and execution. In: Greco, S., Bouchon-Meunier, B., Coletti, G., Fedrizzi, M., Matarazzo, B., Yager, R. (eds.) IPMU 2012, Part II. CCIS, vol. 298, pp. 570–579. Springer, Heidelberg (2012). https://doi.org/10.1007/978-3-642-31715-6_60

25. Szałas, A.: Revisiting object-rule fusion in query languages. In: Cristani, M., Toro, C., Zanni-Merk, C., Howlett, R., Jain, L. (eds.) Proceedings of the 24th International Conference on KES 2020, Knowledge-Based and Intelligent Information and Engineering Systems. Procedia Computer Science, vol. 176, pp. 50–59. Elsevier, Amsterdam (2020)

26. Urquhart, A.: Many-valued logic. In: Gabbay, D., Guenthner, F. (eds.) Handbook of Philosophical Logic, vol. 3, pp. 71–116. Reidel, Dordrecht (1986)
27. Verna, D.: Method combinators. In: Cooper, D. (ed.) Proceedinds of 11th European Lisp Symposium ELS 2018, pp. 32–41. ELSAA (2018)
28. Vitória, A., Małuszyński, J., Szałas, A.: Modeling and reasoning with paraconsistent rough sets. Fundam. Inform. **97**(4), 405–438 (2009)
29. Yao, Y., Wong, S., Lin, T.: A review of rough set models. In: Lin, T., Cercone, N. (eds.) Rough Sets and Data Mining, pp. 47–75. Springer, Boston (1997). https://doi.org/10.1007/978-1-4613-1461-5_3
30. Zhang, Q., Xie, Q., Wang, G.: A survey on rough set theory and its applications. CAAI Trans. Intell. Technol. **1**(4), 323–333 (2016)

Related Methods and Hybridization

Minimizing Depth of Decision Trees
with Hypotheses

Mohammad Azad[1] [iD], Igor Chikalov[2] [iD], Shahid Hussain[3] [iD],
and Mikhail Moshkov[4(✉)] [iD]

[1] College of Computer and Information Sciences, Department of Computer Science,
Jouf University, Sakaka 72441, Saudi Arabia
mmazad@ju.edu.sa
[2] Intel Corporation, 5000 W Chandler Blvd, Chandler, AZ 85226, USA
[3] Dhanani School of Science and Engineering, Computer Science Program,
Habib University, Karachi 75290, Pakistan
[4] Computer, Electrical and Mathematical Sciences and Engineering Division,
King Abdullah University of Science and Technology (KAUST),
Thuwal 23955-6900, Saudi Arabia
mikhail.moshkov@kaust.edu.sa

Abstract. In this paper, we consider decision trees that use both conventional queries based on one attribute each and queries based on hypotheses about values of all attributes. Such decision trees are similar to ones studied in exact learning, where membership and equivalence queries are allowed. We present dynamic programming algorithms for minimization of the depth of above decision trees and discuss results of computer experiments on various data sets and randomly generated Boolean functions.

Keywords: Decision tree · Hypothesis · Depth

1 Introduction

Decision trees are widely used in many areas of computer science and related fields, for example, test theory (initiated by Chegis and Yablonskii [6]), rough set theory (initiated by Pawlak [8–10]), and exact learning (initiated by Angluin [4,5]). These theories are closely related. In particular, attributes from rough set theory and test theory correspond to membership queries from exact learning. Exact learning considers additionally the so-called equivalence queries. The notion of "minimally adequate teacher" that allows both membership and equivalence queries was discussed by Angluin in [3]. Relations between exact learning and PAC learning proposed by Valiant [11] were discussed in [4]. In this paper, we add the notion of a hypothesis to the model that has been considered in rough set theory as well in test theory. This model allows us to use an analog of equivalence queries.

Let T be a decision table with n conditional attributes $f_1, \ldots, f_n$ having values from the set $\omega = \{0, 1, 2, \ldots\}$ in which rows are pairwise different and

© Springer Nature Switzerland AG 2021
S. Ramanna et al. (Eds.): IJCRS 2021, LNAI 12872, pp. 123–133, 2021.
https://doi.org/10.1007/978-3-030-87334-9_11

each row is labeled with a decision from ω. For a given row of T, we should recognize the decision attached to this row. To this end, we can use decision trees based on two types of queries. We can ask about the value of an attribute $f_i \in \{f_1, \ldots, f_n\}$ on the given row. We will obtain an answer of the kind $f_i = \delta$, where δ is the number in the intersection of the given row and the column f_i. We can also ask if a hypothesis $f_1 = \delta_1, \ldots, f_n = \delta_n$ is true, where $\delta_1, \ldots, \delta_n$ are numbers from the columns $f_1, \ldots, f_n$, respectively. Either this hypothesis will be confirmed or we obtain a counterexample in the form $f_i = \sigma$, where $f_i \in \{f_1, \ldots, f_n\}$ and σ is a number from the column f_i different from δ_i. The considered hypothesis is called proper if $(\delta_1, \ldots, \delta_n)$ is a row of the table T. We consider the depth of a decision tree as its time complexity, which is equal to the maximum number of queries in a path from the root to a terminal node of the tree.

Decision trees using hypotheses can be essentially more efficient than the decision trees using only attributes. Let us consider an example, the problem of computation of the conjunction $x_1 \wedge \cdots \wedge x_n$. The minimum depth of a decision tree solving this problem using the attributes $x_1, \ldots, x_n$ is equal to n. However, the minimum depth of a decision tree solving this problem using proper hypotheses is equal to 1: it is enough to ask only about the hypothesis $x_1 = 1, \ldots, x_n = 1$. If it is true, then the considered conjunction is equal to 1. Otherwise, it is equal to 0.

We consider the following five types of decision trees:

1. Decision trees that use only attributes.
2. Decision trees that use only hypotheses.
3. Decision trees that use both attributes and hypotheses.
4. Decision trees that use only proper hypotheses.
5. Decision trees that use both attributes and proper hypotheses.

For each type of decision trees, we design a dynamic programming algorithm that, for a given decision table, finds the minimum depth of a decision tree of the considered type for this table. Note that algorithms for the minimization of the depth for decision trees of type 1 were considered in [1] for decision tables with one-valued decisions and in [2] for decision tables with many-valued decisions.

For the conjunction of n variables, the considered algorithms construct a decision tree of type 1 with the depth equal to n and the decision trees of types 2, 3, 4, and 5 with the depth equal to 1.

It is interesting to study not only specially chosen examples as the conjunction of n variables. We compute the minimum depth of a decision tree for each of the considered five types for eight decision tables from the UCI ML Repository [7]. We do the same for randomly generated Boolean functions with n variables, where $n = 3, \ldots, 6$.

In particular, from the results obtained for Boolean functions it follows that, in general case, the decision trees of types 2 and 4 are better than the decision trees of type 1, and the decision trees of types 3 and 5 are better than the decision trees of types 2 and 4.

The rest of the paper is organized as follows. In Sects. 2 and 3, we consider main notions. In Sects. 4 and 5 – dynamic programming algorithms for the depth minimization. Section 6 contains results of computer experiments and Sect. 7 – short conclusions.

2   Decision Tables

A decision table is a rectangular table T with $n \geq 1$ columns filled with numbers from the set $\omega = \{0, 1, 2, \ldots\}$ of nonnegative integers. Columns of this table are labeled with the conditional attributes $f_1, \ldots, f_n$. Rows of the table are pairwise different. Each row is labeled with a number from ω that is interpreted as a decision. Rows of the table are interpreted as tuples of values of the conditional attributes.

A decision table can be represented by a word over the alphabet $\{0, 1, ;, |\}$ in which numbers from ω are in binary representation (are represented by words over the alphabet $\{0, 1\}$), the symbol ";" is used to separate two numbers from ω, and the symbol "|" is used to separate two rows (we add to each row corresponding decision as the last number in the row). The length of this word will be called the size of the decision table.

A decision table T is called empty if it has no rows. The table T is called degenerate if it is empty or all rows of T are labeled with the same decision.

We denote $F(T) = \{f_1, \ldots, f_n\}$ and denote by $D(T)$ the set of decisions attached to the rows of T. For any conditional attribute $f_i \in F(T)$, we denote by $E(T, f_i)$ the set of values of the attribute f_i in the table T. We denote by $E(T)$ the set of conditional attributes of T for which $|E(T, f_i)| \geq 2$.

A system of equations over T is an arbitrary equation system of the kind

$$\{f_{i_1} = \delta_1, \ldots, f_{i_m} = \delta_m\},$$

where $m \in \omega$, $f_{i_1}, \ldots, f_{i_m} \in F(T)$, and $\delta_1 \in E(T, f_{i_1}), \ldots, \delta_m \in E(T, f_{i_m})$ (if $m = 0$, then the considered equation system is empty).

Let T be a nonempty table. A subtable of T is a table obtained from T by removal of some rows. We correspond to each equation system S over T a subtable TS of the table T. If the system S is empty, then $TS = T$. Let S be nonempty and $S = \{f_{i_1} = \delta_1, \ldots, f_{i_m} = \delta_m\}$. Then TS is the subtable of the table T containing the rows from T, which in the intersection with the columns $f_{i_1}, \ldots, f_{i_m}$ have numbers $\delta_1, \ldots, \delta_m$, respectively. Such nonempty subtables, including the table T, are called separable subtables of T. We denote by $SEP(T)$ the set of separable subtables of the table T.

3   Decision Trees

Let T be a nonempty decision table with n conditional attributes $f_1, \ldots, f_n$. We consider the decision trees with two types of queries. We can choose an attribute $f_i \in F(T) = \{f_1, \ldots, f_n\}$ and ask about its value. This query has the following possible answers: $\{f_i = \delta\}$, where $\delta \in E(T, f_i)$. We can formulate a hypothesis over T in the form of $H = \{f_1 = \delta_1, \ldots, f_n = \delta_n\}$, where $\delta_1 \in E(T, f_1), \ldots, \delta_n \in E(T, f_n)$, and ask about this hypothesis. This query has the following possible answers: $H, \{f_1 = \sigma_1\}, \sigma_1 \in E(T, f_1)\backslash\{\delta_1\}, \ldots, \{f_n = \sigma_n\}, \sigma_n \in E(T, f_n)\backslash\{\delta_n\}$. The first answer means that the hypothesis is true. Other answers are counterexamples. The hypothesis H is called proper for T if $(\delta_1, \ldots, \delta_n)$ is a row of the table T.

A decision tree over T is a marked finite directed tree with the root in which

- Each terminal node is labeled with a number from the set $D(T) \cup \{0\}$.
- Each node, which is not terminal (such nodes are called working), is labeled with an attribute from the set $F(T)$ or with a hypothesis over T.
- If a working node is labeled with an attribute f_i from $F(T)$, then, for each possible answer $\{f_i(x) = \delta\}$, $\delta \in E(T, f_i)$, there is exactly one edge labeled with this answer, which leave this node and there are no any other edges leaving this node.
- If a working node is labeled with a hypothesis $H = \{f_1 = \delta_1, \ldots, f_n = \delta_n\}$ over T, then, for each possible answer $H, \{f_1 = \sigma_1\}, \sigma_1 \in E(T, f_1)\backslash\{\delta_1\}, \ldots, \{f_n = \sigma_n\}, \sigma_n \in E(T, f_n)\backslash\{\delta_n\}$, there is exactly one edge labeled with this answer, which leaves this node and there are no any other edges leaving this node.

Let Γ be a decision tree over T and v be a node of Γ. We now define an equation system $S(\Gamma, v)$ over T associated with the node v. We denote by ξ the directed path from the root of Γ to the node v. If there are no working nodes in ξ, then $S(\Gamma, v)$ is the empty system. Otherwise, $S(\Gamma, v)$ is the union of equation systems attached to the edges of the path ξ.

A decision tree Γ over T is called a decision tree for T if, for any node v of Γ,

- The node v is terminal if and only if the subtable $TS(\Gamma, v)$ is degenerate.
- If v is a terminal node and the subtable $TS(\Gamma, v)$ is empty, then the node v is labeled with the decision 0.
- If v is a terminal node and the subtable $TS(\Gamma, v)$ is nonempty, then the node v is labeled with the decision attached to all rows of $TS(\Gamma, v)$.

A complete path in Γ is an arbitrary directed path from the root to a terminal node in Γ. As the time complexity of a decision tree, we consider its depth that is the maximum number of working nodes in a complete path in the tree or, which is the same, the maximum length of a complete path in the tree. We denote by $h(\Gamma)$ the depth of a decision tree Γ.

We will use the following notation:

- $h^{(1)}(T)$ is the minimum depth of a decision tree for T, which uses only attributes from $F(T)$.
- $h^{(2)}(T)$ is the minimum depth of a decision tree for T, which uses only hypotheses over T.
- $h^{(3)}(T)$ is the minimum depth of a decision tree for T, which uses both attributes from $F(T)$ and hypotheses over T.

- $h^{(4)}(T)$ is the minimum depth of a decision tree for T, which uses only proper hypotheses over T.
- $h^{(5)}(T)$ is the minimum depth of a decision tree for T, which uses both attributes from $F(T)$ and proper hypotheses over T.

4   Construction of Directed Acyclic Graph $\Delta(T)$

Let T be a nonempty decision table with n conditional attributes $f_1, \ldots, f_n$. We now consider an algorithm $\mathcal{A}_0$ for the construction of a directed acyclic graph (DAG) $\Delta(T)$, which will be used for the study of decision trees. Nodes of this graph are some separable subtables of the table T. During each iteration we process one node. We start with the graph that consists of one node T, which is not processed and finish when all nodes of the graph are processed.

Algorithm $\mathcal{A}_0$ (construction of DAG $\Delta(T)$).

Input: A nonempty decision table T with n conditional attributes $f_1, \ldots, f_n$.
Output: Directed acyclic graph $\Delta(T)$.

1. Construct the graph that consists of one node T, which is not marked as processed.
2. If all nodes of the graph are processed, then the algorithm halts and returns the resulting graph as $\Delta(T)$. Otherwise, choose a node (table) Θ that has not been processed yet.
3. If Θ is degenerate, then mark the node Θ as processed and proceed to step 2.
4. If Θ is not degenerate, then, for each $f_i \in E(\Theta)$, draw a bundle of edges from the node Θ. Let $E(\Theta, f_i) = \{a_1, \ldots, a_k\}$. Then draw k edges from Θ and label these edges with systems of equations $\{f_i = a_1\}, \ldots, \{f_i = a_k\}$. These edges enter nodes $\Theta\{f_i = a_1\}, \ldots, \Theta\{f_i = a_k\}$, respectively. If some of the nodes $\Theta\{f_i = a_1\}, \ldots, \Theta\{f_i = a_k\}$ are not present in the graph, then add these nodes to the graph. Mark the node Θ as processed and return to step 2.

The following statement about time complexity of the algorithm $\mathcal{A}_0$ follows immediately from Proposition 3.3 [1].

Proposition 1. *The time complexity of the algorithm $\mathcal{A}_0$ is bounded from above by a polynomial on the size of the input table T and the number $|SEP(T)|$ of different separable subtables of T.*

In general case, the time complexity of the algorithm $\mathcal{A}_0$ is exponential depending on the size of the input decision tables. Note that, in Sect. 3.4 of the book [1], classes of decision tables are described for each of which the number of separable subtables of decision tables from the class is bounded from above by a polynomial on the number of columns in the tables. For each of these classes, the time complexity of the algorithm $\mathcal{A}_0$ is polynomial depending on the size of the input decision tables.

5   Minimizing the Depth of Decision Trees

Let T be a nonempty decision table with n conditional attributes $f_1, \ldots, f_n$. We can use the DAG $\Delta(T)$ to compute values $h^{(1)}(T), \ldots, h^{(5)}(T)$. Let $t \in \{1, \ldots, 5\}$. To find the value $h^{(t)}(T)$, for each node Θ of the DAG $\Delta(T)$, we compute the value $h^{(t)}(\Theta)$. It will be convenient for us to consider not only subtables that are nodes of $\Delta(T)$ but also empty subtable Λ of T and subtables T_r that contain only one row r of T and are not nodes of $\Delta(T)$. We begin with these special subtables and terminal nodes of $\Delta(T)$ (nodes without leaving edges) that are degenerate separable subtables of T and step-by-step move to the table T.

Let Θ be a terminal node of $\Delta(T)$ or $\Theta = T_r$ for some row r of T. Then $h^{(t)}(\Theta) = 0$: the decision tree that contains only one node labeled with the decision attached to all rows of Θ is a decision tree for Θ. If $\Theta = \Lambda$, then $h^{(t)}(\Theta) = 0$: the decision tree that contains only one node labeled with 0 will be considered as a decision tree for Λ.

Let Θ be a nonterminal node of $\Delta(T)$ such that, for each child Θ' of Θ, we already know the value $h^{(t)}(\Theta')$. Based on this information, we can find the minimum depth of a decision tree for Θ, which uses for the subtables corresponding to children of the root decision trees of the type t and in which the root is labeled

- With an attribute from $F(T)$ (we denote by $h_a^{(t)}(\Theta)$ the minimum depth of such a decision tree).
- With a hypothesis over T (we denote by $h_h^{(t)}(\Theta)$ the minimum depth of such a decision tree).
- With a proper hypothesis over T (we denote by $h_p^{(t)}(\Theta)$ the minimum depth of such a decision tree).

Since Θ is nondegenerate, the set $E(\Theta)$ is nonempty. We now describe three procedures for computing the values $h_a^{(t)}(\Theta)$, $h_h^{(t)}(\Theta)$, and $h_p^{(t)}(\Theta)$, respectively.

Let us consider a decision tree $\Gamma(f_i)$ for Θ in which the root is labeled with an attribute $f_i \in E(\Theta)$. For each $\delta \in E(T, f_i)$, there is an edge that leaves the root and enters a node $v(\delta)$. This edge is labeled with the equation system $\{f_i = \delta\}$. The node $v(\delta)$ is the root of a decision tree of the type t for $\Theta\{f_i = \delta\}$ for which the depth is equal to $h^{(t)}(\Theta\{f_i = \delta\})$. It is clear that

$$h(\Gamma(f_i)) = 1 + \max\{h^{(t)}(\Theta\{f_i = \delta\}) : \delta \in E(T, f_i)\}.$$

Since $h^{(t)}(\Theta\{f_i = \delta\}) = h^{(t)}(\Lambda) = 0$ for any $\delta \in E(T, f_i) \backslash E(\Theta, f_i)$,

$$h(\Gamma(f_i)) = 1 + \max\{h^{(t)}(\Theta\{f_i = \delta\}) : \delta \in E(\Theta, f_i)\}. \tag{1}$$

It is clear that, for any $\delta \in E(\Theta, f_i)$, the subtable $\Theta\{f_i = \delta\}$ is a child of Θ in the DAG $\Delta(T)$, i.e., we know the value $h^{(t)}(\Theta\{f_i = \delta\})$.

One can show that $h(\Gamma(f_i))$ is the minimum depth of a decision tree for Θ in which the root is labeled with the attribute f_i and which uses for the subtables corresponding to children of the root decision trees of the type t.

We should not consider attributes $f_i \in F(T) \backslash E(\Theta)$ since, for each such attribute, there is $\delta \in E(T, f_i)$ with $\Theta\{f_i = \delta\} = \Theta$, i.e., based on this attribute we cannot construct an optimal decision tree for Θ. As a result, we have

$$h_a^{(t)}(\Theta) = \min\{h(\Gamma(f_i)) : f_i \in E(\Theta)\}. \tag{2}$$

Computation of $h_a^{(t)}(\Theta)$. Construct the set of attributes $E(\Theta)$. For each attribute $f_i \in E(\Theta)$, compute the value $h(\Gamma(f_i))$ using (1). Compute the value $h_a^{(t)}(\Theta)$ using (2).

Remark 1. Let Θ be a nonterminal node of the DAG $\Delta(T)$ such that, for each child Θ' of Θ, we already know the value $h^{(t)}(\Theta')$. Then the procedure of computation of the value $h_a^{(t)}(\Theta)$ has polynomial time complexity depending on the size of decision table T.

A hypothesis $H = \{f_1 = \delta_1, \ldots, f_n = \delta_n\}$ over T is called admissible for Θ and an attribute $f_i \in F(T) = \{f_1, \ldots, f_n\}$ if, for any $\sigma \in E(T, f_i) \backslash \{\delta_i\}$, $\Theta\{f_i = \sigma\} \neq \Theta$. The hypothesis H is not admissible for Θ and an attribute $f_i \in F(T)$ if and only if $|E(\Theta, f_i)| = 1$ and $\delta_i \notin E(\Theta, f_i)$. The hypothesis H is called admissible for Θ if it is admissible for Θ and any attribute $f_i \in F(T)$.

Let us consider a decision tree $\Gamma(H)$ for Θ in which the root is labeled with an admissible for Θ hypothesis $H = \{f_1 = \delta_1, \ldots, f_n = \delta_n\}$. The set of answers for the query corresponding to the hypothesis H is equal to $A(H) = \{H, \{f_1 = \sigma_1\}, \ldots, \{f_n = \sigma_n\} : \sigma_1 \in E(T, f_1) \backslash \{\delta_1\}, \ldots, \sigma_n \in E(T, f_n) \backslash \{\delta_n\}\}$. For each $S \in A(H)$, there is an edge that leaves the root of $\Gamma(H)$ and enters a node $v(S)$. This edge is labeled with the equation system S. The node $v(S)$ is the root of a decision tree of the type t for ΘS, which depth is equal to $h^{(t)}(\Theta S)$. It is clear that

$$h(\Gamma(H)) = 1 + \max\{h^{(t)}(\Theta S) : S \in A(H)\}.$$

We have $\Theta H = \Lambda$ or $\Theta H = T_r$ for some row r of T. Therefore $h^{(t)}(\Theta H) = 0$. Since H is admissible for Θ, $E(\Theta, f_i) \backslash \{\delta_i\} = \emptyset$ for any attribute $f \in F(T) \backslash E(\Theta)$. It is clear that $\Theta\{f_i = \sigma\} = \Lambda$ and $h^{(t)}(\Theta\{f_i = \sigma\}) = 0$ for any attribute $f_i \in E(\Theta)$ and any $\sigma \in E(T, f_i) \backslash \{\delta_i\}$ such that $\sigma \notin E(\Theta, f_i)$. Therefore

$$h(\Gamma(H)) = 1 + \max\{h^{(t)}(\Theta\{f_i = \sigma\}) : f_i \in E(\Theta), \sigma \in E(\Theta, f_i) \backslash \{\delta_i\}\}. \tag{3}$$

It is clear that, for any $f_i \in E(\Theta)$ and any $\sigma \in E(\Theta, f_i) \backslash \{\delta_i\}$, the subtable $\Theta\{f_i = \sigma\}$ is a child of Θ in the DAG $\Delta(T)$, i.e., we know the value $h^{(t)}(\Theta\{f_i = \sigma\})$.

One can show that $h(\Gamma(H))$ is the minimum depth of a decision tree for Θ in which the root is labeled with the hypothesis H and which uses for the subtables corresponding to children of the root decision trees of the type t.

We should not consider hypotheses that are not admissible for Θ since, for each such hypothesis H for corresponding query, there is an answer $S \in A(H)$ with $\Theta S = \Theta$, i.e., based on this hypothesis we cannot construct an optimal decision tree for Θ.

Computation of $h_h^{(t)}(\Theta)$. First, we construct a hypothesis

$$H_\Theta = \{f_1 = \delta_1, \ldots, f_n = \delta_n\}$$

for Θ. Let $f_i \in F(T) \backslash E(\Theta)$. Then δ_i is equal to the only number in the set $E(\Theta, f_i)$. Let $f_i \in E(\Theta)$. Then δ_i is the minimum number from $E(\Theta, f_i)$ for which $h^{(t)}(\Theta\{f_i = \delta_i\}) = \max\{h^{(t)}(\Theta\{f_i = \sigma\}) : \sigma \in E(\Theta, f_i)\}$. It is clear that H_Θ is admissible for Θ. Compute the value $h(\Gamma(H_\Theta))$ using (3). Simple analysis of (3) shows that $h(\Gamma(H_\Theta)) = h_h^{(t)}(\Theta)$.

Remark 2. Let Θ be a nonterminal node of the DAG $\Delta(T)$ such that, for each child Θ' of Θ, we already know the value $h^{(t)}(\Theta')$. Then the procedure of computation of the value $h_h^{(t)}(\Theta)$ has polynomial time complexity depending on the size of decision table T.

Computation of $h_p^{(t)}(\Theta)$. For each row $r = (\delta_1, \ldots, \delta_n)$ of the decision table T, we check if the corresponding proper hypothesis $H_r = \{f_1 = \delta_1, \ldots, f_n = \delta_n\}$ is admissible for Θ. For each admissible for Θ proper hypothesis $H_r = \{f_1 = \delta_1, \ldots, f_n = \delta_n\}$, we compute the value $h(\Gamma(H_r))$ using (3). One can show that the minimum among the obtained numbers is equal to $h_p^{(t)}(\Theta)$.

Remark 3. Let Θ be a nonterminal node of the DAG $\Delta(T)$ such that, for each child Θ' of Θ, we already know the value $h^{(t)}(\Theta')$. Then the procedure of computation of the value $h_p^{(t)}(\Theta)$ has polynomial time complexity depending on the size of decision table T.

For $t = 1, \ldots, 5$, we describe an algorithm $\mathcal{A}_t$ that, for a given decision table T, calculates the value $h^{(t)}(T)$, which is the minimum depth of a decision tree of the type t for the table T. During the work of this algorithm, we find for each node Θ of the DAG $\Delta(T)$ the value $h^{(t)}(\Theta)$.

Algorithm $\mathcal{A}_t$ (computation of $h^{(t)}(T)$).

Input: A nonempty decision table T and the directed acyclic graph $\Delta(T)$.
Output: The value $h^{(t)}(T)$.

1. If a number is attached to each node of the DAG $\Delta(T)$, then return the number attached to the node T as $h^{(t)}(T)$ and halt the algorithm. Otherwise, choose a node Θ of the graph $\Delta(T)$ without attached number, which is either a terminal node of $\Delta(T)$ or a nonterminal node of $\Delta(T)$ for which all children have attached numbers.
2. If Θ is a terminal node, then attach to it the number $h^{(t)}(\Theta) = 0$ and proceed to step 1.
3. If Θ is not a terminal node, then depending on the value t do the following:

   – In the case $t = 1$, compute the value $h_a^{(1)}(\Theta)$ and attach to Θ the value $h^{(1)}(\Theta) = h_a^{(1)}(\Theta)$.

- In the case $t = 2$, compute the value $h_h^{(2)}(\Theta)$ and attach to Θ the value $h^{(2)}(\Theta) = h_h^{(2)}(\Theta)$.
- In the case $t = 3$, compute the values $h_a^{(3)}(\Theta)$ and $h_h^{(3)}(\Theta)$, and attach to Θ the value $h^{(3)}(\Theta) = \min\{h_a^{(3)}(\Theta), h_h^{(3)}(\Theta)\}$.
- In the case $t = 4$, compute the value $h_p^{(4)}(\Theta)$ and attach to Θ the value $h^{(4)}(\Theta) = h_p^{(4)}(\Theta)$.
- In the case $t = 5$, compute the values $h_a^{(5)}(\Theta)$ and $h_p^{(5)}(\Theta)$, and attach to Θ the value $h^{(5)}(\Theta) = \min\{h_a^{(5)}(\Theta), h_p^{(5)}(\Theta)\}$.

Proceed to step 1.

Using Remarks 1–3 one can prove the following statement.

Proposition 2. *For $t = 1, \ldots, 5$, the time complexity of the algorithm $\mathcal{A}_t$ is bounded from above by a polynomial on the size of the input table T and the number $|SEP(T)|$ of different separable subtables of T.*

6 Results of Experiments

We make experiments with eight decision tables from the UCI ML Repository [7]. Results are represented in Table 1. The first three columns contain the information about the considered decision table T: its name, the number of rows, and the number of conditional attributes. The last five columns contain values $h^{(1)}(T), \ldots, h^{(5)}(T)$ (minimum values for each decision table are in bold).

Decision trees with the minimum depth using attributes (type 1) are optimal for 5 decision tables, using hypotheses (type 2) are optimal for 4 tables, using attributes and hypotheses (type 3) are optimal for 8 tables, using proper hypotheses (type 4) are optimal for 3 tables, using attributes and proper hypotheses (type 5) are optimal for 7 tables.

For the decision table SOYBEAN-SMALL, we must use attributes to construct an optimal decision tree. For this table, it is enough to use only attributes. For the decision tables BREAST-CANCER and NURSERY, we must use both attributes

Table 1. Experimental results for decision tables from [7]

Decision table T	Number of rows	Number of attributes	$h^{(1)}(T)$	$h^{(2)}(T)$	$h^{(3)}(T)$	$h^{(4)}(T)$	$h^{(5)}(T)$
BALANCE-SCALE	625	5	**4**	**4**	**4**	**4**	**4**
BREAST-CANCER	266	10	6	6	**5**	6	**5**
CARS	1728	7	**6**	**6**	**6**	**6**	**6**
HAYES-ROTH-DATA	69	5	**4**	**4**	**4**	**4**	**4**
NURSERY	12960	9	8	8	**7**	8	**7**
SOYBEAN-SMALL	47	36	**2**	4	**2**	6	**2**
TIC-TAC-TOE	958	10	6	6	**5**	8	6
ZOO-DATA	59	17	**4**	**4**	**4**	5	**4**
Average			5.00	5.25	4.63	5.88	4.75

and hypotheses to construct optimal decision trees. For these tables, it is enough to use attributes and proper hypotheses. For the decision table TIC-TAC-TOE, we must use both attributes and hypotheses to construct optimal decision trees. For this table, it is not enough to use attributes and proper hypotheses.

For $n = 3, \ldots, 6$, we generate randomly 100 Boolean functions with n variables. We represent each Boolean function with n variables as a decision table with n columns labeled with these variables and with 2^n rows that are all possible n-tuples of values of the variables. Each row is labeled with the decision that is the value of the function on the corresponding n-tuple. For each function, using its decision table representation, we find the minimum depth of a decision tree of the type t computing this function, $t = 1, \ldots, 5$. For each Boolean function, each hypothesis over the decision table representing it is proper. Therefore, for each Boolean function, $h^{(2)} = h^{(4)}$ and $h^{(3)} = h^{(5)}$.

Results of experiments are represented in Table 2. The first column contains the number of variables in the considered Boolean functions. The last five columns contain information about values $h^{(1)}, \ldots, h^{(5)}$ in the format $_{min}Avg_{max}$.

Table 2. Experimental results for Boolean functions

Number of variables n	$h^{(1)}$	$h^{(2)}$	$h^{(3)}$	$h^{(4)}$	$h^{(5)}$
3	$_2 2.8163_3$	$_1 2.0612_3$	$_1 1.8878_2$	$_1 2.0612_3$	$_1 1.8878_2$
4	$_3 3.9400_4$	$_2 3.0500_4$	$_2 2.9700_3$	$_2 3.0500_4$	$_2 2.9700_3$
5	$_4 4.9500_5$	$_4 4.0800_5$	$_3 3.9900_4$	$_4 4.0800_5$	$_3 3.9900_4$
6	$_5 5.9900_6$	$_5 5.0100_6$	$_5 5.0000_5$	$_5 5.0100_6$	$_5 5.0000_5$

From the obtained results it follows that, in general case, the decision trees of types 2 and 4 are better than the decision trees of type 1, and the decision trees of types 3 and 5 are better than the decision trees of types 2 and 4.

7 Conclusions

In this paper, we studied modified decision trees that use both queries based on one attribute each and queries based on hypotheses about values of all attributes. We designed dynamic programming algorithms for minimization of the depth of such decision trees and considered results of computer experiments. In the future, we are planning to study the number of nodes in the modified decision trees.

Acknowledgments. Research reported in this publication was supported by King Abdullah University of Science and Technology (KAUST). The authors are greatly indebted to anonymous reviewers for useful comments and suggestions.

References

1. AbouEisha, H., Amin, T., Chikalov, I., Hussain, S., Moshkov, M.: Extensions of Dynamic Programming for Combinatorial Optimization and Data Mining. Intelligent Systems Reference Library, vol. 146. Springer, Cham (2019). https://doi.org/10.1007/978-3-319-91839-6
2. Alsolami, F., Azad, M., Chikalov, I., Moshkov, M.: Decision and Inhibitory Trees and Rules for Decision Tables with Many-valued Decisions. Intelligent Systems Reference Library, vol. 156. Springer, Cham (2020). https://doi.org/10.1007/978-3-030-12854-8
3. Angluin, D.: Learning regular sets from queries and counterexamples. Inf. Comput. **75**(2), 87–106 (1987)
4. Angluin, D.: Queries and concept learning. Mach. Learn. **2**(4), 319–342 (1988)
5. Angluin, D.: Queries revisited. Theor. Comput. Sci. **313**(2), 175–194 (2004)
6. Chegis, I.A., Yablonskii, S.V.: Logical methods of control of work of electric schemes. Trudy Mat. Inst. Steklov (in Russian) **51**, 270–360 (1958)
7. Dua, D., Graff, C.: UCI machine learning repository. University of California, Irvine, School of Information and Computer Sciences (2017). http://archive.ics.uci.edu/ml
8. Pawlak, Z.: Rough sets. Int. J. Parallel Program. **11**(5), 341–356 (1982)
9. Pawlak, Z.: Rough Sets - Theoretical Aspects of Reasoning about Data, Theory and Decision Library: Series D, vol. 9. Kluwer (1991)
10. Pawlak, Z., Skowron, A.: Rudiments of rough sets. Inf. Sci. **177**(1), 3–27 (2007)
11. Valiant, L.G.: A theory of the learnable. Commun. ACM **27**(11), 1134–1142 (1984)

The Influence of Fuzzy Expectations on Triples of Triangular Norms in the Weighted Fuzzy Petri Net for the Subject Area of Passenger Transport Logistics

Yurii Bloshko[1]([⊠]), Zbigniew Suraj[1], and Oksana Olar[2]

[1] Institute of Computer Science, Rzeszów University, Rzeszów, Poland
zbigniew.suraj@ur.edu.pl
[2] Yuriy Fedkovych Chernivtsi National University, Chernivtsi, Ukraine
o.olar@chnu.edu.ua

Abstract. This paper continues the analysis of the application of different triples of t-/s-norms and their results in the weighted fuzzy Petri nets for the subject area of passenger transport logistics. The analysis applies the range of 27 different triples of functions which are located in-between minimal (LtN, LtN, ZsN) and maximal (optimized) (ZtN, ZtN, LsN) triples. It also includes classical triple (ZtN, GtN, ZsN) which is located exactly in the middle of this range and remains a good starting point in the comparison of the achieved results. This paper includes a deeper look on the already achieved numerical values as well as decisions and proposes a new approach which will unleash the full potential of the net and applied triples of functions. The idea includes the conception of application of user's expectation. Therefore, the decision-support system provides the results based not only on the input values which were previously filled by the experts in the corresponding subject area, but also on the expectations which can be either met or rejected in the process of calculation.

Keywords: Decision-making system · Intelligent computational techniques · Weighted fuzzy Petri net · Triangular norms · Knowledge representation · Transport logistic problem

1 Introduction

The conception of application of weighted fuzzy Petri nets (wFPN) in the subject area of passenger transport logistics (PTL) was described by the authors in papers [1–6]. In order to analyze the development of wFPN, different approaches and applications of numerous combinations of triples of functions were considered. Moreover, the research of influence of triples on the numerical results and thereby decisions was conducted. Every research starts from the creation of the tables of type "Object-property" as well as establishing connections between them [7, 8]. These connections provide a switch (transfer) between objects and properties in the connected tables. Each table contains

© Springer Nature Switzerland AG 2021
S. Ramanna et al. (Eds.): IJCRS 2021, LNAI 12872, pp. 134–148, 2021.
https://doi.org/10.1007/978-3-030-87334-9_12

some numbers of columns which are considered as properties and some number of rows which are considered as objects. Additionally, each intersection may include some fuzzy value in the range [0, 1] which describes the strength of the connection for some property with the object. As the weight is getting closer to 1 as stronger the connection is. Every table is filled with the knowledge provided by the experts in the relevant field of studies. The following step is to create production rules on the basis of existing tables. Production rules are created in accordance to the formula: IF r_{i1} AND (OR)... AND (OR) r_{in} THEN d_j, where r_{ik} ($k = 1,..., n$) – property, and d_j – object. Production rules serve as the basis for creating wFPN, where input places – properties, output places – objects. Also, knowledge table of type "Object-property" allows creation of a knowledge flow: fuzzy value of the object achieved in the production rule becomes an input fuzzy value associated with a property in the following table which was previously connected to the current one. Thus, an object from the previous level becomes a property in the following table. Moreover, every wFPN includes weights which describe the strength of the connection between property and object [9, 10]. In the knowledge table, the weight is represented at the intersection of the respective property and object, while in wFPN the numerical value of the weight is assigned to the arc connecting the input place with the transition. In this manner, the input value that is given to the transition is the value of input place multiplied with the value (weight) of the arc. Each transition represents a production rule. Therefore, based on the logical operator in the production rule, the corresponding combination of t-/s-norms is used to form a triple of functions (In, Out_1, Out_2), where In – input operator, Out_1 – first output operator, Out_2 - second output operator. Additionally, each transition includes beta $\beta(t)$ and gamma $\gamma(t)$ [11]. The truth degree function beta was proposed to be calculated by the following formula $\beta(t) = k/(k + 1)$ for the research of wFPN in the subject area of PTL (where k is the number of input places connected to this transition) [12]. Yet, it is allowed to set the value for beta $\beta(t)$ by the experts in the given subject area or generated from data tables [13]. The value for the threshold function gamma $\gamma(t)$ is set by the experts. This function describes the minimal threshold value which should be achieved by the input function in the triple in order to fire transition t:

$$In(w_{i1} \cdot M(p_{i1}), w_{i2} \cdot M(p_{i2}), \ldots w_{ik} \cdot M(p_{ik})) \geq \gamma(t) > 0 \qquad (1)$$

where: (a) In is an input operator instantiated with some t-/s-norm; (b) w_{ij} ($j = 1,..., k$) is a weight which is connected with the corresponding input place; (c) $M(p_{ij})$ is a marking of an input place of transition t.

All wFPN model are simulated in a special software PNeS® which was created for such kinds of research [14]. Paper [15] introduced graphical cube for logical AND which describes position of every possible triple in accordance to these sequences:

1. for t-norm: LtN ≤ EtN ≤ GtN ≤ HtN ≤ ZtN;
2. for s-norm: ZsN ≤ HsN ≤ GsN ≤ EsN ≤ LsN [11], where:

- ZtN(a ,b) = min(a, b), ZsN(a,b) = max(a,b) (Zadeh t-/s-norm);

- $HtN(a,b) = \begin{cases} 0 \text{ for } a = b = 0 \\ \frac{ab}{a+b-ab} \text{ otherwise} \end{cases}$, $HsN\ (a,b) = \begin{cases} 1 \text{ for } a = b = 1 \\ \frac{a+b-2ab}{1-ab} \text{ otherwise} \end{cases}$
  (Hamacher t-/s-norm);
- $GtN(a, b) = ab$, $GsN(a,b) = a + b - ab$ (Goguen t-/s-norm);
- $EtN(a, b) = \frac{ab}{2-(a+b-ab)}$, $EsN(a,b) = \frac{a+b}{1+ab}$ (Einstein t-/s-norm);
- $LtN(a, b) = \max(0, a + b - 1)$, $LsN(a,b) = \min(1, a + b)$ (Lukasiewicz t-/s-norm).

These sequences were graphically represented in a form of a cube (Fig. 1).

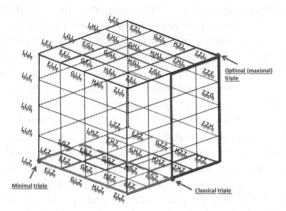

Fig. 1. Cube with 125 combinations of triples of functions (case of logical AND).

Authors' research of triples of function in the subject area of PTL mostly covered the application of logical AND in the production rules [3]. Papers [3, 5] presented a research of triples from the blue rectangle, while paper [6] presented a research of the triples from the green rectangle. A conception of expectations is introduced in this paper to analyze the influence of previously set output value on the real result. The main purpose of activating the third element of the triple is to take the value from the output place and thus correlate the previously obtained calculations.

Figure 2 presents two rectangles highlighted in green and blue colors in the cube (Fig. 1) with a detailed list and locations of triples of functions. They are presented in a range from the minimal (LtN, LtN, ZsN) to the maximal one (ZtN, ZtN, LsN) with a connecting classical triple (ZtN, GtN, ZsN) in the middle. Green rectangle is located on the bottom plane of the cube and describes the list of triples from the minimal to the classical one, while blue rectangle is associated with a side plane and describes triples in a range from classical to the maximal one. As far as green rectangle is located on the bottom plane (Fig. 1), the third function of the triple (Zadeh s-norm) remains the same, because the location of the function is associated with the bottom plane of the cube. Yet, there are 5 alternatives of functions to choose from the horizontal lines (LtN, EtN, GtN, HtN, ZtN) in the first place of the triple and 3 alternatives to choose from the vertical lines (LtN, EtN, GtN) in the second place of the triple.

Blue rectangle can be described in the same manner. The difference with the green rectangle is that it is located on the other (side) plane of the cube (Fig. 1). Thus, the first function (Zadeh t-norm) is stable. Also, there are 3 alternatives to choose from the vertical lines (GtN, HtN, ZsN) in the second place of the triple and 5 alternatives from the horizontal lines (ZsN, HsN, GsN, EsN, LsN) to choose from in the third place.

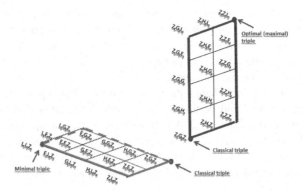

Fig. 2. The detailed representation of planes of the cube with triangular norms.

The aim of the research is to analyze the groups of triples that lie in these rectangles because they form a chain from minimum (LtN, LtN, ZsN) to maximum (ZtN, ZtN, LsN) through the classics (ZtN, GtN, ZsN). Thus, the growth and relationship of the obtained results is the purpose of the analysis in the context of the decision-making process.

2   Fuzzy Expectations for wFPN

The research covers the disclosure of application of different combination of t-/s-norms in the subject area of PTL. From the very beginning, authors presented the idea, where the decision-support system wFPN was giving some decisions based on the user's features. Yet, there was missed a condition, where the decision may vary from the user's expectations. Suppose the client is afraid to fly (hence it is better not to use the aviation branch). Based on his preferences presented in the properties list, there may occur a circumstance when the model suggests aviation type of transportation as the best one. Here comes the situation when client's personal fears do not meet model's decision causing dissonance (i.e.: a proper suggestion by the model is different from the expected one).

The conception of user's expectation implies the idea, where output places of wFPN are also filled with some fuzzy value in the range [0, 1]. This approach allows to consider user's expectation in the decision-making system. Moreover, it allows to see the impact of the third function of the triple on the calculations in the wFPN, since it will apply the value from the output place in the calculations to form the final output of the system. Also, expectations of the result will be included: low expectations will lead to the decrease of the output value, high expectation will increase the output value.

This paper aims to test two approaches: a) only final level of output is filled with expected fuzzy values; b) all of output places are filled with expected fuzzy values. First approach implies the influence only on the resulting output (one time influence), while the second approach covers the analysis of all output objects in the net and setting expectation for each of them.

3   The Review of wFPN Model for the Experiment on Triples of Functions

Paper [5] presented calculations and analysis of combinations of triples from the blue rectangle in Fig. 1. The peculiarity of the wFPN model in the subject area of PTL lies in the circumstances, where only knowledge about input features and their truth degree are available. Thus, in accordance to the description of activation of each element of the triple – the last element was always neglected. The last function in the triple takes as input: the results of the second function and the value on the output place, which is empty and therefore does not change the final result [3]. Only 3 possible combinations from the blue rectangle could be tested. As a result, different outputs were obtained that could be compared and analyzed. Additionally, two different approaches of analysis were presented and tested which also led to different results. Therefore, the analysis of the green rectangle (Fig. 1) was performed for the same input values in order to extend the range of values and decisions with the results obtained on the basis of previously considered approaches.

The benefit of applied triples from the green rectangle is in their location on a different plane of the cube. As it was mentioned before, the third element of the triple is neglected because of the wFPN structure in this area of research. Thus, blue rectangle neglects 5 possible alternatives of the third element reducing the number of possible combination to be tested from 15 to just 3. Yet, only one element of the triple is neglected in a green rectangle leading to the possibility of testing and analyzing 15 combinations of triples.

This paper is a continuation of research on the use of various triples of functions, their impact on the numerical result representing the degree of confidence in the proposed decision and the effectiveness of the selected combination.

The visualization of the wFPN model for the experiment is presented in Fig. 3.

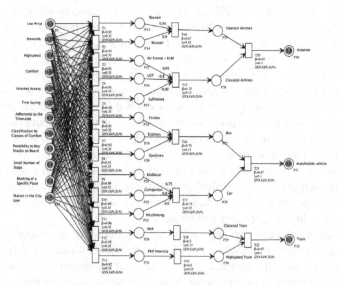

Fig. 3. wFPN model for the best type of transport with fuzzy expectations.

The input values for the net are considered from the customer's perspective and are generalized for the output objects. Detailed properties for each branch are described in the hierarchy [2, 4, 5]. The model itself includes 4 levels of places and 3 levels of transitions. The first level of places represents properties and their fuzzy values from the knowledge table of the lowest level, while the last level of places represents objects from the top-level knowledge table. Remark: first level of nets (between first level of places and first level of transitions) formally includes different weights, but they could not be depicted in the Fig. 3 due to the oversaturation of their number in the representation of the wFPN model. Additionally, there may occur a situation when the initially established connection will be deleted. Even if the previously established weight has some fuzzy value (up to 1.0), the input place with value equal to 0.0 achieved in the process of the calculation on the previous level of transitions will neglect the connection on the following level of transitions.

Figure 3 already includes expecting fuzzy values at the resulting output places: expectation for the "Aviation" object is equal to 0.8, for the "Automobile vehicle" object is equal to 0.6 and for the "Train" object is equal to 0.7. After making the calculations, the decision suggests disclosing the means of transport. Each branch has its own input values and fuzzy expectations (FE).

Figure 3 describes the first approach of application of fuzzy expectations, where only the final output places are filled with some values. The second approach of the analysis of fuzzy expectations implies the full acknowledgment of objects presented in the wFPN model. In this case, the second and the third levels of places in Fig. 3 will be also filled with fuzzy values. Fuzzy values for objects on the second, third and fourth levels are presented in Table 1.

Table 1. Fuzzy values for objects on the second, third, and fourth levels.

Transport companies	FE	Transport kinds	FE	Types of transport	FE
Ryanair	0,9	Lowcost airlines	0,9	Aviation	0,8
Wizzair	0,85				
Air France – KLM	0,7	Classical airlines	0,8		
LOT	0,75				
Lufthansa	0,8				
Flixbus	0,8	Bus	0,5	Automobile vehicle	0,6
Ecolines	0,85				
Eurolines	0,85				
Blablacar	0,4	Car	0,65		
Companion	0,5				
Hitchhiking	0,1				
PKP	0,8	Classical train	0,7	Train	0,7
PKP intercity	0,9	High-speed train	0,8		

The next step is to apply the same input values presented in Fig. 3 for the wFPN model as the first approach and fill in the same table with values taken from Table 1 as the second approach.

4 The Influence of Fuzzy Expectations on the Results of the wFPN Model

Table 2 presents outputs achieved with 29 triples of functions for the same input values (Fig. 3). The highest achieved output values among objects for the same triple of functions are marked in bold. The first observation is that the output values became higher in comparison to the previous researches [5, 6]. It can be explained with the addition of expectation fuzzy values which were also high and made an impact on the last level of calculations. Also, it can be mentioned that some decisions were changed for some triples after application of fuzzy expectations [5, 6]. Secondly, the existence of LtN function in the triples again led to the situation that transitions were not fired, because input value did not meet the condition for the firing (1). These outputs are marked in italic. The only exception is triple (ZtN, LtN, ZsN) which has a possibility reach the end of calculations.

Note to Table 2: during the operation of the net, triple (ZtN, LtN, ZsN) gave a value of 0 for one of the sub-objects. Therefore, the transition associated with this object was removed from the net as it did not provide any additional information in the decision making process. Also, it necessitates to mention that outcomes for objects "Train" and "Aviation" in case of application of a maximal triple (ZtN, ZtN, LsN) are equal, since they have reached their maximum which is equal to 1.0. From the observational point of view, they are treated equally best, because they cannot be compared otherwise.

The following step is to apply two strategies to find out the proper decision (object) and triples which led to this decision [3]. Strategy 1 implies the approach on finding out the number of triples that achieved the same output object which is most often as a result. Strategy 2 is a mathematical approach in accordance to the following formula:

$$Res(Obj_i) = \frac{Res(LtN, LtN, ZsN)_{Obj_i} + \cdots + Res(ZtN, GtN, ZsN)_{Obj_i} + \cdots + Res(ZtN, ZtN, LsN)_{Obj_i}}{Num\_of\_triples}$$

(2)

Strategy 1: It is clearly shown in the Table 2 that the object "Aviation" achieves the highest output result by the majority of triples (16 in total). Moreover, object "Train" also achieves the highest possible outcome value equal to 1.0 after application of the triple (ZtN, ZtN, LsN). Yet, objects "Aviation" and "Train" are treated equally only after calculations of the maximal triple and at the same time, it is the only triple where they cannot be compared. Remark: triple (ZtN, LtN, ZsN) is not considered because of the possible changes in the outputs.

Strategy 2:

- "Aviation" $= \frac{4 \cdot 0.8 + 0.809 + \cdots + 0.983 + 1 + 1}{16} = \frac{13.857}{16} = \mathbf{0.8660625}$.
- "Automobile vehicle" $= \frac{4 \cdot 0.6 + 0.617 + \cdots + 0.697 + 0.78 + 0.825}{16} = \frac{10.684}{16} = 0.66775$.
- "Train" $= \frac{4 \cdot 0.7 + 0.715 + \cdots + 0.85 + 0.975 + 1}{16} = \frac{12.684}{16} = 0.79275$.

Table 2. Resulting values of 29 triples of function for the output objects (first approach).

Triples/Decisive objects	Aviation	Automobile vehicle	Train
(LtN, LtN, ZsN)	*Undefined*	*Undefined*	*Undefined*
(LtN, EtN, ZsN)	*Undefined*	*Undefined*	*Undefined*
(LtN, GtN, ZsN)	*Undefined*	*Undefined*	*Undefined*
(EtN, LtN, ZsN)	*Undefined*	*Undefined*	*Undefined*
(EtN, EtN, ZsN)	*0,8*	*0,6*	*0,7*
(EtN, GtN, ZsN)	*0,8*	*0,6*	*0,7*
(GtN, LtN, ZsN)	*Undefined*	*Undefined*	*Undefined*
(GtN, EtN, ZsN)	*0,8*	*0,6*	*0,7*
(GtN, GtN, ZsN)	*0,8*	*0,6*	*0,7*
(HtN, LtN, ZsN)	*Undefined*	*Undefined*	*Undefined*
(HtN, EtN, ZsN)	*0,8*	*0,6*	*0,7*
(HtN, GtN, ZsN)	*0,8*	*0,6*	*0,7*
(ZtN, LtN, ZsN)	Undefined *(0,8)*	**0,6 [lost one sub-object]**	Undefined
(ZtN, EtN, ZsN)	**0,8**	0,6	0,7
(ZtN, GtN, ZsN)	**0,8**	0,6	0,7
(ZtN, HtN, ZsN)	**0,8**	0,6	0,7
(ZtN, ZtN, ZsN)	**0,8**	0,6	0,7
(ZtN, GtN, HsN)	**0,809**	0,617	0,715
(ZtN, HtN, HsN)	**0,815**	0,632	0,731
(ZtN, ZtN, HsN)	**0,824**	0,642	0,767
(ZtN, GtN, GsN)	**0,837**	0,639	0,745
(ZtN, HtN, GsN)	**0,858**	0,672	0,783
(ZtN, ZtN, GsN)	**0,88**	0,69	0,846
(ZtN, GtN, EsN)	**0,857**	0,659	0,769
(ZtN, HtN, EsN)	**0,885**	0,704	0,818
(ZtN, ZtN, EsN)	**0,909**	0,727	0,885
(ZtN, GtN, LsN)	**0,983**	0,697	0,85
(ZtN, HtN, LsN)	**1**	0,78	0,975
(ZtN, ZtN, LsN)	**1**	0.825	**1**

In accordance to both strategies, the same object "Aviation" is considered as the best one. Application of the second strategy did not make any influence on the conclusion since object "Aviation" was considered as the best one by all applied triples of functions.

Yet, since both triples: classical (ZtN, GtN, ZsN) and maximal (ZtN, ZtN, LsN) provided the same sequence of decisions as all other triples between them, it is impossible to highlight their benefits. The only exception is two equal decisions achieved by the maximal triple (ZtN, ZtN, LsN), which were discussed above. Moreover, lowering of the threshold function gamma $\gamma(t)$ would not change any decisions since the achieved value there are proportional and correspond to the sequence of decisions that was already achieved. It will only increase the number of triples supporting "Aviation" object as the best decision and at the same time, it will lower the resulting values achieved by the second strategy. It is worth mentioning that values in curly brackets for the triple (ZtN, LtN, ZsN) show alternative result which can be achieved (where applicable) if the threshold functions are lowered.

The only difference which is worth to be highlighted is the numerical outcomes of the triple (ZtN, LtN, ZsN). In case of lowering threshold value, triple (ZtN, LtN, ZsN) has a chance to achieve some numerical value which are highlighted in brackets. Yet, the decision for the object "Train" remained stable "Undefined". It can be explained with the fact that objects which led to the decision "Train" were earlier zeroed on the previous level of calculation due to the LtN function.

The most important observation of the results is that the rise of numerical values directly corresponds to their location in the cube (Figs. 1, 2). Thus, as the triple is located closer to the maximum – the higher outcome will be achieved. Additionally, the value of expectations influenced the outcomes. Since the expectations were high, they increased results in Table 2 compared to results presented in papers [5, 6].

The next approach to be presented is filling of all output places with fuzzy values (Table 1) in the wFPN model (Fig. 3) which forms Table 3. In the same manner, the highest results are marked in bold and results which were achieved after lowering gamma are in italic.

The notable difference between Tables 2 and 3 and therefore between approaches of applying expectation fuzzy values lies in the observation of influence of LtN function and avoidance of "Undefined" result. Even if the value was dropped to 0 in the process of calculation of the second element of the triple, the third function of the triple could make an influence and avoid "Undefined" output. Additionally, it correlates the low values achieved by the triples in a green rectangle (and some in a blue one) to a higher output numerical values. Yet, these values totally correspond to the fuzzy expectations and therefore neglect decision-making process of the net. On the other hand, if values from the blue rectangle are being compared in Tables 2 And 3, it is possible to estimate the increase of resulting numerical values in the Table 3, which applied a different approach (fuzzy expectations at every output place). Such an increase can be explained in the following way: at each level of places (except for the very first one), fuzzy expectations were keeping the output value high and such tendency was kept until the very last level of output places. At the same time, the first approach (Table 2) increases only the last level of output places, while every level of places between the very first and the last one may have a decrease, since the third function of the triple is neglected (note: output place is empty at the moment of firing of a transition).

Moreover, it is worth noting that last three triples of functions in the Table 3 received the highest possible fuzzy value which is equal to 1.0 for each object at the output. In

Table 3. Resulting values of 29 triples of function for three output objects (second approach).

Triples/Decisive objects	Aviation	Automobile vehicle	Train
(LtN, LtN, ZsN)	*Undefined*	*Undefined*	*Undefined*
(LtN, EtN, ZsN)	*Undefined*	*Undefined*	*Undefined*
(LtN, GtN, ZsN)	*Undefined*	*Undefined*	*Undefined*
(EtN, LtN, ZsN)	*0,8*	*0,6*	*0,7*
(EtN, EtN, ZsN)	*0,8*	*0,6*	*0,7*
(EtN, GtN, ZsN)	*0,8*	*0,6*	*0,7*
(GtN, LtN, ZsN)	*0,8*	*0,6*	*0,7*
(GtN, EtN, ZsN)	*0,8*	*0,6*	*0,7*
(GtN, GtN, ZsN)	*0,8*	*0,6*	*0,7*
(HtN, LtN, ZsN)	*0,8*	*0,6*	*0,7*
(HtN, EtN, ZsN)	*0,8*	*0,6*	*0,7*
(HtN, GtN, ZsN)	*0,8*	*0,6*	*0,7*
<u>(ZtN, LtN, ZsN)</u>	**0,8**	0,6 [one sub-object didn't reach the threshold without gamma lowering]	0,7
(ZtN, EtN, ZsN)	**0,8**	0,6	0,7
(ZtN, GtN, ZsN)	**0,8**	0,6	0,7
(ZtN, HtN, ZsN)	**0,8**	0,6	0,7
(ZtN, ZtN, ZsN)	**0,8**	0,65	0,7
(ZtN, GtN, HsN)	**0,84**	0,699	0,77
(ZtN, HtN, HsN)	**0,845**	0,718	0,781
(ZtN, ZtN, HsN)	**0,858**	0,779	0,814
(ZtN, GtN, GsN)	**0,921**	0,791	0,868
(ZtN, HtN, GsN)	**0,926**	0,815	0,878
(ZtN, ZtN, GsN)	**0,934**	0,868	0,901
(ZtN, GtN, EsN)	**0,950**	0,846	0,912
(ZtN, HtN, EsN)	**0,953**	0,868	0,919
(ZtN, ZtN, EsN)	**0,957**	0,906	0,933
(ZtN, GtN, LsN)	**1**	1	1
(ZtN, HtN, LsN)	**1**	1	1
(ZtN, ZtN, LsN)	**1**	1	1

this case, objects cannot be compared between each other, since there is no numerical difference between them. Therefore, it can be considered as a disadvantage. Yet, if results from these three triples are compared with the corresponding results in Table 2, it can be estimated that it is easier to reach the maximum output after application of fuzzy expectations for every output place.

Additional interesting observation concerns triple (ZtN, LtN, ZsN). In the process of calculations of the triple (ZtN, LtN, ZsN) for the object "Automobile vehicle", one of the sub-objects "Car" did not receive its value since condition for firing a transition was not satisfied earlier. Yet, it is worth highlighting that results for (ZtN, LtN, ZsN) in the Table 3 compared with results in Table 2 have a difference. In Table 2, object "Automobile vehicle" is losing subject "Car" in any case: either threshold function is high and transition is not fired either it is fired but LtN function zeroed output for the object "Car". Table 3 introduces a difference with the application of fuzzy expectations. In case the threshold function gamma $\gamma(t)$ is initially set to 0.15 for the corresponding output object (as it is described above), the initial condition for firing a transition would not be satisfied and object will be lost, but the remaining object allows the computation to be completed under these conditions. This is marked in Table 3 in the same way as in Table 2. However, in the case of lowering the value of the threshold function, the object remains in relation to the fuzzy expectations: LtN function lowers the result of calculations to 0, but fuzzy expectation is considered in the calculations and therefore object "Automobile vehicle" receives two inputs for the calculations. Yet, in accordance to calculations, another sub-object "Bus" takes an advantage over sub-object "Car" with an influence on the resulting value of the final object "Automobile vehicle". To sum it up: in Table 2 object "Automobile vehicle" always receives only one input "Bus", since object "Car" is always getting 0 value or may not be even achieved at all (failure to comply condition (1)), while in the Table 3, the dependence of achieving output results lies exceptionally on the threshold function value (i.e. meeting the condition for firing the transition) and the output as follows will be mostly dependent on the output value which in turn is the fuzzy expectation. Thus, fuzzy expectations may serve as the saving tool for objects that apply functions which return extremely low or zero outputs.

Moreover, in case of application of LtN function in a first place of a triple, the results tend to be "Undefined" in every situation (Tables 2, 3). It usually happens for transitions which have a big number of input nodes. It can be explained in the following way: each transition on the first level has many inputs compared to 2 or 3 inputs on the second and third levels. Thus, the peculiarity of the formula of LtN function: $LtN(a, b) = \max(a + b - 1, 0)$ tends to lower the output value at each iteration. Therefore, if input values are already low, then this function may result in a 0 value after first round of iteration. Yet, even if input values are relatively high, they tend to get lower after each iteration. As follows, a high number of iterations tend to decrease the output value until it gets 0. In this manner, LtN function is highly ineffective with a big number of inputs since it will directly correspond to the same number of iterations minus one leading to the zero-output value.

Then, it necessitates to apply two strategies to reach some conclusions about results in the Table 3.

Strategy 1: "Aviation" object remains the best one for every triple of function. "Train" object is the second-best option and "Automobile vehicle" object is the last suggested option.

Remark: triple (ZtN, LtN, ZsN) does not make any influential changes in the sequence of decisions in strategy 1.

Strategy 2: Only results achieved with the initially set threshold function gamma $\gamma(t)$ are included. Therefore, in the range of triple from the classical to the maximal, fuzzy expectations correlate the decision-making process of the net, while the decisive part remains by first two functions of the triple. In case of applying triples in the range from the minimal to the classical one, fuzzy expectation makes a higher influence neglecting output values:

- "Aviation" $= \frac{4\cdot0.8+0.84+\cdots+0.957+3\cdot1}{16} = \frac{14.384}{16} = \mathbf{0.899};$
- "Automobile vehicle" $= \frac{3\cdot0.6+0.65+\cdots+0.906+3\cdot1}{16} = \frac{12.74}{16} = 0.79625;$
- "Train" $= \frac{4\cdot0.7+0.77+\cdots+0.933+3\cdot1}{16} = \frac{13.576}{16} = 0.8485.$

First observations are as follows. The sequence of proposed decisions in both strategies 1 and 2 is equal in the same manner as it was for the Table 2. Additionally, lowering of the threshold function will lead to the decrease of the resulting values for each object but their sequence will remain the same since results from the green rectangle are located in the same correspondence as results from the blue rectangle. The only notable difference which can be observed between strategies 2 for Tables 2 And 3 is in the increase of numerical values for each object after application of the second strategy (where each output place is previously filled with fuzzy expectations). Thus, the next step is to compare two strategies with results achieved in Tables 2 And 3 after lowering the threshold function gamma $\gamma(t)$. Case 1 describes the situation, where fuzzy expectations are set only for the final level of output places and Case 2 describes the situations, where fuzzy expectations are set for every output place. Remark: triple (ZtN, LtN, ZsN) is not taken into consideration:

- Case 1 – "Aviation" $= \frac{13.857+6\cdot0.8}{16+6} = \frac{18.657}{22} = \mathbf{0.84804545454}.$
- Case 2 – "Aviation" $= \frac{14.384+9\cdot0.8}{16+9} = \frac{21.584}{25} = \mathbf{0.86336}.$
- Case 1 – "Automobile vehicle" $= \frac{10.684+6\cdot0.6}{16+6} = \frac{14.284}{22} = 0.64927272727.$
- Case 2 – "Automobile vehicle" $= \frac{12.74+9\cdot0.6}{16+9} = \frac{18.14}{25} = 0.7256.$
- Case 1 – "Train" $= \frac{12.684+6\cdot0.7}{16+6} = \frac{16.884}{22} = 0.76745454545.$
- Case 2 – "Train" $= \frac{13.576+9\cdot0.7}{16+9} = \frac{19.876}{25} = 0.79504.$

As it was expected, with the addition of results from triples in the green rectangle, the average mathematical average meaning was also decreased, because the resulting values there are smaller than in the blue rectangle. Yet, it was proven that the sequence of decisions remained the same with the same sequence of outputs taken from another group of triples. Still, it is worth noting that the final calculations showed one notable thing: the application of the second case (when every output place is previously filled with fuzzy values) leads to higher output result compared to the first case (when only the final level of output places is filled with fuzzy values). It can be explained in a way that the second case keeps values at a high level at each level of transitions, while the first case influences only the last level of (transitions and) places.

5 Conclusions

This paper presented the conception of application of fuzzy expectations and their influence on wFPN. Additionally, different triples of functions in the range from the minimal triple (LtN, LtN, ZsN) to the maximal one (ZtN, ZtN, LsN) were tested. The following observations were achieved:

- the practical proof of achieved resulting values for different triples of function and their correspondence to the location in a cube from minimal to maximal one (Fig. 1). The case where fuzzy expectations are not applied, the decision-making process totally relies on first two functions of the triple in a step-by-step firing sequence of wFPN model;
- there were tested two different strategies for verification of the achieved results (objects) and triples which provided the highest probability of its correctness between minimal, maximal and classical triples;
- classical triple of functions (ZtN, GtN, ZsN) is proposed as the best option at this moment. With the addition of fuzzy expectations, the same object "Aviation" was concluded as the best one with a verification by both strategies and cases (whether only the decisive output places are filled with fuzzy values (Table 2) or every output place is filled with them (Table 3));
- the influence of the number of inputs for the triples from green rectangle plays a vital role in calculations (in case fuzzy expectations are not applied): depending on the chosen combination of norms, there exists a risk of dependence of output numerical value to the number of inputs. If the function applies iteration for each additional connection (arc) then there may occur a situation when the output value decreases with each iteration, since functions in a green rectangle tend to achieve low outputs. Moreover, the output value can drop to 0 as it occurs with the application of the LtN function which is recommended not to be applied in accordance to the current research;
- additionally, there may be a dependency leading to a different order of decisions described in the previous point. Therefore, there is a need to correlate the outputs (with the activation of the third element of the triple). LtN function is the function which gives the lowest possible numeric value and may lead to the 0 value at some level of places leading to the "Undefined" state at the last level of places.

In case of application of triples in the range between minimal and classical triples, the threshold function gamma $\gamma(t)$ plays a critical role: if it is too high, there is a high probability not to reach the end of calculations, since triples in this range operates with low numbers. The problem of application of triples from a green rectangle lies in achieving low values. Thus, there exists a problem of not achieving any decisions if the condition for firing a transition (1) is not satisfied and the transition does not fire leading to the "Undefined" state at the last level of places or (either) calculations achieved zero value at some level of places (incl. the last one). Fuzzy expectations enable the third element of the triple and increases the output value in any case, because of the formulas which are described for every s-norm (yet, the transition should remain initially fireable). The problem of applying fuzzy expectations for triples that can approach the maximum output value. There exists a possibility of achieving the maximal value being equal to 1.0 for each object. Thus, there is no possibility to estimate the difference between fuzzy outputs for different objects (Tables 2, 3). Fuzzy expectations may also correlate the problems described above, but they will withdraw decision-making process in case of applying triples from the green rectangle, since the output will be equal to the expectations (Tables 2, 3).

The following conclusion for the given experiment was achieved: the influence of fuzzy expectations grows as the chosen triple of function is located closer to the minimal triple (LtN, LtN, ZsN), while the influence of decision-making process raises as the chosen triple is located to the maximal one (ZtN, ZtN, LsN). Thus, there is additional advantage of the classical triple (ZtN, GtN, ZsN), since it is located exactly in the middle and may balance the internal decision-making process with the outer expectations. The next topic for disclosure is how to establish fuzzy expectations so that they correlate precisely with the decision-making process. Moreover, other classes of wFPN, such as T2GFPN [16], PFPN [17], FGFPN [18], can be used to improve the calculations with the analysis of the effectiveness of fuzzy expectations in their application.

References

1. Suraj, Z., Olar, O., Bloshko, Y.: Optimized fuzzy petri nets and their application for transport logistics problem. In: Proceedings of International Workshop on Concurrency, Specification and Programming (CS&P 2019), Olsztyn, Poland, 24–26 September 2019 (2019). http://ceur-ws.org/Vol-2571/CSP2019_paper_5.pdf
2. Suraj, Z., Olar, O., Bloshko, Y.: Hierarchical weighted fuzzy Petri nets and their application for transport logistics problem. In: Proceedings of International Conference on Intelligent Systems and Knowledge Engineering (FLINS/ISKE 2020), Cologne, Germany, 18–21 August 2020, World Scientific Proceedings Series on Computer Engineering and Information Science, Volume 12: Developments of Artificial Intelligence Technologies in Computation and Robotics, pp. 404–411. World Scientific, Singapore (2020)
3. Suraj, Z., Olar, O., Bloshko, Y.: The analysis of human oriented system of weighted fuzzy Petri nets for passenger transport logistics problem. In: Kahraman, C., Cevik, O.S., Oztaysi, B., Sari, I., Cebi, S., Tolga, A. (eds.) Intelligent and Fuzzy Systems (INFUS 2020). Advances in Intelligent Systems and Computing, vol. 1197, pp. 1580–1588. Springer, Cham (2021). https://doi.org/10.1007/978-3-030-51156-2_184
4. Bloshko, Y., Suraj, Z., Olar, O.: Towards optimization of triangular norms in weighted fuzzy petri nets for hierarchical applications in subject area of passenger transport logistics. Approved at the FUZZ-IEEE 2021, Luxembourg, 11–14 July 2021 (2021, in print)
5. Suraj, Z., Olar, O., Bloshko, Y.: Modeling of passenger transport logistics based on intelligent computational techniques. Int. J. Comput. Intell. Syst. (Submitted)

6. Suraj, Z., Olar, O., Bloshko, Y.: The analysis of triples of triangular norms for the subject area of passenger transport logistics. In: Rocha, Á., Adeli, H., Dzemyda, G., Moreira, F., Ramalho Correia, A.M. (eds.) Trends and Applications in Information Systems and Technologies. WorldCIST 2021. Advances in Intelligent Systems and Computing, vol. 1365, pp. 29–38. Springer, Cham (2021).https://doi.org/10.1007/978-3-030-72657-7_3

7. Lyashkevych, V., Olar, O., Lyashkevych, M.: Software ontology subject domain intelligence diagnostics of computer means. In: Proceedings of 7th IEEE International Conference on Intelligent Data Acquisition and Advanced Computing Systems: Technology and Applications (IDAACS-2013), Berlin, Germany, September 2013, pp. 12–14 (2013)

8. Lokazyuk, V.: Software for creating knowledge base of intelligent systems of diagnosing process. In: Lokazyuk, V., Olar, O., Lyaskevych, V. (eds.) Advanced Computer System and Networks: Design and Application, ACSN 2009, Lviv, pp. 140–145 (2009)

9. Chen, S.-M.: Weighted fuzzy reasoning using weighted fuzzy Petri nets. IEEE Trans. Knowl. Data Eng. **14**(2), 386–397 (2002)

10. Suraj, Z., Hassanien, A.E., Bandyopadhyay, S.: Weighted generalized fuzzy petri nets and rough sets for knowledge representation and reasoning. In: Bello, R., Miao, D., Falcon, R., Nakata, M., Rosete, A., Ciucci, D. (eds.) Rough Sets (IJCRS 2020). Lecture Notes in Computer Science, vol. 12179, pp. 61–77. Springer, Cham (2020). https://doi.org/10.1007/978-3-030-52705-1_5

11. Klement, E.P., Mesiar, R., Pap, E.: Triangular Norms. Kluwer, Dordrecht (2000)

12. Suraj, Z., Olar, O., Bloshko, Y.: Conception of fuzzy Petri net to solve transport logistics problems. Curr. Res. Math. Comput. Sci. **II**, 303–313 (2018)

13. Skowron, A.: Boolean reasoning for decision rules generation. In: Komorowski, J., Raś, Z.W. (eds.) Methodologies for Intelligent Systems. ISMIS 1993. Lecture Notes in Computer Science (Lecture Notes in Artificial Intelligence), vol. 689, pp. 295–305. Springer, Berlin, Heidelberg (1993). https://doi.org/10.1007/3-540-56804-2_28

14. Suraj, Z., Grochowalski, P.: Petri nets and PNeS in modeling and analysis of concurrent systems. In: Proceedings International Workshop on Concurrency, Specification and Programming (CS&P 2017), Warsaw, Poland, 25–27 September 2017, pp. 1–12 (2017)

15. Suraj, Z.: Toward optimization of reasoning using generalized fuzzy petri nets. In: Nguyen, H., Ha, Q.T., Li, T., Przybyła-Kasperek, M. (eds.) Rough Sets. IJCRS 2018. Lecture Notes in Computer Science, vol. 11103, pp. 294–308. Springer, Cham (2018). https://doi.org/10.1007/978-3-319-99368-3_23

16. Suraj, Z., Grochowalski, P.: Fuzzy petri nets with linear orders for intervals. In: Martín-Vide, C., Neruda, R., Vega-Rodríguez, M.A. (eds.) TPNC 2017. LNCS, vol. 10687, pp. 150–161. Springer, Cham (2017). https://doi.org/10.1007/978-3-319-71069-3_12

17. Suraj, Z.: Parameterised fuzzy petri nets for approximate reasoning in decision support systems. In: Hassanien, A.E., Salem, A.-B., Ramadan, R., Kim, T.-H. (eds.) AMLTA 2012. CCIS, vol. 322, pp. 33–42. Springer, Heidelberg (2012). https://doi.org/10.1007/978-3-642-35326-0_4

18. Suraj, Z., Grochowalski, P., Bandyopadhyay, S.: Flexible generalized fuzzy petri nets for rule-based systems. In: Martín-Vide, C., Mizuki, T., Vega-Rodríguez, M.A. (eds.) TPNC 2016. LNCS, vol. 10071, pp. 196–207. Springer, Cham (2016). https://doi.org/10.1007/978-3-319-49001-4_16

Possibility Distributions Generated by Intuitionistic L-Fuzzy Sets

Stefania Boffa$^{(\boxtimes)}$ (ID) and Davide Ciucci (ID)

Dipartimento di Informatica, Sistemistica e Comunicazione, Università degli Studi di Milano-Bicocca, Viale Sarca 336, 20126 Milano, Italy
{stefania.boffa,davide.ciucci}@unimib.it

Abstract. In this work, we bridge possibility theory with intuitionistic L-fuzzy sets, by identifying a special class of possibility distributions corresponding to intuitionistic L-fuzzy sets based on a complete residuated lattice with an involution. Moreover, taking the Łukasiewicz n-chains as structures of truth degrees, we propose an algorithm to compute the intuitionistic L-fuzzy set corresponding to a given possibility distribution, in case it exists.

Keywords: Possibility distributions · Intuitionistic L- fuzzy sets · Complete residuated lattices · Łukasiewicz n-chains

1 Introduction

Possibility distributions are the building blocks of *possibility theory* [15]. The concept of *possibility* was investigated by several scholars, expecially by Shackle [32], Lewis [27], Cohen [10], and Zadeh [33]. Moreover, possibility theory and its applications were widely explored by Dubois, Prade and colleagues in many works [12,13,16].

A possibility distribution π_x is a map associated to a variable x, from a universe U to a totally ordered scale L with a top and bottom, such as the unit interval $[0, 1]$. Depending on the interpretations, $\pi_x(u)$ estimates the degree of ease, the degree of unsurprizingness or of expectedness, the degree of acceptability or of preference related to the proposition "the value of x is u" [14]. Here, we focus on possibility distributions arising when a degree of plausibility needs to be assigned to an L-set as in the following example[1].

Suppose that $\mathcal{V}$ is a collection of features regarding a flat (for instance *small size* and *low price*). Then, each specific flat F is associated to an L-set $\omega_F : \mathcal{V} \to L$, where $\omega_F(v)$ is the truth degree to which F has the attribute v of $\mathcal{V}$. Therefore, we could consider a possibility distribution π such that $\pi(\omega_F)$ expresses the degree of possibility that a given customer prefers a given flat F (described by ω_F) before he/she knows it in advance.

---

[1] L-sets were introduced by Goguen [21] as generalizations of fuzzy sets.

© Springer Nature Switzerland AG 2021
S. Ramanna et al. (Eds.): IJCRS 2021, LNAI 12872, pp. 149–163, 2021.
https://doi.org/10.1007/978-3-030-87334-9_13

Specifically, we deal with possibility distributions whose domain is composed of all L-sets on a given universe assuming that L is a complete residuated lattice with an involution [24]. Besides, they are interpreted as *preference functions*, thus standing for a counterpart to utility functions [11,17].

Mainly, we aim to discover the existing connections between this type of possibility distributions and intuitionistic L-fuzzy sets, a generalization of intuitionistic fuzzy sets, based on a lattice L instead of [0, 1] as the set of truth-values. To this purpose, we view intuitionistic L-fuzzy sets as generalizations of *orthopairs*, which are pairs of disjoint subsets of a universe used to model uncertainty [7]. Given a set of propositional variables $\mathcal{V}$, an orthopair (P, N) on $\mathcal{V}$ has an epistemic meaning: P is the set of variables *known to be true*, N is the set of variables *known to be false*, and $\mathcal{V} \setminus (P \cup N)$ is the set of *unknown variables* by a given agent. In [8], the authors provided the following correspondence between orthopairs and Boolean possibility distributions: An orthopair on $\mathcal{V}$ generates a Boolean possibility distribution whose domain Ω is made of all evaluation functions on $\mathcal{V}$. On the other hand, not all Boolean possibility distributions having Ω as domain are generated by an orthopair on $\mathcal{V}$. Consequently, orthopairs on $\mathcal{V}$ individuate a special class of Boolean possibility distributions on Ω. In this article, we intend to extend this correspondence by using fuzzy logic. More precisely, we identify intuitionistic L-fuzzy sets of a given universe $\mathcal{V}$ with particular possibility distributions, which assign a degree of L to each L-set of $\mathcal{V}$.

In providing theoretical results, we suppose that *complete residuated lattices with an involution* are our algebraic structures of truth values [19]. However, examples and algorithms are confined to finite substructures of the *standard Łukasiewicz MV-algebra* [6,28]. Our choice depends on that Łukasiewicz implication is usually used for fuzzy logic applications because it is the only plausible continuous implication operation on [0, 1] [31].

The article is organized as follows. The next section reviews some basic notions and results regarding residuated lattices and intuitionistic fuzzy sets. In Sect. 3, we firstly assign a special possibility distribution to each intuitionistic L-fuzzy set. Then, we prove that possibility distributions corresponding to intuitionistic L-fuzzy sets are normal. After that, confining to $IMTL$-algebras, we establish under what conditions a possibility distribution assumes value $\mathbf{0}$. In Sect. 4, we firstly show that there exist normal possibility distributions not generated by an intuitionistic L-fuzzy set. Then, we find the intuitionistic L-fuzzy set that generates a given possibility distribution, in case it exists. Moreover, in Subsect. 4.1, considering the Łukasiewicz n-chains as algebraic structures of truth degrees, we provide procedures to compute the possibility distribution generated by a given intuitionistic L-fuzzy set, and vice-versa, the intuitionistic L-fuzzy set generating a given possibility distribution. Finally, in the last section, we briefly discuss the potential developments of our results.

2   Preliminaries

This section describes some notations, preliminary notions and results, which will be used in this article.

2.1   Algebraic Structures of Truth Values

As basic structures of truth degrees, we choose complete residuated lattices, which are widely adopted for applications of fuzzy logic [22,23,26].

Definition 1 [25]. *A residuated lattice is an algebra $\langle L, \wedge, \vee, \otimes, \rightarrow, 0, 1 \rangle$, where*

(i) $\langle L, \wedge, \vee, 0, 1 \rangle$ is a bounded lattice;

(ii) $\langle L, \otimes, 1 \rangle$ is a commutative monoid, i.e. $\otimes$ is a binary operation that is commutative, associative, and $a \otimes 1 = a$ for each $a \in L$;

(iii) $a \otimes b \le c$ if and only if $a \le b \rightarrow c$, for each $a, b, c \in L$ (adjunction property).

A residuated lattice $(L, \wedge, \vee, \otimes, \rightarrow, 0, 1)$ is complete if its reduct $(L, \wedge, \vee)$ is a complete lattice. In a residuated lattice, a unary operation named *negation* is derived as follows: $\neg x = x \rightarrow 0$, for each $x \in L$. In this paper, we deal with residuated lattices where the negation is an *involution*, namely the so-called *double negation law* holds: $\neg\neg x = x$ for each $x \in L$. The following proposition lists some properties satisfied by every residuated lattice with an involution.

Proposition 1. *Let $\langle L, \wedge, \vee, \otimes, \rightarrow, 0, 1 \rangle$ be a residuated lattice, then the followings hold for each $a, b, c \in L$:*

(i) $a \wedge b \le a$ and $a \wedge b \le b$;

(ii) if $a \le b$ and $a \le c$ then $a \le b \wedge c$;

(iii) $a \vee b = 1$ if and only if $a = 1$ or $b = 1$;

(iv) $a \wedge b = 1$ if and only if $a = 1$ and $b = 1$;

(v) $a \wedge b = 0$ if and only if $a = 0$ or $b = 0$;

(vi) $a \otimes b = 1$ if and only if $a = 1$ and $b = 1$;

(vii) $a \rightarrow b = 1$ if and only if $a \le b$;

(viii) if $\neg$ is an involution, then $a \rightarrow b = \neg b \rightarrow \neg a$.

Special residuated lattices with an involution are the so-called IMTL-algebras, which are the algebraic structures for monoidal t-norm based logic with an involutive negation.

Definition 2 [18]. *A residuated lattice with an involution $\langle L, \wedge, \vee, \otimes, \rightarrow, 0, 1 \rangle$ is an IMTL-algebra if and only if it satisfies the pre-linearity axiom:*

$$(a \rightarrow b) \vee (b \rightarrow a) = 1 \text{ for each } a, b \in L. \tag{1}$$

In providing examples and algorithms, we must restrict to a class of finite substructures

$$\{\langle Ł_n, \wedge, \vee, \otimes, \rightarrow, 0, 1 \rangle \text{ with } n \in \mathbb{N}\} \tag{2}$$

of the standard Łukasiewicz MV-algebra [6], where $Ł_n$ is the *n-element Łukasiewicz chain* given by $Ł_n = \{k/n \mid 0 \le k \le n \text{ and } n \in \mathbb{Z}\}$, and the operations in (2) are defined as follows: let $a, b \in Ł_n$, then $a \wedge b = \min(a, b)$, $a \vee b = \max(a, b)$, $a \otimes b = \max(0, a + b - 1)$ (*Łukasiewicz conjunction*), and $a \rightarrow b = \min(1, 1 - a + b)$ (*Łukasiewicz implication*). Moreover, $\neg a = 1 - a$ for each $a \in Ł_n$. These structures also satisfy the *pre-linearity axiom* defined by (1). For convenience, we will indicate a residuated lattice $(L, \wedge, \vee, \otimes, \rightarrow, 0, 1)$ with its support L.

2.2   Intuitionistic Fuzzy Sets and Intuitionistic L-fuzzy Sets

Intuitionistic fuzzy sets (IF sets for short) were introduced by Atanassov in [1,2] to generalize the concept of *fuzzy sets* in order to explicitly take into account the *non-belongingness* to a set. More formally:

Definition 3. *Let X be a universe such that $X \neq \emptyset$. An* intuitionistic fuzzy set *A of X is defined as $A = \{(x, \mu(x), \nu(x)) \mid x \in X\}$, where the maps $\mu : X \to [0,1]$ and $\nu : X \to [0,1]$ satisfy the condition $\mu(x) + \nu(x) \leq 1$, for each $x \in X$.*

The values $\mu(x)$ and $\nu(x)$ are respectively called degree of membership *and* non-membership *of x to A, and $1 - (\mu(x) + \nu(x))$ is called* hesitation margin *of x to A.*

Let us observe that an IF set coincides with a fuzzy set when the hesitation margin of each element of the starting universe is equal to 0. In this work, we look at IF sets as generalizations of orthopairs by using fuzzy logic. Given a universe X, (P, N) is an *orthopair* on X if and only if $P, N \subseteq X$ and $P \cap N = \emptyset$ [7]. It is easy to understand that (P, N) can be identified with a particular intuitionistic fuzzy set $\{(x, \mu(x), \nu(x)) \mid x \in X\}$, where μ and ν coincides with the characteristic functions of P and N, respectively. That is, orthopairs coincide with the Boolean sub-collection of IF sets.

IF sets were extended to *intuitionistic L-fuzzy sets* (ILF sets for short) considering an appropriate lattice L instead of $[0,1]$ as the set of truth-values [3,4,20]. Our results are based on *intuitionistic L-fuzzy sets* valued on a complete residuated lattice satisfying the double negation law.

Definition 4. *Let $\langle L, \wedge, \vee, \otimes, \to, 0, 1 \rangle$ be a complete residuated lattice having an involution $\neg$, and let X be a non-empty set. An intuitionistic L-fuzzy set A of X is defined by $A = \{(x, \mu(x), \nu(x)) \mid x \in X\}$, where $\mu : X \to L$ and $\nu : X \to L$ satisfy the condition $\mu(x) \leq \neg\nu(x)$, for each $x \in X^2$.*

The components of an intuitionistic L-fuzzy set (μ, ν) of X satisfy the identity $\mu(x) \otimes \nu(x) = 0$ for each $x \in X$. Thus, if $\langle [0,1], \wedge, \vee, \otimes, \to, 0, 1 \rangle$ is the standard Łukasiewicz MV-algebra, they represent *contrary properties* [5].

For convenience, in the sequel, we briefly write (μ, ν) to denote the intuitionistic L-set $\{(\mu(x), \nu(x)) \mid x \in X\}$ when X is clear from the context.

3   From Intuitionistic L-Fuzzy Sets to Possibility Distributions

In this section, we firstly assign a particular possibility distribution to each intuitionistic L-fuzzy set. Then, we prove that possibility distributions corresponding to ILF sets are normal. After that, confining to complete $IMTL$-algebras, we establish under what conditions a possibility distribution assumes value 0.

---

[2] We notice that, as in Definition 3, μ and ν have a symmetrical role, in the sense that $\mu(x) \leq \neg\nu(x)$ is equivalent to $\nu(x) \leq \neg\mu(x)$.

3.1   Possibility Distributions

A *possibility distribution* is a map from a universe X to a totally ordered scale L equipped with a top, a bottom, and an order-reversing map (such as the unit interval $[0, 1]$ with the function assigning $1 - \lambda$ to each $\lambda \in [0, 1]$). The universe of discourse can be an attribute domain, a set of interpretation of a propositional language, etc. In this work, we focus on possibility distributions having the following form

$$\pi : \mathsf{L}^{\mathcal{V}} \to \mathsf{L}, \tag{3}$$

where L is a complete residuated lattice with an involution, and $\mathsf{L}^{\mathcal{V}}$ is the set of all L-sets of a non-empty universe $\mathcal{V}$, i.e., $\mathsf{L}^{\mathcal{V}} = \{\omega \mid \omega : \mathcal{V} \to \mathsf{L}\}$. Of course, since we choose complete residuated lattices with an involution as algebraic structures of truth degrees, our results also hold for the standard definition of possibility distribution, where L is a totally ordered scale. We use the symbol Π to denote the set of all possibility distributions given by (3), i.e., $\Pi = \{\pi \mid \pi : \mathsf{L}^{\mathcal{V}} \to \mathsf{L}\}$.

In possibility theory, a very important role is played by *normal possibility distributions* [15].

Definition 5. *A possibility distribution $\pi \in \Pi$ is* normal *if and only if there exists $\omega \in L^{\mathcal{V}}$ such that $\pi(\omega) = \mathbf{1}$. Moreover, given $\pi \in \Pi$, we put $\mathcal{K}(\pi) = \{\omega \in L^{\mathcal{V}} \mid \pi(\omega) = \mathbf{1}\}$, and we call $\mathcal{K}(\pi)$ the* kernel *of π.*

3.2   Possibility Distributions Generated by Intuitionistic L-fuzzy Sets

Every intuitionistic L-fuzzy set (μ, ν) determines a possibility distribution $\pi_{(\mu,\nu)} \in \Pi$.

Definition 6. *Let $\omega \in L^{\mathcal{V}}$, then*

$$\pi_{(\mu,\nu)}(\omega) = \bigwedge_{v \in \mathcal{V}} (\mu(v) \to \omega(v)) \otimes (\nu(v) \to \neg\omega(v)). \tag{4}$$

We call $\pi_{(\mu,\nu)}$ the possibility distribution generated by (μ, ν).

Let us point out that our possibility distributions play a different role from those based on rough set theory [9,29,30]. Indeed, a possibility distribution, defined by (4), is viewed as a preference function that arises by aggregating the mappings of an intuitionistic L-fuzzy set, which are interpreted as preference functions too[3]. This interpretation can be better understood from the following illustrative example, where a possibility distribution is generated by an intuitionistic L-set in a concrete situation.

---

[3] Additionally, given a intuitionistic L-fuzzy set (μ, ν), the value $\pi_{(\mu,\nu)}(\omega)$ can be also understood as an answer to a bipolar fuzzy query given by (μ, ν), where μ and ν respectively express positive and negative elastic constraints.

Example 1. Imagine that a real estate agent wants to discover the degree of possibility to which a given client C prefers a given flat F that he/she does not know in advance, starting from a pair of specific preference functions, expressed by the client on a set of features concerning apartments.

Then, let $\mathcal{V} = \{v_1, \ldots, v_{10}\}$ be a collection of features regarding flats (for instance *small size* and *low price*), and let $Ł_5 = \{0, 0.25, 0.5, 0.75, 1\}$ be the 5-element Łukasiewicz chain (see Subsect. 2.1). We suppose that

- each flat F is described by an $Ł_5$-set $\omega_F : \mathcal{V} \to Ł_5$, where $\omega_F(v)$ is the truth degree to which F has the attribute v of $\mathcal{V}$;
- the preferences of a given customer C on the attributes of $\mathcal{V}$ are described by an intuitionistic $Ł_5$-fuzzy set (μ_C, ν_C) of $\mathcal{V}$. This means that given $v \in \mathcal{V}$, C prefers apartments having the attribute v with a degree at least $\mu_C(v)$ and at most $\neg\nu_C(v)$ (i.e., C prefers apartments that do not have v with a degree at least $\nu(v)$) in the scale $Ł_5$. For example, if v is the attribute *small size*, then $\mu(v) = 0.5$ and $\nu(v) = 0.25$ respectively mean that the customer prefers flats being *small* at least 0.5 and not more than 0.75 in the scale $Ł_5$, since $\neg\nu(v) = 0.75$.
  Let us notice that μ and ν are also fuzzy constraints: given $v \in \mathcal{V}$, $\mu(v)$ and $\neg\nu(v)$ represent degrees of priority, namely a suitable flat must have the attribute v with a degree between $\mu(v)$ and $\neg\nu(v)$, according to the preference of C. Moreover, we say that μ and ν are respectively positive and negative preference functions because their interpretation is based on the customer preferences about the presence or the absence of certain properties in its ideal apartment.

Hence, $\pi_{(\mu_C, \nu_C)} : Ł_5^{\mathcal{V}} \to Ł_5$ given by (4), is a new preference function, where $\pi_{(\mu_C, \nu_C)}(\omega_F)$ is the degree of possibility that customer C prefers apartment ω_F, and it is computed by aggregating μ_C and ν_C that capture the preferences expressed by C on the attributes of $\mathcal{V}$. In other words, $\pi_{(\mu_C, \nu_C)}$ is a possibility distribution prescribing to what extent a flat is judged to be suitable for C according to the constraints given by (μ_C, ν_C). For example, let F_i and F_j be flats represented by ω_{F_i} and ω_{F_j}, and let C be a customer whose preferences are represented by (μ_C, ν_C) (see Table 1). By Eq. 4, $\pi_{(\mu_C, \nu_C)}(\omega_{F_i}) = 1$ and $\pi_{(\mu_C, \nu_C)}(\omega_{F_j}) = 0.25$. This

Table 1. Values assumed by $\omega_{F_i}, \omega_{F_j}, \mu_C$ and ν_C on $\mathcal{V}$

	v_1	v_2	v_3	v_4	v_5	v_6	v_7	v_8	v_9	v_{10}
ω_{F_i}	0.5	0.5	0.75	1	0	0.5	0.25	0.5	0.5	0.25
ω_{F_j}	0.75	0.5	0.5	0.25	0.25	0.5	0.75	1	0	0.25

	v_1	v_2	v_3	v_4	v_5	v_6	v_7	v_8	v_9	v_{10}
μ_C	0.25	0.25	0.75	1	0	0	0	0.25	0.25	0.25
ν_C	0.5	0.5	0.25	0	0.25	0.5	0.75	0.25	0	0.5

means that we can believe that C prefers ω_{F_i} more than ω_{F_j}. Consequently, the real estate agent could propose ω_{F_i} to C directly and exclude ω_{F_j}.

In the following example, we find the possibility distribution generated by a given ILF set.

Example 2. Let $\mathcal{V} = \{a, b\}$, and let $\mathsf{L}_5 = \{0, 0.25, 0.5, 0.75, 1\}$ be the 5-element Łukasiewicz chain (see Subsect. 2.1). Then, Π is composed of the L_5-sets $\omega_1, \ldots, \omega_{25} : \{a, b\} \to \mathsf{L}_5$ defined by Table 2. We consider the ILF set (μ, ν) given by $\{(a, 0.25, 0.25), (b, 0.5, 0.5)\}$ (i.e. $\mu(a) = 0.25$, $\mu(b) = 0.5$, $\nu(a) = 0.25$, and $\nu(b) = 0.5$).

Then, by (4), the possibility distribution generated by (μ, ν) is given by

$$\pi_{(\mu,\nu)}(\omega_i) = \begin{cases} 1 & \text{if } i \in \{1, 2, 3\}, \\ 0.5 & \text{if } i \in \{4, 8, 9, 12, 13, 16, 17, 20, 21, 25\}, \\ 0.75 & \text{otherwise.} \end{cases} \quad (5)$$

Table 2. Values assumed by $\omega_1, \ldots, \omega_{25}$ on $\{a, b\}$

	ω_1	ω_2	ω_3	ω_4	ω_5	ω_6	ω_7	ω_8	ω_9	ω_{10}	ω_{11}	ω_{12}	ω_{13}
a	0.25	0.5	0.75	0	0	0	0	0	0.25	0.25	0.25	0.25	0.5
b	0.5	0.5	0.5	0	0.25	0.5	0.75	1	0	0.25	0.75	1	0

	ω_{14}	ω_{15}	ω_{16}	ω_{17}	ω_{18}	ω_{19}	ω_{20}	ω_{21}	ω_{22}	ω_{23}	ω_{24}	ω_{25}
a	0.5	0.5	0.5	0.75	0.75	0.75	0.75	1	1	1	1	1
b	0.25	0.75	1	0	0.25	0.75	1	0	0.25	0.5	0.75	1

Remark 1. When $\mathsf{L} = \{0, 1\}$ and $\mathcal{V}$ is a set of propositional variables, Eq. 4 provides the following correspondence between Boolean possibility distributions and orthopairs, which has already been shown in [8]. Given an ILF set (μ, ν), then $\mu : \mathcal{V} \to \{0, 1\}$ and $\nu : \mathcal{V} \to \{0, 1\}$ are respectively the characteristic functions of the sets O_μ and O_ν that form an orthopair on $\mathcal{V}$. Furthermore, $\{0, 1\}^{\mathcal{V}}$ consists of all Boolean evaluation functions on $\mathcal{V}$. Hence, it is easy to check that given $\pi \in \Pi$ and $\omega \in \{0, 1\}^{\mathcal{V}}$, $\pi(\omega) = 1$ (according to Eq. 4) if and only if ω is a model of the propositional formula $\phi_\mu \wedge \phi_\nu$ such that

$$\phi_\mu := \begin{cases} \bigwedge_{v \in O_\mu} v & \text{if } O_\mu \neq \emptyset \\ \top & \text{otherwise} \end{cases} \quad \text{and} \quad \phi_\nu := \begin{cases} \bigwedge_{v \in O_\nu} \neg v & \text{if } O_\nu \neq \emptyset \\ \top & \text{otherwise} \end{cases}$$

where $\wedge, \neg$, and $\top$ are respectively interpreted with the conjunction, the negation, and the top of a Boolean algebra.

An intuitionistic L-fuzzy set (μ, ν) determines also a collection of L-sets $\mathcal{I}_{(\mu,\nu)}$:

$$\mathcal{I}_{(\mu,\nu)} = \{\omega : \mathcal{V} \to L \text{ such that } \mu(v) \leq \omega(v) \leq \neg\nu(v) \text{ for each } v \in \mathcal{V}\}. \qquad (6)$$

Remark 2. $\mathcal{I}_{(\mu,\nu)}$ is a non-empty set since $\mu \in \mathcal{I}_{(\mu,\nu)}$.

The following theorem states that $\mathcal{I}_{(\mu,\nu)}$ coincides with the kernel of $\pi_{(\mu,\nu)}$, and so with $\mathcal{K}(\pi_{(\mu,\nu)}) = \{\omega \in L^{\mathcal{V}} \mid \pi_{(\mu,\nu)}(\omega) = 1\}$.

Theorem 1. *Let (μ, ν) be an intuitionistic L-fuzzy set, and let $\omega \in L^{\mathcal{V}}$. Then, $\pi_{(\mu,\nu)}(\omega) = \mathbf{1}$ if and only if $\omega \in \mathcal{I}_{(\mu,\nu)}$.*

Proof. Let $\omega \in L^{\mathcal{V}}$ such that $\pi_{(\mu,\nu)}(\omega) = 1$. Then, by (4), $\bigwedge_{v \in \mathcal{V}}(\mu(v) \to \omega(v)) \otimes (\nu(v) \to \neg\omega(v)) = 1$, for each $v \in \mathcal{V}$.

Using Proposition 1 (items (iii), (vi), and (vii)), we have that $\mu(v) \leq \omega(v)$ and $\nu(v) \leq \neg\omega(v)$, for each $v \in \mathcal{V}$.

Moreover, by Proposition 1(viii), $\nu(v) \leq \neg\omega(v)$ implies $\neg\neg\omega(v) \leq \neg\nu(v)$ for each $v \in \mathcal{V}$, and since $\neg$ is an involution, we finally get $\omega(v) \leq \neg\nu(v)$ for each $v \in \mathcal{V}$. Hence, $\mu(v) \leq \omega(v) \leq \neg\nu(v)$ for each $v \in \mathcal{V}$, and so, we can conclude that ω belongs to $\mathcal{I}_{(\mu,\nu)}$ (see (6)).

Analogously, we can prove that if $\omega \in \mathcal{I}_{(\mu,\nu)}$ then $\pi_{(\mu,\nu)}(\omega) = 1$.

Example 3. Consider Example 2, then $\mathcal{I}_{(\mu,\nu)} = \{\omega_1, \omega_2, \omega_3\}$, which is also equal to $\mathcal{K}(\pi_{(\mu,\nu)})$.

Therefore, as an immediate consequence of Theorem 1 and Remark 2, we have that possibility distributions generated by an intuitionistic L-set are always normal.

Corollary 1. *Let $\pi \in \Pi$. If π is generated by an intuitionistic L-fuzzy set, then π is normal.*

3.3 Possibility Distributions Generated by Intuitionistic L-fuzzy Sets Based on an IMTL-algebra

In this subsection, confining to complete IMTL-algebras, we discover when a possibility distribution (generated by an ILF set) assumes value $\mathbf{0}$.

At first, let us prove the following lemma.

Lemma 1. *Let $\langle L, \wedge, \vee, \otimes, \to, \mathbf{0}, \mathbf{1} \rangle$ be a complete IMTL-algebra, and let (μ, ν) be an intuitionistic L-fuzzy set. Then, $(\mu(v) \to \omega(v)) \vee (\nu(v) \to \neg\omega(v)) = \mathbf{1}$, for each $v \in \mathcal{V}$.*

Proof. Let $v \in \mathcal{V}$ such that $\mu(v) \to \omega(v) \neq \mathbf{1}$. Since the pre-linearity axiom holds, we get $\omega(v) \to \mu(v) = \mathbf{1}$. By Proposition 1(vii), $\omega(v) \leq \mu(v)$. Moreover, by Definition 4, $\mu(v) \leq \neg\nu(v)$. Hence, $\omega(v) \leq \neg\nu(v)$ that is equivalent to $\omega(v) \to \neg\nu(v) = \mathbf{1}$. Finally, by Proposition 1 (viii), we get $\nu(v) \to \neg\omega(v) = \mathbf{1}$.

Analogously, let $v \in \mathcal{V}$, if $\nu(v) \to \neg\omega(v) \neq \mathbf{1}$, then we can prove that $\mu(v) \to \omega(v) = \mathbf{1}$.

By Proposition 1(iv), we conclude that $(\mu(v) \to \omega(v)) \vee (\nu(v) \to \neg\omega(v)) = \mathbf{1}$.

A possibility distribution generated by an ILF set (valued on a complete IMTL-algebra) is equal to 0 only in some particular cases. More precisely, the next theorem holds.

Theorem 2. *Let* $\langle L, \wedge, \vee, \otimes, \rightarrow, 0, 1 \rangle$ *be a complete IMTL-algebra, let* (μ, ν) *be an intuitionistic L-fuzzy set, and let* $\omega \in L^{\mathcal{V}}$. *Then,* $\pi_{(\mu,\nu)}(\omega) = 0$ *if and only if there exists* $v \in \mathcal{V}$ *such that* $\mu(v) = 1$ *and* $\omega(v) = 0$, *or* $\mu(v) = 0$ *and* $\omega(v) = 1$.

Proof. ($\Leftarrow$) This implication is trivial.
($\Rightarrow$) Let $\omega \in L^{\mathcal{V}}$ such that $\pi_{(\mu,\nu)}(\omega) = 0$. Then, by Proposition 1 (v), there exists $v \in \mathcal{V}$ such that

$$(\mu(v) \rightarrow \omega(v)) \otimes (\nu(v) \rightarrow \neg\omega(v)) = 0. \tag{7}$$

By Lemma 1,
$$\mu(v) \rightarrow \omega(v) = 1 \text{ or } \nu(v) \rightarrow \neg\omega(v) = 1. \tag{8}$$

Eventually, by Definition 1 ($a \otimes 1 = a$ for each $a \in L$), Eqs. (7) and (8) imply that $\mu(v) = 1$ and $\omega(v) = 0$, or $\mu(v) = 0$ and $\omega(v) = 1$.

Example 4. Consider Example 2. Then, $\pi_{(\mu,\nu)}(\omega) \neq 0$ for each $\omega \in \mathsf{L}_5^{\{a,b\}}$. In fact, $\mu(a), \mu(b) \notin \{0, 1\}$.

4   From Possibility Distributions to Intuitionistic Fuzzy Sets

This section mainly aims to find the intuitionistic L-fuzzy set that generates a given possibility distribution $\pi : L^{\mathcal{V}} \rightarrow L$ by means of Eq. 4.

Let us recall that Eq. 4 leads to define a normal possibility distribution for each intuitionistic L-set. On the other hand, it is not always possible to do the opposite. Namely, there exist normal possibility distributions from $L^{\mathcal{V}}$ to L that do not correspond to any intuitionistic L-fuzzy set by means of Eq. 4. The following is an example.

Example 5. Consider Example 2, we can prove that no intuitionistic L_5-fuzzy set generates the possibility distribution $\pi : \mathsf{L}_5^{\{a,b\}} \rightarrow \mathsf{L}_5$ defined by the following formula: let $\omega_i \in \mathsf{L}_5^{\{a,b\}}$,

$$\pi(\omega_i) = \begin{cases} 1 & \text{if } i \in \{1, 2, 3\}, \\ 0 & \text{otherwise.} \end{cases} \tag{9}$$

Since the pre-linearity axiom holds in $\langle \mathsf{L}_5, \wedge, \vee, \otimes, \rightarrow, 0, 1 \rangle$, we can apply Theorem 2. Consequently, in case π is generated by an intuitionistic L_5-fuzzy set, it must be $\omega_i(a) \in \{0, 1\}$ or $\omega_i(b) \in \{0, 1\}$, for each $i \in \{4, \ldots, 25\}$. But, this contradicts Table 2, where $\omega_i(a)$ and $\omega_i(b)$ do not belong to $\{0, 1\}$ for each $i \in \{10, 11, 14, 15, 18, 19\}$.

In the sequel, we write Π^* to indicate the set of all possibility distributions of Π that are generated by an intuitionistic L-fuzzy set.

Now, we want to establish when a given possibility distribution belongs to Π^*. In order to do this, we firstly associate an intuitionistic L-fuzzy set to every possibility distribution starting from its kernel.

Definition 7. *Given $\pi \in \Pi$ and $v \in \mathcal{V}$, then $\mu_\pi, \nu_\pi : \mathcal{V} \to L$ are defined as follows:*

$$\mu_\pi(v) = \bigwedge_{\omega \in \mathcal{K}(\pi)} \omega(v) \quad and \quad \nu_\pi(v) = \bigwedge_{\omega \in \mathcal{K}(\pi)} \neg\omega(v). \tag{10}$$

We can prove that the functions given by (10) form an intuitionistic L-fuzzy set.

Proposition 2. *Let $\pi \in \Pi$, then (μ_π, ν_π) is an intuitionistic L-fuzzy set.*

Proof. By Proposition 1(i), we get

$$\bigwedge_{\omega \in \mathcal{K}(\pi)} \omega(v) \leq \omega(v) \quad and \quad \bigwedge_{\omega \in \mathcal{K}(\pi)} \neg\omega(v) \leq \neg\omega(v), \quad \text{for each } v \in \mathcal{V}. \tag{11}$$

Moreover, by Proposition 1 (vii, viii),

$$\bigwedge_{\omega \in \mathcal{K}(\pi)} \neg\omega(v) \leq \neg\omega(v), \text{ implies that } \omega(v) \leq \neg \bigwedge_{\omega \in \mathcal{K}(\pi)} \neg\omega(v) \quad \text{for each } v \in \mathcal{V}.$$
$$\tag{12}$$

Hence, by (10), we can conclude that $\mu_\pi(v) \leq \neg\nu_\pi(v)$, for each $v \in \mathcal{V}$.

Since (μ_π, ν_π) is an intuitionistic L-fuzzy set, it generates a new possibility distribution (by means of Eq. 4) that we denote with π^*. In general, π^* does not coincide with π. For example, it is easy to verify that if π is given by (5) then π^* is given by (9). Consequently, $\pi^* \neq \pi$. Of course, $\pi = \pi^*$ implies that $\pi \in \Pi^*$, and more precisely that π is generated by the intuitionistic L-fuzzy set (μ_π, ν_π). Furthermore, the following theorem shows that (μ_π, ν_π) is the only intuitionistic L-fuzzy set that can generate π.

Theorem 3. *Let $\pi \in \Pi$. If $\pi \in \Pi^*$, then π is generated by (μ_π, ν_π).*

Proof. Let $\pi \in \Pi^*$, there exists (μ, ν) that generates π. So, we want to prove that $(\mu, \nu) = (\mu_\pi, \nu_\pi)$.

First of all, we show that $\mu = \mu_\pi$. Let $v \in \mathcal{V}$. Then, by Theorem 1, $\mu(v) \leq \omega(v)$ for each $\omega \in \mathcal{K}(\pi)$. Consequently, by Proposition 1(ii), $\mu(v) \leq \bigwedge_{\omega \in \mathcal{K}(\pi)} \omega(v)$. Namely, $\mu(v) \leq \mu_\pi(v)$.

Moreover, if $\mu_\pi(v) < \mu(v)$, then there exists $\omega \in \mathcal{K}(\pi)$ such that $\omega(v) < \mu(v)$, but it contradicts Theorem 1. Then, $\mu_\pi(v) \leq \mu(v)$.

Analogously, we can prove that $\nu = \nu_\pi$.

Example 6. Consider the possibility distribution $\pi_{(\mu,\nu)}$ given by (5). For convenience, we indicate $\pi_{(\mu,\nu)}$ with π. Example 2 shows that π is generated by (μ, ν), which is $\{(a, 0.25, 0, 25), (b, 0.5, 0.5)\}$. Consequently, $\pi \in \Pi^*$. Moreover, Theorem 3 assures us that $(\mu, \nu) = (\mu_\pi, \nu_\pi)$. Indeed, $\mu_{\pi_{(\mu,\nu)}}(a) = \omega_1(a) \wedge \omega_2(a) \wedge \omega_3(a) = 0.25 \wedge 0.5 \wedge 0.75 = 0.25$, $\mu_{\pi_{(\mu,\nu)}}(b) = \omega_1(b) \wedge \omega_2(b) \wedge \omega_3(b) = 0.5 \wedge 0.5 \wedge 0.5 = 0.5$, $\nu_{\pi_{(\mu,\nu)}}(a) = \neg\omega_1(a) \wedge \neg\omega_2(a) \wedge \neg\omega_3(a) = 0.75 \wedge 0.5 \wedge 0.25 = 0.25$, and $\nu_{\pi_{(\mu,\nu)}}(b) = \neg\omega_1(b) \wedge \neg\omega_2(b) \wedge \neg\omega_3(b) = 0.5 \wedge 0.5 \wedge 0.5 = 0.5$.

Using Theorem 3, we provide a necessary and sufficient condition for a possibility distribution to be generated by an ILF set.

Corollary 2. *Let $\pi \in \Pi$. Then, $\pi \in \Pi^*$ if and only if $\pi = \pi^*$, namely*

$$\pi(\omega) = \bigwedge_{v \in \mathcal{V}} (\mu_\pi(v) \to \omega(v)) \otimes (\nu_\pi(v) \to \neg\omega(v)), \quad \text{for each } \omega \in L^\mathcal{V}.$$

The following proposition will be used in the next subsection. It shows that the kernel of π^* always includes that of π.

Proposition 3. *Let $\pi \in \Pi$. Then, $\mathcal{K}(\pi) \subseteq \mathcal{K}(\pi^*)$.*

Proof. Let $\omega \in \mathcal{K}(\pi)$. Then, by (11) and (12), we get $\mu_\pi(v) \leq \omega(v) \leq \neg\nu_\pi(v)$ for each $v \in \mathcal{V}$. Thus, by Theorem 1, $\omega \in \mathcal{K}(\pi^*)$.

4.1   An Algorithm to Find the Intuitionistic L-fuzzy Set Generating a Given Possibility Distribution

In this subsection, assuming that our structures of truth degrees are the Łukasiewicz n-chains defined by (2), we propose three algorithms to achieve the following goals.

(i) Compute the intuitionistic $Ł_n$-fuzzy set corresponding to a given possibility distribution by means of (10).
(ii) Find the values assumed by the possibility distribution generated by a given intuitionistic $Ł_n$-fuzzy set.
(iii) Establish whether or not a given possibility distribution is generated by an intuitionistic $Ł_n$-fuzzy set.

Firstly, we propose the procedure INT-L-SET (see Algorithm 1) based on Eq. 10. Its input consists of a finite set $\mathcal{V}$, a positive integer n (to determine the corresponding Łukasiewicz n-chain), and a possibility distribution π from $Ł_n^\mathcal{V}$ to $Ł_n$. Its output is a pair (μ, ν) of mappings from $\mathcal{V}$ to $Ł_n$. By Proposition 2, (μ, ν)

is an intuitionistic $Ł_n$-fuzzy set, and by Theorem 3, if $\pi \in \Pi^*$ then it generates π.

Algorithm 1: The algorithm for finding the intuitionistic $Ł_n$-fuzzy set corresponding to a given possibility distribution by means of (10).

---

procedure INT-L-SET $(\mathcal{V}, n, \pi)$
foreach $v \in \mathcal{V}$ **do**
$\quad$ $\mu(v), \nu(v) \to 1$;
$\quad$ **foreach** $\omega \in Ł_n^{\mathcal{V}}$ *such that* $\pi(\omega) = 1$ **do**
$\quad\quad$ $\mu(v) \leftarrow \min\{\mu(v), \omega(v)\}$;
$\quad\quad$ $\nu(v) \leftarrow \min\{\nu(v), 1 - \omega(v)\}$;

return (μ, ν);
end procedure

The next procedure (see Algorithm 2) is constructed by using the following proposition, where Eq. 4 is rewritten for all possibility distributions generated by an intuitionistic $Ł_n$-fuzzy set[4].

Proposition 4. *Let π be a possibility distribution generated by an intuitionistic $Ł_n$-fuzzy set (μ, ν), and let $\omega \in Ł_n^{\mathcal{V}}$. Then, $\pi(\omega) = \bigwedge_{v \in \mathcal{V}} \alpha_\omega(v)$, where*

$$\alpha_\omega(v) = \begin{cases} \mu(v) \to \omega(v) & \text{if } \omega(v) \leq \mu(v), \\ \nu(v) \to \neg\omega(v) & \text{if } \omega(v) \geq \neg\nu(v), \\ 1 & \text{otherwise.} \end{cases} \quad (13)$$

Proof. Let $v \in \mathcal{V}$ such that $\omega(v) \leq \mu(v)$. By Definition 4, $\mu(v) \leq \neg\nu(v)$. Hence, $\omega(v) \leq \neg\nu(v)$. By Proposition 1(vii,viii), $\nu(v) \leq \neg\omega(v)$, and so $\nu(v) \to \neg\omega(v) = 1$. Finally, using Eq. 4, $\alpha_\omega(v) = (\mu(v) \to \omega(v)) \otimes 1$, hence $\alpha_\omega(v) = (\mu(v) \to \omega(v))$ from Definition 1(ii).

Analogously, given $v \in \mathcal{V}$ such that $\omega(v) > \mu(v)$, we can prove that $\alpha_\omega(v)$ is given by (13).

Proposition 4 leads to the procedure VALUE (just apply the Lukasiewicz operations to (13)) taking as input a finite set $\mathcal{V}$, a function ω from $\mathcal{V}$ to $Ł_n$, and an intuitionistic $Ł_n$-fuzzy set (μ, ν), and producing as output the value $\pi_{(\mu,\nu)}(\omega)$.

---

[4] More in general, Proposition 4 holds when we consider complete residuated lattices with an involution and $[0, 1]$ as support.

Algorithm 2: The algorithm to find the values assumed by the possibility distribution generated by the intuitionistic $Ł_n$-fuzzy set (μ, ν).

---

procedure VALUE $(\mathcal{V}, \omega, (\mu, \nu))$
$m \leftarrow 1;$
foreach $v \in \mathcal{V}$ **do**
    if $\omega(v) < \mu(v)$ **then**
        $\alpha_v(\omega) \leftarrow 1 - \mu(v) + \omega(v);$
    else
        if $\omega(v) > 1 - \nu(v)$ **then**
            $\alpha_v(\omega) \leftarrow 2 - \omega(v) + \nu(v);$
        else
            $\alpha_v(\omega) \leftarrow 1;$
    $m \leftarrow \min\{m, \alpha_v(\omega)\};$
return $m;$
end procedure

Finally, we present the procedure DISTRIBUTION to establish whether or not a given possibility distribution $\pi : Ł_n^{\mathcal{V}} \mapsto Ł_n$ is generated by the intuitionistic $Ł_n$-fuzzy set (μ_π, ν_π) (see Algorithm 3). In detail, firstly, the intuitionistic $Ł_n$-fuzzy set (μ_π, ν_π) is computed by INT-L-SET. Then, using the procedure VALUE, it is checked whether or not $\pi = \pi^*$, where π^* is the possibility distribution generated by (μ_π, ν_π). Eventually, if $\pi = \pi^*$, then π is generated by (μ_π, ν_π). Otherwise, the answer is that $\pi \notin \Pi^*$ (from Theorem 3 and Corollary 2). Moreover, by Proposition 3, we know that $\pi(\omega) = \pi^*(\omega)$ for each $\omega \in \mathcal{K}(\pi)$. Hence, we must apply the procedure VALUE only for each $\omega \in Ł_n^{\mathcal{V}} \setminus \mathcal{K}(\pi)$.

Algorithm 3: The algorithm to establish whether or not a given possibility distribution is generated by an intuitionistic $Ł_n$-fuzzy set.

---

procedure DISTRIBUTION $(\mathcal{V}, n, \pi)$
$i \leftarrow 0;$
$(\mu, \nu) \leftarrow$ INT-L-SET$(\mathcal{V}, n, \pi);$
foreach $\omega \in Ł_n$ **such that** $\pi(\omega) \neq 1$ **do**
    if $\pi(\omega) \neq$ VALUE$(\mathcal{V}, \omega, (\mu, \nu))$ **then**
        $i \leftarrow 1;$
        break;
if $i = 1$ **then**
    print π is not generated by an intuitionistic $Ł_n$-fuzzy set;
else
    print π is generated by the intuitionistic $Ł_n$-fuzzy set (μ, ν);
return;
end procedure

---

5   Conclusions and Future Directions

In this article, we identified each intuitionistic L-fuzzy set with a special normal possibility distribution. On the other hand, we showed that not all normal possibility distributions can be identified with an intuitionistic L-fuzzy set.

In the future, we intend to explore the connection between possibility theory and intuitionistic L-fuzzy sets in more detail. As an example, we would like to discover other properties (in addition to normality) characterizing possibility distributions generated by ILF sets. Also, we could associate a collection of ILF sets to each possibility distribution, and hence, generalize by using fuzzy logic, the correspondence between Boolean possibility distributions and sets of orthopairs [8].

On a longer term, the link between the two theories could be used in applications, by applying techniques developed for IFS to Possibility Theory and, whenever possible, the other way round.

References

1. Atanassov, K.: Review and new results on intuitionistic fuzzy sets. preprint Im-MFAIS-1-88, Sofia 5, 1 (1988)
2. Atanassov, K.: Intuitionistic fuzzy sets. Int. J. Bioautom. **20**, 1 (2016)
3. Atanassov, K.T.: Intuitionistic Fuzzy Sets: Theory and Applications (studies in fuzziness and soft computing), vol. 35. Physica-Verlag, Heidelberg (1999)
4. Atanassov, K.T., Stoeva, S.: Intuitionistic l-fuzzy sets. In: Cybernetics and Systems Research 2, vol. 23, pp. 539–540, Trappl R. (ed.) Elsevier Scientific Publications, Amsterdam (1984)
5. Boffa, S., Murinová, P., Novák, V.: Graded decagon of opposition with fuzzy quantifier-based concept-forming operators. In: Lesot, M.J., et al. (eds.) IPMU 2020. CCIS, vol. 1239, pp. 131–144. Springer, Cham (2020). https://doi.org/10.1007/978-3-030-50153-2_10
6. Chang, C.C.: Algebraic analysis of many valued logics. Trans. Am. Math. Soc. **88**(2), 467–490 (1958)
7. Ciucci, D.: Orthopairs and granular computing. Granul. Comput. **1**(3), 159–170 (2016)
8. Ciucci, D., Dubois, D., Lawry, J.: Borderline vs. unknown: comparing three-valued representations of imperfect information. Int. J. Approx. Reason. **55**(9), 1866–1889 (2014)
9. Ciucci, D., Forcati, I.: Certainty-based rough sets. In: Polkowski, L., et al. (eds.) IJCRS 2017. LNCS (LNAI), vol. 10314, pp. 43–55. Springer, Cham (2017). https://doi.org/10.1007/978-3-319-60840-2_3
10. Cohen, L.J.: The probable and the provable (1977)
11. Dubois, D., Fargier, H., Prade, H.: Possibility theory in constraint satisfaction problems: handling priority, preference and uncertainty. Appl. Intell. **6**(4), 287–309 (1996)
12. Dubois, D., Moral, S., Prade, H.: A semantics for possibility theory based on likelihoods. J. Math. Anal. Appl. **205**(2), 359–380 (1997)
13. Dubois, D., Prade, H.: Ranking fuzzy numbers in the setting of possibility theory. Inf. Sci. **30**(3), 183–224 (1983)

14. Dubois, D., Prade, H.: Fuzzy rules in knowledge-based systems. In: Yager, R.R., Zadeh, L.A. (eds.) An Introduction to Fuzzy Logic Applications in Intelligent Systems. The Springer International Series in Engineering and Computer Science (Knowledge Representation, Learning and Expert Systems), vol. 165, pp. 45–68. Springer, Boston, MA (1992). https://doi.org/10.1007/978-1-4615-3640-6_3

15. Dubois, D., Prade, H.: Possibility theory: qualitative and quantitative aspects. In: Smets, P. (ed.) Quantified Representation of Uncertainty and Imprecision. HDRUMS, vol. 1, pp. 169–226. Springer, Dordrecht (1998). https://doi.org/10.1007/978-94-017-1735-9_6

16. Dubois, D., Prade, H.: Possibility Theory: An Approach to Computerized Processing of Uncertainty. Springer Science & Business Media, Heidelberg (2012)

17. Dubois, D., Prade, H.: Possibility theory and its applications: where do we stand? In: Kacprzyk, J., Pedrycz, W. (eds.) Springer Handbook of Computational Intelligence, pp. 31–60. Springer, Heidelberg (2015). https://doi.org/10.1007/978-3-662-43505-2_3

18. Esteva, F., Godo, L.: Monoidal t-norm based logic: towards a logic for left-continuous t-norms. Fuzzy Sets Syst. **124**(3), 271–288 (2001)

19. Galatos, N., Raftery, J.G.: Adding involution to residuated structures. Stud. Log. **77**(2), 181–207 (2004)

20. Gerstenkorn, T., Tepavĉević, A.: Lattice valued intuitionistic fuzzy sets. Open Math. **2**(3), 388–398 (2004)

21. Goguen, J.A.: L-fuzzy sets. J. Math. Anal. Appl. **18**(1), 145–174 (1967)

22. Goguen, J.A.: The logic of inexact concepts. Synthese, pp. 325–373 (1969)

23. Gottwald, S., Gottwald, P.S.: A Treatise on Many-Valued Logics, vol. 3. Research Studies Press Baldock, Boston (2001)

24. Hájek, P.: Basic fuzzy logic and bl-algebras. Soft. Comput. **2**(3), 124–128 (1998)

25. Hájek, P.: Metamathematics of Fuzzy Logic, vol. 4. Springer Science & Business Media, Heidelberg (2013)

26. Höhle, U.: On the fundamentals of fuzzy set theory. J. Math. Anal. Appl. **201**(3), 786–826 (1996)

27. Lewis, D.: Counterfactuals. John Wiley & Sons, Hoboken (2013)

28. Lukasiewicz, J.: Untersuchungen uber den aussagenkalkul. CR des seances de la Societe des Sciences et des Letters de Varsovie, cl. III 23 (1930)

29. Nakata, M., Sakai, H.: Lower and upper approximations in data tables containing possibilistic information. In: Peters, J.F., Skowron, A., Marek, V.W., Orłowska, E., Słowiński, R., Ziarko, W. (eds.) Transactions on Rough Sets VII. LNCS, vol. 4400, pp. 170–189. Springer, Heidelberg (2007). https://doi.org/10.1007/978-3-540-71663-1_11

30. Nakata, M., Sakai, H.: Rough sets by indiscernibility relations in data sets containing possibilistic information. In: Flores, V., et al. (eds.) IJCRS 2016. LNCS (LNAI), vol. 9920, pp. 187–196. Springer, Cham (2016). https://doi.org/10.1007/978-3-319-47160-0_17

31. Novák, V., Perfilieva, I., Dvorak, A.: Insight into Fuzzy Modeling. John Wiley & Sons, Hoboken (2016)

32. Shackle, G.L.S.: Decision Order and Time in Human Affairs. Cambridge University Press, Cambridge (2010)

33. Zadeh, L.A.: Fuzzy sets as a basis for a theory of possibility. Fuzzy Sets Syst. **1**(1), 3–28 (1978)

Feature Selection and Disambiguation in Learning from Fuzzy Labels Using Rough Sets

Andrea Campagner[(✉)] and Davide Ciucci[ID]

Dipartimento di Informatica, Sistemistica e Comunicazione,
University of Milano–Bicocca, viale Sarca 336, 20126 Milano, Italy
a.campagner@campus.unimib.it

Abstract. In this article, we study the setting of *learning from fuzzy labels*, a generalization of supervised learning in which instances are assumed to be labeled with a fuzzy set, interpreted as an epistemic possibility distribution. We tackle the problem of feature selection in such task, in the context of *rough set theory* (RST). More specifically, we consider the problem of RST-based feature selection as a means for *data disambiguation*: that is, retrieving the most plausible precise instantiation of the imprecise training data. We define generalizations of decision tables and reducts, using tools from generalized information theory and belief function theory. We study the computational complexity and theoretical properties of the associated computational problems.

Keywords: Fuzzy labels · Rough sets · Feature selection · Belief functions · Entropy

1 Introduction

Weakly supervised learning [34] refers to Machine Learning tasks in which training instances are not required to be associated with a precise target label: the annotations can be either imprecise or partial. Such tasks could be a consequence of certain data pre-processing operations such as anonymization [24]; could be due to imprecise measurements or expert opinions; or to limit data annotation costs [20]. Some particularly relevant instances of weakly supervised learning are *superset learning* [15] (i.e. instances are associated with sets of candidate labels), *learning from evidential labels* [6,9] (i.e., instances are associated with belief functions over the labels) and *learning from fuzzy labels* [10,13]. In this latter setting, which is the focus of this article, each instance x is annotated with a fuzzy set μ of candidate labels. These fuzzy sets have an epistemic semantics and represent possibility distributions π_μ: only one of the labels is the correct one and the fuzzy membership degrees, then, describe the possibility degree of the labels. For example, an image could be tagged with {horse : 1, pony : 0.8, zebra : 0.5, dog : 0.0}, suggesting that the animal shown

© Springer Nature Switzerland AG 2021
S. Ramanna et al. (Eds.): IJCRS 2021, LNAI 12872, pp. 164–179, 2021.
https://doi.org/10.1007/978-3-030-87334-9_14

on the picture is one among {horse, pony, zebra}: though it is not exactly known which of them, it is known that *horse* is deemed more plausible than *pony*, which in turn is deemed more plausible than *zebra*.[1]

While in recent years the superset learning task has been widely investigated [4,16,17], also using approaches based on Rough Set theory [4,25], the *learning from fuzzy labels* task has, comparatively, been given less attention mainly due to the high computational complexity of the problem [13] and to the difficulty of acquiring such data [14]: most works have focused on the problem of classification [6,9,10,13,23] (in particular in those tasks where the acquisition of such fuzzy labels is easier, e.g. multi-rater learning and self-regularized learning [11]), while other tasks such as *feature selection*, despite their importance, have mostly been ignored.

In this article, drawing from our previous work on superset feature selection [4], we attempt to close this gap by proposing methods, based on Rough Set Theory (RST), Belief Function Theory (BFT) and possibility theory, to address the problem of feature selection. Remarkably, in line with the generalized risk minimization paradigm [13], we consider this task as a means for *data disambiguation*, i.e., for the purpose of figuring out the most plausible precise instantiation of the imprecise training data. For this purpose we propose a generalization of standard Decision Tables and we describe different definitions of reducts. In particular, in Sect. 2 we provide the necessary background knowledge on possibility theory, Rough Set theory and Belief Function theory; in Sect. 3.1 we define a generalization of decision tables to the learning from fuzzy label settings; in Sect. 3.2 we introduce several notions of reducts and study their relationships and computational complexity properties; in Sect. 3.3 we propose a generalization of entropy reducts, in order to provide an approach for performing feature selection which is more apt at the design of heuristics or approximation algorithms; finally, in Sect. 4, we summarize our results and describe some open problems.

2 Background

In this section, we recall basic notions of rough set theory (RST) and evidence theory, which will be used in the main part of the article.

2.1 Possibility Theory

Possibility theory is a theory of uncertainty, alternative to probability theory, which allows for the quantification of degrees of possibility on the basis of a fuzzy set [33]. We recall that a fuzzy set (equivalently, a possibility distribution)

---

[1] We note that in the *learning from fuzzy labels* setting, the set of candidate labels (that is, the labels with a membership degree greater than 0) is given a disjunctive interpretation: only one of those labels is correct, but we don't precisely know which one, and the membership degrees represent *degrees of belief*. Thus, in this article, we do not consider the conjunctive interpretation, in which the membership degrees are *degrees of truth* (and, thus, could be seen as a generalization of multi-label learning).

F can be seen as a function $F : X \mapsto [0, 1]$, that is, a generalization of the characteristic function representation of classical sets. A possibility measure is a function $pos_F : 2^X \mapsto [0, 1]$ such that

1. $pos_F(\emptyset) = 0$ and $pos_F(X) = 1,;$
2. if $A \cap B = \emptyset$ then $pos_F(A \cup B) = max(pos_F(A), pos_F(B))$.

It can be easily seen that every possibility measure is induced by a fuzzy set F as $pos_F(A) = max_{x \in A} F(x)$: in this case we say that F is the possibility distribution corresponding to the possibility measure pos_F.

A possibility distribution F is *normal* if $\exists x \in X. F(x) = 1$: in this article we will focus on normal possibility distributions. Given $\alpha \in [0, 1]$, the *alpha*-cut of F is defined as $F^\alpha = \{x \in X : F(x) \geq \alpha\}$, while the *strong* α-cut is defined as $F^{\alpha+} = \{x \in X : F(x) > \alpha\}$: we recall that the collection of α-cuts of F is sufficient to determine F [19].

The epistemic view [8] of possibility distributions refers to the common interpretation under which a possibility distribution represents the degrees of belief (of an agent) towards a set of possible alternatives. We refer the reader to [7,13] for a discussion of epistemic possibility distributions in Machine Learning.

2.2   Rough Set Theory

Rough set theory has been proposed by Pawlak [22] as a framework for representing and managing uncertain data, and has since been widely applied for various problems in the ML domain (see [2] for a recent overview and survey). We briefly recall the main notions of RST, especially regarding its applications to feature selection.

A decision table (DT) is a triple $DT = \langle U, Att, t \rangle$ such that U is a universe of objects and Att is a set of *attributes* employed to represent objects in U. Formally, each attribute $a \in Att$ is a function $a : U \rightarrow V_a$, where V_a is the domain of values of a. Moreover, $t \notin Att$ is a distinguished *decision* attribute, which represents the target decision (also labeling or annotation) associated with each object in the universe. We say that DT is *inconsistent* if the following holds: $\exists x_1, x_2 \in U, \forall a \in Att, a(x_1) = a(x_2)$ and $t(x_1) \neq t(x_2)$.

Given $B \subseteq Att$, we can define the *indiscernibility relation* with respect to B as $x I_B x'$ iff $\forall a \in B, a(x') = a(x)$. Clearly, it is an equivalence relation that partitions the universe U in equivalence classes, also called *granules of information*, $[x]_B$. Then, the *indiscernibility partition* is denoted as $\pi_B = \{[x]_B \mid x \in U\}$.

We say that $B \subseteq Att$ is a *decision reduct* for DT if $\pi_B \leq \pi_t$ (where the order $\leq$ is the refinement order for partitions, that is, π_t is a coarsening of π_B) and there is no $C \subsetneq B$ such that $\pi_C \leq \pi_t$. Then, evidently, a reduct of a decision table DT represents a set of non-redundant and necessary features to represent the information in DT. We say that a reduct R is *minimal* if it is among the smallest (with respect to cardinality) reducts.

Given $B \subseteq Att$ and a set $S \subseteq U$, a *rough approximation* of S (with respect to B) is defined as the pair $B(S) = \langle l_B(S), u_B(S) \rangle$, where $l_B(S) = \bigcup \{[x]_B \mid [x]_B \subseteq$

$S\}$ is the *lower approximation* of S, and $u_B(s) = \bigcup\{[x]_B \mid [x]_B \cap S \neq \emptyset\}$ is the corresponding *upper approximation*.

Finally, given $B \subseteq Att$, the *generalized decision* with respect to B for an object $x \in U$ is defined as $\delta_B(x) = \{t(x') \mid x' \in [x]_B\}$. Notably, if DT is not inconsistent and B is a reduct, then $\delta_B(x) = \{t(x)\}$ for all $x \in U$.

We note that in the RST literature, there exist several definitions of reduct that, while equivalent on consistent Decision Tables, are generally non-equivalent for inconsistent ones. We refer the reader to [28] for an overview of such a list and a study of their dependencies. Moreover, we remark that, given a decision table, the problem of finding the minimal reduct is in general Σ_2^P-complete (by reduction to the *Shortest Implicant* problem [31]). We recall that Σ_2^P is the complexity class defined by problems that can be verified in polynomial time given access to an oracle for an NP-complete problem [1].

Finally, we recall that some previous works have investigated the generalization of Rough Set Theory to the case of imprecise data, both in the case of set-valued data [4,21,25] and in the case of possibility distributions [5], or more general uncertainty representations [30]. Nakata et al. [18] discuss a generalization of Rough Set Theory to the case where every attribute value is expressed as a possibility distribution and study generalized notions of rough approximations: though this approach uses a cut-based approach similar to the one we adopt in this paper, the authors do not study generalizations of reducts to this setting. Ciucci et al. [5] focus on a specific type of possibility distribution (certainty distributions) and study different notions for both rough approximations and reducts: in our work we consider the case of general possibility distributions, but only for the decision attribute. Also, we note that both articles [5,18] do not consider applications to the learning from fuzzy labels setting. Finally, Trabelsi et al. [30] considered the generalization of RST to account for evidential data in the decision attribute and proposed a definition of reducts in that setting: while the approach adopted by the authors shares some similarities with the approach we propose, the former does not agree with the generalized risk minimization principle [13] and hence cannot be applied to the task of data disambiguation.

2.3 Belief Function Theory

Belief Function theory (BFT), also known as Dempster-Shafer theory or evidence theory, has been introduced by Dempster and subsequently formalized by Shafer in [26]. Given a *frame of discernment* X, which represents all possible states of a system under study, a *basic belief assignment* (bba) is a function $m : 2^X \to [0,1]$, such that $m(\emptyset) = 0$ and $\sum_{A \in 2^X} m(A) = 1$. The *support* of m is defined as $supp(m) = \{A \subseteq X : m(A) > 0\}$.

From a bba, a pair of functions, called respectively *belief* and *plausibility*, can be defined as follows:

$$Bel_m(A) = \sum_{B:B \subseteq A} m(B) \qquad Pl_m(A) = \sum_{B:B \cap A \neq \emptyset} m(B) \tag{1}$$

As can be seen from these definitions, there is a clear correspondence between BFT and, respectively, RST and possibility theory. In the first case, it is easy to note that belief functions (resp., plausibility functions) correspond to lower approximations (resp., upper approximations) in RST whenever the support m is a partition of X; we refer the reader to [32] for further connections between the two theories. In the case of possibility theory, any possibility measure (resp. necessity measure) is a plausibility (resp. belief) function: indeed, it can be shown that possibility theory can be seen as a special case of BFT where we require that m is *consonant* [26], that is $\forall A_1, A_2 \in supp(m) . A_1 \subseteq A_2 \vee A_2 \subseteq A_1$ (i.e., $supp(m)$ with the order given by $\subseteq$ is a linear order).

Finally, we recall that several generalizations of information-theoretic concepts, specifically the concept of *entropy* (which was also proposed to generalize the definition of reducts in RST [27]), have been defined for BFT. Most relevantly, we recall the definition of *optimistic aggregate uncertainty* [3,4]:

$$OAU(m) = \min_{p \in \mathcal{P}(m)} H(p), \tag{2}$$

where $\mathcal{P}(m)$ is the set of probability distributions p such that $Bel_m \leq p \leq Pl_m$ and $H(p) = -\sum_{x \in X} p(x)log_2 p(x)$ is the Shannon entropy of p.

3   Possibilistic Decision Tables and Reducts

In this section, we extend some key concepts of rough set theory to the setting of learning from fuzzy labels.

3.1   Possibilistic Decision Tables

In the *learning from fuzzy labels* setting, an object $x \in U$ is not necessarily assigned a single annotation $t(x) \in V_t$, but may instead be associated with an epistemic statement (elicited by an agent, human or computational) encoding the relative plausibility of a set S of candidate annotations, one of which is assumed to be the true annotation associated with x. The relative plausibility of the candidate annotations is expressed as a possibility distribution (or, equivalently, as a fuzzy set) over the label set. To model this idea in terms of RST, we generalize the definition of a decision table as follows.

Definition 1. *A* possibilistic decision table *(PDT) is a tuple $P = \langle U, Att, t, d \rangle$, where $\langle U, Att, t \rangle$ is a decision table, i.e.:*

- *U is a universe of objects of interest;*
- *Att is a set of attributes (or features);*
- *t is the (real) decision attribute (whose value, in general, is not known);*
- *$d \notin Att$ is a map from objects to possibility distributions over V_t, $d : U \rightarrow \mathcal{F}(V_t)$ such that the weak superset property holds: $d(x)_{t(x)} > 0$ for all $x \in U$.*

Remark 1. By $d(x)_y$ we denote the possibility degree assigned to class label y for object x. We adopt this convention (over the alternative $d(x)(y)$) in order to simplify the notation.

The intuitive meaning of the possibility distribution d is that, if $|d(x)^{0+}| > 1$ for some $x \in U$, then the real decision associated with x (i.e. $t(x)$) is not known precisely, but is known to be in $d(x)^{0+}$. Furthermore, if $d(x)_a > d(x)_b$ then the decision a is considered more plausible than decision b for object x. Nonetheless, an alternative *preferential* interpretation can also be considered (similarly to the superset learning setting [4,16]): in this context, the inequality $d(x)_y \leq d(x)_{y'}$ would mean that, for object x, the label y' is preferred to y. Interestingly, while in the superset learning setting the two interpretations coincide (in the sense that they define the same notion of reducts), this is not the case in the learning from fuzzy labels setting. In the following, we will mainly focus on the epistemic interpretation, though we will occasionally make reference also to the preferential one when the two differ. First, we note that Definition 1 is a proper generalization of both standard and superset decision tables (SDT) [4]: indeed, if $d(x)^{0+} = d(x)^1$ for all $x \in U$, then we have a superset decision table; and, in the particular case where it also holds that $|d(x)^{0+}| = 1$ for all $x \in U$, then we have a standard decision table. We remark that the *weak superset property* forbids the real decision $t(x)$, for any object x, to be considered impossible (that is, we assume that there are no labeling errors) but nothing more is assumed: in particular, the stronger requirement that $d(x)_{t(x)} = 1$ (which means that $t(x)$ is considered fully plausible) is not guaranteed to hold. We call this latter requirement the *strong superset property*.

While both conditions can be seen as proper generalizations of the *superset property* in superset learning [16,17], we argue that under the epistemic interpretation of a PDT, the strong superset property is, in a specific sense, trivial: indeed, were this property be satisfied, then the PDT P would be equivalent to a SDT (specifically, the SDT $S = \langle U, Att, t, d_S \rangle$ s.t. $\forall x \in U.d_S(x) = d(x)^1$) as under the strong superset condition (i.e. $d(x)_{t(x)} = 1$) the real annotation $t(x)$ is guaranteed to lie among those with an associated possibility degree equal to 1.

By contrast, in the preferential interpretation, the strong superset property only implies that $t(x)$ is the most preferred annotation for x: this, in general, does not imply that other possible annotations should not be considered.

A PDT can be associated with a collection of compatible (standard) decision tables, which we call instantiations of the PDT:

Definition 2. *An* instantiation *of a PDT $P = \langle U, Att, t, d \rangle$ is a standard decision table $T = \langle U, Att, t' \rangle$ such that $d(x)_{t'(x)} > 0$ for all $x \in U$. The collection of* instantiations *of P is denoted $\mathcal{I}(P)$.*

We note that the collection $\mathcal{I}(P)$ inherits a ranking of the instantiations from the definition of the possibilistic decision attribute d:

Definition 3. *Let $I_1, I_2 \in \mathcal{I}(P)$ be two instantiations of a PDT P. Then we say that I_1 is (conservatively) less possible than I_2, denoted $I_1 \leq_C I_2$, if:*

$$min_{x \in U} d(x)_{t'}^{I_1} \leq min_{x \in U} d(x)_{t'}^{I_2} \tag{3}$$

We say that I_1 is dominated in possibility by I_2, denoted $I_1 \leq_D I_2$, if:

$$\forall x \in U. \ d(x)_{t'}^{I_1} \leq d(x)_{t'}^{I_2} \tag{4}$$

where, in both definitions $d(x)_{t'}^{I_i}$ refers to the value of the decision attribute d (in P) on the label $t'(x)$ in the instantiation I_i.

It is easy to observe that the following result holds:

Proposition 1. *The order $\leq_C$ determines a possibility distribution (equivalently, a fuzzy set) $\mu_{\mathcal{I}(P)}$ on the collection $\mathcal{I}(P)$ where, for each $I \in \mathcal{I}(P)$:*

$$\mu_{\mathcal{I}(P)}(I) = min_{x \in U} d(x)_{t'}^{I} \tag{5}$$

Proof. The result easily follows from the observation that $\leq_C$ is a weak ordering on $\mathcal{I}(P)$. Using the product fuzzy set construction [19], it is then easy to see that we can associate with $\leq_C$ a possibility distribution which is equivalent to $\mu_{\mathcal{I}(P)}$. □

The order $\leq_D$, on the other hand, cannot be directly associated with a (standard) possibility distribution on $\mathcal{I}(P)$, as it only defines a partial order: thus, it defines an L-fuzzy set over the set of instantiations where, in general, $L \neq ([0,1], min, max)$. Interestingly, the $\leq_D$ order is equivalent to the notion of dominance [12] in multi-criteria decision making: this could suggest that this ordering over instantiations (and the corresponding definitions of reducts) could be of particular interest in the *preferential* interpretation of the *learning from fuzzy-label* setting.

The following definition generalizes the notion of inconsistency for a PDT:

Definition 4. *For $B \subset Att$ and $\alpha \in [0,1)$ the PDT P is (α, B)-inconsistent if*

$$\exists x_1, x_2 \in U, \forall a \in B, a(x_1) = a(x_2) \ and \ d(x_1)^{\alpha+} \cap d(x_2)^{\alpha+} = \emptyset . \tag{6}$$

We call such a pair x_1, x_2 (α, B)-inconsistent. If condition (6) is not satisfied, then P is (α, B)-consistent. In particular, we say that P is weakly B-consistent if it is $(0, B)$-consistent; while we say that P is B-consistent when it is (α, B)-consistent for every α.

From the definition, we see that the notion of consistency (dually, inconsistency) for a PDT is richer than its classical counterpart and, in general, implies the non-existence of indiscernible objects with non-overlapping decisions, at any given α-cut of the possibility distribution defined by d. We say that an instantiation I is α-consistent with a PDT P if the following holds for all x_1, x_2: if x_1, x_2 are (α, Att)-consistent in S, then they are consistent in I.

3.2 Possibilistic Reducts

Learning from fuzzy labels, as a proper generalization of superset learning, encompasses the idea of *data disambiguation*: the goal of such a task is to jointly learn a function, mapping novel objects to the corresponding correct decision, and figuring out the most plausible instantation of the available data.

In the case of superset learning the notion of *plausibility* of an instantation can be entirely captured through the principle of simplicity [13] as any two instantiations are, a priori, equally plausible as they are both associated a possibility degree equal to 1: Thus, an instantation that can be explained by a simple model is more plausible than an instantation that requires a more complex one (this approach is, in turn, inspired by the *Occam's razor* principle).

In Rough Set Theory, the most natural measure of model complexity is the size of a reduct: indeed, this approach has been applied, in superset learning, to define so-called Minimum Description Length (MDL) reducts [4] which refer to the minimal reducts among all reducts of all possible instantiations. The most natural generalization of this notion to the setting of learning from fuzzy labels leads to the following definition:

Definition 5. *A set of attributes $R \subseteq Att$ is a* possibilistic reduct *of a PDT P if there exists an instantation $I \in \mathcal{I}(P)$ s.t. R is a reduct for I. A minimum description length (MDL) instantation is one of the instantiations of P admitting a reduct of minimum size (compared to all the reducts of all possible instantiations). We call the corresponding reducts* possibilistic MDL reducts.

While meaningful from a conceptual perspective, it is easy to observe that this definition of reducts disregards the most important difference between the superset learning and learning from fuzzy label settings: that is, the instantiations can be associated with an inherent measure of plausibility, given by the orders $\leq_C, \leq_D$. Indeed, the following result trivially holds:

Proposition 2. *Let P be a PDT, and let $\mathcal{S}(P) = \langle U, Att, t, d_S \rangle$ be the SDT defined from P s.t. $\forall x. \ d_S(x) = d(x)^{0+}$. Then, R is a possibilistic reduct (resp. possibilistic MDL reduct) of P iff it is a superset reduct (resp. MDL reduct) of $\mathcal{S}(P)$.*

Proposition 2 shows that the notion of a possibilistic reduct discards the epistemic information expressed by the decision attribute, and is thus equivalent to the notion of a superset reduct. In order to capture the richer semantics of PDTs, we argue that any proper definition of reduct should take into account not only the simplicity of the induced model (that is, the size of the reducts) but also the epistemic information encoded by the (possibilistic) decision attribute d. For this reason, we consider the following definitions of reducts:

Definition 6. *For each $\alpha \in (0, 1]$, let $\mathcal{S}(P)^\alpha$ be the SDT defined from P s.t. $\forall x. \ d_S^\alpha(x) = d(x)^\alpha$. For each possibilistic reduct R, denote by $\mathcal{I}(R) \subseteq \mathcal{I}(P)$ the collection of instantiations of P for which R is a reduct. Then, R:*

- *Is an α-possibilistic reduct if it is a superset reduct of $\mathcal{S}(P)^\alpha$, and an α-MDL reduct if it is also a MDL reduct of $\mathcal{S}(P)^\alpha$;*
- *Is a C-reduct if it is a possibilistic reduct and $\nexists R' \subseteq Att$ s.t. both $|R'| \leq |R|$ and $\exists I_1 \in sup_{\leq_C}\mathcal{I}(R), I_2 \in sup_{\leq_C}\mathcal{I}(R'). \ I_1 <_C I_2^2$;*

---

[2] Here $sup_{\leq_C}\mathcal{I}(R) = \{I \in \mathcal{I}(R) : \nexists I' \in \mathcal{I}(R) \text{ s.t. } I <_C I'\}$.

- *Is a λ-reduct, with $\lambda \in [0,1]$, if it is a possibilistic reduct and $sup_{I \in \mathcal{I}(R)}(1 - \lambda)\mu_{\mathcal{I}(P)}(I) - \lambda\frac{|R|}{|Att|}$ is maximal among all possibilistic reducts;*
- *Is a D-reduct if it is a possibilistic reduct and there is no $R' \subseteq Att$ s.t. both $|R'| \leq |R|$ and $\exists I_1 \in sup_{\leq_D}\mathcal{I}(R), I_2 \in sup_{\leq_D}\mathcal{I}(R')$. $I_1 <_D I_2$;*

Given a possibilistic reduct R of a given PDT P, we denote by α^R the maximum α s.t. R is an α-possibilistic reduct of P. We note the following basic properties:

Theorem 1. *The problem of finding all possibilistic reducts (resp. all C-reducts, all λ-reducts for any given value of λ) can be polynomially reduced to the problem of finding all α-possibilistic reducts and α-MDL reducts. In particular:*

- *R is a 0-possibilistic reduct iff it is a possibilistic reduct iff it is a λ-reduct ($\lambda = 1$);*
- *R is a C-reduct iff $\nexists R'$ s.t. both $|R'| \leq |R|$ and $\alpha^{R'} \geq \alpha^R$.*

Proof. As regards possibilistic reducts, it is trivial to show that the collection of possibilistic reducts is the same as the collection of 0-possibilistic reducts. For all other types of reducts, the proof is constructive: we describe an algorithm that finds all α-possibilistic and α-MDL reducts (see Algorithm 1), and show that this procedure can be effectively used (see Algorithms 2, 3) for finding all other types of reducts with no more than polynomial (in the number of reducts) overhead. For a PDT P let $\alpha(P) = \{\alpha' \in (0,1] : \exists x \in U, y \in V_t \text{ s.t. } d(x)_t = \alpha'\}$. The overhead for Algorithm 2 is $O(n^2)$ and for Algorithm 3 is $\Theta(n)$ (where n is the number of reducts). Thus, the main statement of the theorem holds. The other statements can be easily proved. $\qquad \square$

Algorithm 1. The brute-force algorithm for finding the α-possibilistic and α-MDL reducts of a possibilistic decision table P.

---

    procedure α-POSSIBILISTIC-REDUCTS(P: possibilistic decision table)
        for all $\alpha \in \alpha(P)$ in decreasing order **do**
            $poss\text{-}reds_\alpha \leftarrow Superset\text{-}Reducts(\mathcal{S}(P)^\alpha)$
            $MDL\text{-}reds_\alpha \leftarrow Find\text{-}Shortest(poss\text{-}reds_\alpha)$
        end for
        return $poss\text{-}reds_\alpha$, $MDL\text{-}reds_\alpha$ ▷ The collections of α-possibilistic and α-MDL reducts
    end procedure

We do not know if a similar technique could also be applied to compute the D-reducts: we leave this as open problem.

As a direct consequence of Theorem 1, we can see that the problem of finding all α-possibilistic (resp. α-MDL) reducts is not harder than finding all superset (resp. MDL) reducts of a given SDT.

Theorem 2. *The problem of finding all α-possibilistic (resp. α-MDL) reducts is no computationally harder than the problem of finding all superset (resp. MDL) reducts. Thus, in particular the problem of deciding whether, given a PDT P and $k \in \mathbb{N}^+$, the α-MDL reducts of P are of size $\leq k$ is Σ_2^P complete.*

Algorithm 2. The algorithm for finding the C-reducts of a possibilistic decision table P.

procedure C-REDUCTS(P: possibilistic decision table)
    $MDL\text{-}reds_\alpha \leftarrow \alpha\text{-}Possibilistic\text{-}Reducts(P)$
    $C\text{-}reds \leftarrow MDL\text{-}reds_1$
    for all $\alpha \in \alpha(P) \setminus \{1\}$ **do**
        for all $r \in MDL\text{-}reds_\alpha$ **do**
            if $\nexists r' \in C\text{-}reds$ s.t. $|r'| < |r|$ **then**
                $C\text{-}reds.append(r)$
            end if
        end for
    end for
    return $C\text{-}reds$ ▷ The set of C-reducts
end procedure

Algorithm 3. The algorithm for finding the λ-reducts of a possibilistic decision table P.

procedure λ-REDUCTS(P: possibilistic decision table, $\lambda \in [0,1]$)
    $poss\text{-}reds_\alpha \leftarrow \alpha\text{-}Possibilistic\text{-}Reducts(P)$
    $\lambda\text{-}reds \leftarrow \emptyset$
    $\theta \leftarrow 0$
    $map \leftarrow \emptyset$
    for all $\alpha \in \alpha(P)$ in decreasing order **do**
        for all $r \in poss\text{-}reds_\alpha$ **do**
            $\theta\text{-}temp \leftarrow (1-\lambda)\alpha - \lambda\frac{|r|}{|Att|}$
            $map.append(\langle r, \theta\text{-}temp\rangle)$
            if $\theta\text{-}temp \geq \theta$ **then**
                $\theta \leftarrow \theta\text{-}temp$
            end if
        end for
    end for
    $lambda\text{-}reds \leftarrow$ all $r \in map$
    return $\lambda\text{-}reds$ ▷ The set of λ-reducts
end procedure

Proof. For each α the reduction is trivial, as the problem of finding the α-MDL reducts of P is equivalent to finding the MDL reducts of $\mathcal{S}(P)^\alpha$. Note also that $|\alpha(P)| \leq |U||V_t|$: this implies that the problem α-MDL $Reduct$ requires, in the worst case, a polynomial (in the size of the PDT P) number of calls to a procedure for checking MDL Reducts. This can also be easily seen from Algorithm 1. □

Despite this result, showing that finding minimal reducts (that is, α-MDL, C-reducts or λ-reducts) for a PDT is not harder than finding MDL reducts for a SDT (which, in turn, is no harder than finding minimal reducts for a classical DT), all the reduct search problems considered require worst-case exponential time (in the size of the PDT). Indeed, while heuristics could be applied to speed up the computation of reducts [29] (specifically, to reduce the complexity of the *find-shortest-reducts* step in Algorithm 1) the proposed algorithms still require enumerating all the possible instantiations. Therefore, in the following section, we propose an alternative definition of reducts in order to reduce the computational costs.

3.3   Entropy Reducts

Following [4] we discuss an alternative definition of reduct, based on the notion of entropy [27], which simplifies the complexity of finding a reduct for a SDT. Given a SDT S with decision d, and $W \subseteq V_t$, we can define a basic belief assignment as

$$m(W | [x]_B) = \frac{|\{x' \in [x]_B : d(x') = W\}|}{|[x]_B|}. \tag{7}$$

Let P be a PDT, $\alpha \in [0,1]$, $B \subseteq Att$ be a set of attributes and denote by $IND_B = \{[x]_B\}$ the equivalence classes (granules) with respect to B. Let $d^\alpha_{[x]_B}$ be the restriction of d on the equivalence class $[x]_B$ for the derived SDT $S(P)^\alpha$, and let $m(\cdot | [x]_B^\alpha)$ the corresponding bba. Then, we define the OAU entropy of d, conditional on B and possibility degree α, as:

$$OAU(d|B, \alpha) = \sum_{[x]_B \in IND_B} \frac{|[x]_B|}{|U|} OAU(m(\cdot | [x]_B^\alpha)) \tag{8}$$

That is, the OAU entropy of a PDT (conditional on a set of attributes B and a possibility degree α) is obtained by first computing the derived SDT $S(P)^\alpha$, and then computing the (weighted) average of the OAU entropies of the bbas (see Eq. 2) determined by the granules $\{[x]_B : x \in U\}$. Based on the OAU entropy of a PDT, we can define entropy reducts for PDTs:

Definition 7. *Let $B \subseteq Att$ be a set of attributes, $\alpha \in [0,1]$. Then, we say that B is:*

- *An α-OAU super-reduct if $OAU(d | B, \alpha) \leq OAU(d | Att, \alpha)$;*
- *An α-OAU reduct if no proper subset of B is also a α-OAU super-reduct;*
- *An α-OAU ϵ-approximate super-reduct, with $\epsilon \in [0,1)$, if $OAU(d | B, \alpha) \leq OAU(d | Att, \alpha) - log_2(1 - \epsilon)$;*
- *An α-OAU ϵ-approximate reduct if no proper subset of B is also an α-OAU ϵ-approximate super-reduct.*

Let $[x]_B$ be one of the granules with respect to an α-OAU reduct. Then, the α-OAU *instantiation* with respect to $[x]_B$ is given by

$$dec_{OAU(B,\alpha)}([x]_B) = \arg\max_{v \in V_t} \left\{ p(v) \mid p \in \arg\min_{p \in P_{Bel}} H(p) \right\}, \tag{9}$$

that is, (one of) the most probable among the classes under the probability distributions which corresponds to the minimum value of entropy.

Table 1. An example of possibilistic decision table

	w	x	v	z	d
x_1	0	0	0	0	0
x_2	0	0	0	1	$\{0 : 0.5, 1 : 1.0\}$
x_3	0	1	1	0	0
x_4	0	1	1	1	$\{0 : 1.0, 1 : 0.5\}$
x_5	0	1	0	1	1
x_6	0	1	0	0	$\{0 : 0.5, 1 : 1.0\}$

Example 1. Let $P = \langle U = \{x_1, ..., x_6\}, A = \{w, x, v, z\}, d \rangle$ be the PDT in Table 1. We have that $\alpha(P) = \{0.5, 1\}$, thus in particular $\mathcal{S}(P)^{0.5}$ assigns $\{0, 1\}$ to objects x_2, x_4, x_6; while $\mathcal{S}(P)^1$ assigns 1 to objects x_2, x_6 and 0 to object x_4.

We have $OAU(d\,|\,A, 0.5) = OAU(d\,|\,B, 0.5) = 0$ for $B = \{x, v\}$. Also, it holds that $OAU(d\,|\,\{x\}, 0.5) = OAU(d\,|\,\{v\}, 0.5) > 0$. Thus, B is a 0.5-OAU reduct of SDT. We note that $\{z\}$ is also a 0.5-OAU reduct since, similarly, $OAU(d\,|\,z, 0.5) = 0$.

The 0.5-OAU instantiation given by $\{x, v\}$ is $dec_{x,v}(\{x_1, x_2\}) = 0$, and, similarly, $dec_{x,v}(\{x_3, x_4\}) = 0$ (since for objects $x_1, ..., x_4$ the instantiation with minimal OAU value is the one where all objects are assigned the label 0), while $dec_{x,v}(\{x_5, x_6\}) = 1$. By contrast, 0.5-OAU instantiation given by $\{z\}$ is $dec_z(\{x_1, x_3, x_6\}) = 0$, $dec_z(\{x_2, x_4, x_5\}) = 1$.

There is a single 0.5-MDL instantiation, that is $dec_{MDL}(\{x_1, x_3, x_6\}) = 0$, and $dec_{MDL}(\{x_2, x_4, x_5\}) = 1$, which corresponds to the 0.5-MDL reduct $\{z\}$. Thus, in this case, the 0.5-MDL reduct is equivalent to a 0.5-OAU reduct.

As regards $\mathcal{S}(P)^1$, we note that the decision attribute d is single-valued (hence, there is a single instantiation) and the corresponding DT is consistent. In this case there is a single reduct, namely $C = \{x, v, z\}$: therefore C is the only 1-MDL reduct and the only 1-OAU reduct.

Therefore we have that the set of MDL reducts is equivalent to the set of 0.5-MDL reducts (i.e. $\{\{z\}\}$); while the set of C-reducts is $\{\{z\}, \{x, z, v\}\}$; on the other hand we notice that the set of λ-reducts (for varying λ) is structured as follows:

$$\begin{cases} \{z\} & \lambda \geq 0.5 \\ \{\{z\}, \{x, z, v\}\} & \lambda = 0.5 \\ \{\{x, z, v\}\} & 0 \leq \lambda < 0.5 \end{cases}$$

Note that the set of λ-reducts and C-reducts (and possibilistic reducts, by extension) can include two sets $R, R' \subseteq Att$ s.t. $R \subset R'$ as long as they correspond to two different instantiations of the PDT from which they are derived.

In Example 1, the set of α-MDL reducts was exactly the set of minimal (w.r.t. size) α-OAU reducts: this is not a coincidence, we can show that this is a general property of OAU reducts on consistent PDTs.

Theorem 3. *Let P be a PDT, $\alpha \in (0, 1]$ and assume that $\mathcal{S}(P)^\alpha$ is consistent. Then, the set of consistent α-possibilistic reducts (i.e., the α-possibilistic reducts whose corresponding instantiations are consistent) coincides with the set of α-OAU reducts. Thus, in particular:*

1. *The set of consistent α-MDL reducts coincides with the set of minimal α-OAU reducts;*
2. *Finding the sets of consistent C-reducts and λ-reducts (for all values of λ) can be reduced to finding the set of α-OAU reducts for all values of α;*
3. *Finding the minimal α-OAU reducts is Σ_2^P-complete.*

Proof. We show the proof only for the main statement: the other statements directly follow from the definition of α-MDL reducts and from Theorems 1, 2. Indeed, suppose that R is a consistent α-possibilistic reduct: this means that there exists $I \in \mathcal{I}(R)$, instantiation of $\mathcal{S}(P)^\alpha$ that is consistent. As a consequence $OAU(d|R, \alpha) = 0$ and thus R is a α-OAU super-reduct. Suppose, further, that R were not a α-OAU reduct: then $\exists R' \subset R$ s.t. $OAU(d|R', \alpha) = 0$, but this means that R' is a consistent reduct of $\mathcal{S}(P)^\alpha$ which is a contradiction. Therefore R is an α-OAU reduct and the claim follows.   □

While, as a consequence of Theorem 3, the complexity of finding minimal α-OAU reducts is the same as finding α-MDL ones, even in the approximate case, the former approach to finding reducts is more amenable to optimization as it does not require an explicit enumeration of the instantiations of the PDT. Furthermore, as this approach relies on a quantitative quality measure (i.e., entropy), simple greedy procedures can be implemented with polynomial time complexity (specifically, $O(m^2 \cdot n)$, where m is the number of attributes and n the number of objects), and the guarantee to find an α-OAU reduct (albeit not necessarily minimal w.r.t. size).

4   Conclusion

In this article we studied the problem of feature selection in the learning from fuzzy label setting, and introduced generalized notions of reducts as well as algorithms for feature selection on the basis of this notion. While this paper provides a promising direction for the application of RST-based feature selection in weakly supervised learning, it naturally leaves many questions open. Specifically, we plan to address the following problems in future works:

- In Theorem 1, we showed that most definitions of reducts in a PDT can be derived from the definition of α-possibilistic reducts. Similar characterization also for D-reducts should be investigated in order to better understand the relationship between the latter and other types of reducts;
- In Theorem 3, we showed the equivalence of α-OAU and α-possibilistic reducts in the consistent case. The relation between these two definitions of reduct in the general, non-necessarily consistent case, should also be investigated;

- The definitions of reducts considered in this article, being based on the Pawlak definition of rough approximations, can only be applied to discrete data: thus, the generalization of the proposed approaches to encompass RST techniques that can be applied to continuous data (neighborhood-based or fuzzy-rough approaches) should be investigated.
- We plan to evaluate the performance of the proposed reduct definitions on real PDTs: These, in turn, can be obtained from multi-rater annotations, or through self-labeling techniques [11].

References

1. Arora, S., Barak, B.: Computational Complexity: A modern Approach. Cambridge University Press, Cambridge (2009)
2. Bello, R., Falcon, R.: Rough sets in machine learning: a review. In: Wang, G., Skowron, A., Yao, Y., Ślęzak, D., Polkowski, L. (eds.) Thriving Rough Sets. SCI, vol. 708, pp. 87–118. Springer, Cham (2017). https://doi.org/10.1007/978-3-319-54966-8_5
3. Campagner, A., Ciucci, D.: Orthopartitions and soft clustering: soft mutual information measures for clustering validation. Knowl.-Based Syst. **180**, 51–61 (2019)
4. Campagner, A., Ciucci, D., Hüllermeier, E.: Feature reduction in superset learning using rough sets and evidence theory. In: Lesot, M.J., et al. (eds.) IPMU 2020. CCIS, vol. 1237, pp. 471–484. Springer, Cham (2020). https://doi.org/10.1007/978-3-030-50146-4_35
5. Ciucci, D., Forcati, I.: Certainty-based rough sets. In: Polkowski, L., et al. (eds.) IJCRS 2017. LNCS (LNAI), vol. 10314, pp. 43–55. Springer, Cham (2017). https://doi.org/10.1007/978-3-319-60840-2_3
6. Côme, E., Oukhellou, L., Denoeux, T., Aknin, P.: Learning from partially supervised data using mixture models and belief functions. Pattern Recogn. **42**(3), 334–348 (2009)
7. Couso, I., Borgelt, C., Hullermeier, E., Kruse, R.: Fuzzy sets in data analysis: from statistical foundations to machine learning. IEEE Comput. Intell. Mag. **14**(1), 31–44 (2019)
8. Couso, I., Dubois, D., Sánchez, L.: Random sets and random fuzzy sets as ill-perceived random variables. SpringerBriefs in Computational Intelligence (2014)
9. Denoeux, T.: A k-nearest neighbor classification rule based on dempster-shafer theory. In: Yager, R.R., Liu, L. (eds.) Classic Works of the Dempster-Shafer Theory of Belief Functions. Studies in Fuzziness and Soft Computing, vol. 219, pp. 737–760. Springer, Berlin, Heidelberg (2008). https://doi.org/10.1007/978-3-540-44792-4_29
10. Denœux, T., Zouhal, L.M.: Handling possibilistic labels in pattern classification using evidential reasoning. Fuzzy Sets Syst. **122**(3), 409–424 (2001)
11. El Gayar, N., Schwenker, F., Palm, G.: A study of the robustness of KNN classifiers trained using soft labels. In: Schwenker, F., Marinai, S. (eds.) ANNPR 2006. LNCS (LNAI), vol. 4087, pp. 67–80. Springer, Heidelberg (2006). https://doi.org/10.1007/11829898_7
12. Greco, S., Matarazzo, B., Slowinski, R.: Rough sets theory for multicriteria decision analysis. Eur. J. Oper. Res. **129**(1), 1–47 (2001)

13. Hüllermeier, E.: Learning from imprecise and fuzzy observations: data disambiguation through generalized loss minimization. Int. J. Approx. Reason. **55**(7), 1519–1534 (2014)
14. Hüllermeier, E.: Does machine learning need fuzzy logic? Fuzzy Sets Syst. **281**, 292–299 (2015)
15. Hüllermeier, E., Beringer, J.: Learning from ambiguously labeled examples. Intell. Data Anal. **10**(5), 419–439 (2006)
16. Hüllermeier, E., Cheng, W.: Superset learning based on generalized loss minimization. In: Appice, A., Rodrigues, P.P., Santos Costa, V., Gama, J., Jorge, A., Soares, C. (eds.) ECML PKDD 2015. LNCS (LNAI), vol. 9285, pp. 260–275. Springer, Cham (2015). https://doi.org/10.1007/978-3-319-23525-7_16
17. Liu, L., Dietterich, T.: Learnability of the superset label learning problem. In: ICML, pp. 1629–1637 (2014)
18. Nakata, M., Sakai, H.: An approach based on rough sets to possibilistic information. In: Laurent, A., Strauss, O., Bouchon-Meunier, B., Yager, R.R. (eds.) IPMU 2014. CCIS, vol. 444, pp. 61–70. Springer, Cham (2014). https://doi.org/10.1007/978-3-319-08852-5_7
19. Nguyen, H.T., Walker, C., Walker, E.A.: A First Course in Fuzzy Logic. CRC Press, Boca Raton (2018)
20. Ning, Q., He, H., Fan, C., Roth, D.: Partial or complete, that's the question. arXiv preprint arXiv:1906.04937 (2019)
21. Orlowska, E. (ed.): IncompleteIinformation: Rough Set Analysis. Physica (2013)
22. Pawlak, Z.: Rough sets. Int. J. Comput. Inf. Sci. **11**(5), 341–356 (1982)
23. Quost, B., Denoeux, T.: Clustering and classification of fuzzy data using the fuzzy em algorithm. Fuzzy Sets Syst. **286**, 134–156 (2016)
24. Sakai, H., Liu, C., Nakata, M., Tsumoto, S.: A proposal of a privacy-preserving questionnaire by non-deterministic information and its analysis. In: 2016 IEEE International Conference on Big Data (Big Data), pp. 1956–1965. IEEE (2016)
25. Sakai, H., Nakata, M., Yao, Y.: Pawlak's many valued information system, non-deterministic information system, and a proposal of new topics on information incompleteness toward the actual application. In: Wang, G., Skowron, A., Yao, Y., Ślęzak, D., Polkowski, L. (eds.) Thriving Rough Sets. SCI, vol. 708, pp. 187–204. Springer, Cham (2017). https://doi.org/10.1007/978-3-319-54966-8_9
26. Shafer, G.: A Mathematical Theory of Evidence. Princeton University Press, Princeton (1976)
27. Ślęzak, D.: Approximate entropy reducts. Fundam. Inform. **53**(3–4), 365–390 (2002)
28. Ślęzak, D., Dutta, S.: Dynamic and discernibility characteristics of different attribute reduction criteria. In: Nguyen, H.S., Ha, Q.-T., Li, T., Przybyła-Kasperek, M. (eds.) IJCRS 2018. LNCS (LNAI), vol. 11103, pp. 628–643. Springer, Cham (2018). https://doi.org/10.1007/978-3-319-99368-3_49
29. Thangavel, K., Pethalakshmi, A.: Dimensionality reduction based on rough set theory: a review. Appl. Soft Comput. **9**(1), 1–12 (2009)
30. Trabelsi, S., Elouedi, Z., Lingras, P.: Dynamic reduct from partially uncertain data using rough sets. In: Sakai, H., Chakraborty, M.K., Hassanien, A.E., Ślęzak, D., Zhu, W. (eds.) RSFDGrC 2009. LNCS (LNAI), vol. 5908, pp. 160–167. Springer, Heidelberg (2009). https://doi.org/10.1007/978-3-642-10646-0_19
31. Umans, C.: On the complexity and inapproximability of shortest implicant problems. In: Wiedermann, J., van Emde Boas, P., Nielsen, M. (eds.) ICALP 1999. LNCS, vol. 1644, pp. 687–696. Springer, Heidelberg (1999). https://doi.org/10.1007/3-540-48523-6_65

32. Yao, Y.Y., Lingras, P.J.: Interpretations of belief functions in the theory of rough sets. Inf. Sci. **104**(1–2), 81–106 (1998)
33. Zadeh, L.A.: Fuzzy sets as a basis for a theory of possibility. Fuzzy Sets Syst. **1**(1), 3–28 (1978)
34. Zhou, Z.-H.: A brief introduction to weakly supervised learning. Natl. Sci. Rev. **5**(1), 44–53 (2018)

Right Adjoint Algebras Versus Operator Left Residuated Posets

M. Eugenia Cornejo$^{(\boxtimes)}$ and Jesús Medina

Department of Mathematics, University of Cádiz, Cádiz, Spain
{mariaeugenia.cornejo,jesus.medina}@uca.es

Abstract. Algebraic structures are essential in fuzzy frameworks such as fuzzy formal concept analysis and fuzzy rough set theory. This paper studies two general structures such as right adjoint algebras and operator left residuated posets, introducing several properties which relate them. Different extensions of the operators included in a given operator left residuated poset are presented and a reasoned analysis is shown to guarantee that the equivalence satisfied by the operators in this structure is not a generalization of the usual adjoint property, which is a basic property verified by right adjoint pairs. Operator left residuated posets are also studied in the framework of the Dedekind-MacNeille completion of a poset.

Keywords: Operator left residuated poset · Right adjoint pair · Dedekind-MacNeille completion · Fuzzy modus ponens

1 Introduction

Residuated algebraic structures perform a fundamental role in many areas of mathematics and information sciences, such as many-valued and fuzzy logics, quantum logics and quantum computing, formal concept analysis, rough set theory and fuzzy relation equations, among others [2–8, 13, 14, 19, 24, 25]. In recent years, several logical approaches have used different residuated algebraic operators as logical connectives and have investigated the relationships among these operators. Following this research line, we will establish a comparison between right adjoint pairs and operator left residuated posets.

Right adjoint pairs, which are composed of a conjunctor and implication related by the well-known adjoint property, arise from right multi-adjoint algebras [17]. These pairs provide more flexibility to the frameworks where they are

This work is partially supported by the 2014-2020 ERDF Operational Programme in collaboration with the State Research Agency (AEI) in projects TIN2016-76653-P and PID2019-108991GB-I00, and with the Department of Economy, Knowledge, Business and University of the Regional Government of Andalusia in project FEDER-UCA18-108612, and by the European Cooperation in Science & Technology (COST) Action CA17124.

S. Ramanna et al. (Eds.): IJCRS 2021, LNAI 12872, pp. 180–191, 2021.
https://doi.org/10.1007/978-3-030-87334-9_15

used for making computations, since the commutativity and associativity properties are not required. In the literature, we can find different works where the right adjoint pairs are widely studied from a theoretical point of view [15–18]. From an application perspective, for instance, right adjoint pairs are considered in logic programming for modeling knowledge systems with uncertainty, query procedures and abductive reasoning [21,22,28–31,34–36], in fuzzy relation equations for giving support mechanism for the negotiations of sellers [12] and knowing the main actions that a considered company must perform for increasing their benefits [1,37], in formal concept analysis and rough set theory for establishing different degrees of preference on the attributes/objects of a database [20,32,33].

Operator left residuated posets were presented as general structures for several non-classical logics whose underlying posets do not need to be lattices [9–11]. Indeed, these operators can be used as an algebraic semantics for the quantum mechanics logic in a wide sense. Specifically in [10], it is shown that pseudo-orthomodular, pseudo-Boolean and Boolean posets are a particular case of operators left residuated.

As we mentioned above, this paper relates right adjoint pairs to operator left residuated posets, in order to study in what sense the latter generalizes the well-known adjoint property, which is satisfied by right adjoint pairs. To reach this goal, diverse properties will be included and different proper extensions of the operators considered in a given operator left residuated poset will be defined. In addition, we will study what requirements these extensions must satisfy to ensure that they generalize the fuzzy modus ponens, in the semantical way, following the philosophy proposed by Petr Hájek in [27]. Finally, operator left residuated posets will be studied in the framework of the Dedekind-MacNeille completion of a poset. This fact leads us to an important consequence, that is, the "residuation" property verified by the operators of a given operator left residuated poset is a restriction of the adjoint property satisfied by the operators on the Dedekind-MacNeille completion of a poset.

The paper is organized as follows. Section 2 includes basic preliminary concepts about Dedekind-MacNeille completion and Galois connections for a better understanding. Section 3 introduces proper extensions of the operators considered in a given operator left residuated poset, as well as the required properties for the generalization of the fuzzy modus ponens. Section 4 studies operator left residuated posets defined on the Dedekind-MacNeille completion of a poset. Conclusions and prospects for future work are presented in Sect. 5.

2   Preliminaries

Preliminary notions and results are included in order to make the paper self-contained.

2.1   Dedekind-MacNeille Completion

This section recalls definitions and properties associated with the Dedekind-MacNeille completion of a poset, which can be found for instance in [23]. First of all, the operators involved in the completion of a partial ordered set (poset) are introduced.

Definition 1 ([23]). *Let $(P, \leq)$ be a poset and $A \subseteq P$, the "upper" set and the "lower" set of A are respectively defined as:*

$$U(A) = \{x \in P \mid a \leq x, \text{ for all } a \in A\}$$
$$L(A) = \{x \in P \mid x \leq a, \text{ for all } a \in A\}$$

These operators form an antitone Galois connection, whose definition is recalled next, together with the isotone version.

Definition 2 ([23]). *Let $(P, \leq_P)$ and $(Q, \leq_Q)$ be posets. The pair $(^\uparrow, ^\downarrow)$ of mappings $^\downarrow\colon P \to Q$ and $^\uparrow\colon Q \to P$ is:*

- *An* isotone Galois connection *between P and Q if the next equivalence holds:*

$$p \leq_P q^\uparrow \quad \text{if and only if} \quad p^\downarrow \leq_Q q$$

  for all $p \in P$ and $q \in Q$.
- *An* antitone Galois connection *between P and Q if the next equivalence holds:*

$$p \leq_P q^\uparrow \quad \text{if and only if} \quad q \leq_Q p^\downarrow$$

  for all $p \in P$ and $q \in Q$.

As a consequence, the following properties are satisfied, for all $A, B \subseteq P$.

(i) $A \subseteq LU(A)$ and $A \subseteq UL(A)$
(ii) If $A \subseteq B$ then $U(B) \subseteq U(A)$ and $L(B) \subseteq L(A)$
(iii) If $x_1 \leq x_2$ then $U(x_1) \subseteq U(x_2)$ and $L(x_1) \subseteq L(x_2)$
(iv) $U(A) = ULU(A)$ and $L(A) = LUL(A)$

The Dedekind-MacNeille completion of a poset $(P, \leq)$ is defined, from the operators U and L, as follows.

Definition 3 ([23]). *Let $(P, \leq)$ be a poset. The* Dedekind-MacNeille completion *of P is the set $\mathsf{DM}(P) = \{LU(A) \mid A \subseteq P\}$, which forms a complete lattice with respect to the inclusion ordering.*

It is important to emphasize that Dedekind-MacNeille completion of P forms a closure system in the powerset of P, and therefore the infimum coincides with the intersection and the supremum is the closure of the union.

Now, we present a technical property which will be helpful later.

Proposition 1 ([23]). *Let $(P, \leq)$ be a poset. For all $X \subseteq P$ the following equalities hold in $\mathsf{DM}(P)$:*

$$\bigwedge_{x \in X} L(\{x\}) = L(X) \qquad \bigvee_{x \in X} L(\{x\}) = LU(X)$$

From now on, when the mappings U and L evaluate singleton sets, we will write x instead of $\{x\}$ in order to simplify the notation.

2.2 Algebraic Structures

This section recalls the notions of right adjoint algebras and operator left residuated posets.

Right adjoint algebras are right multi-adjoint algebras [17,18] in which only one right adjoint pair is considered. These algebras generalize residuated lattices, introduced by Dilworth and Ward [26], and neither the commutativity and associativity properties nor the boundary conditions with the top element are required.

Definition 4 ([17,18]). *A right adjoint algebra* $(P_1, \leq_1, P_2, \leq_2, P_3, \leq_3, \&, \swarrow)$ *is a tuple composed of three posets* $(P_1, \leq_1)$, $(P_2, \leq_2)$, $(P_3, \leq_3)$ *and two mappings* $\& \colon P_1 \times P_2 \to P_3$, $\swarrow \colon P_3 \times P_2 \to P_1$ *satisfying the following equivalence*

$$x \, \& \, y \leq_3 z \quad \text{if and only if} \quad x \leq_1 z \swarrow y$$

for all $x \in P_1$, $y \in P_2$ *and* $z \in P_3$. *The pair* $(\&, \swarrow)$ *is called* right adjoint pair *with respect to* P_1, P_2, P_3.

The previous equivalence is called *adjoint property*, and represents a semantics fuzzy extension of modus ponens. The following properties associated with the operators of a right adjoint pair can be obtained from the adjoint property.

Proposition 2 ([17,18]). *Let* $(\&, \swarrow)$ *be a right adjoint pair with respect to the posets* $(P_1, \leq_1)$, $(P_2, \leq_2)$, $(P_3, \leq_3)$, *then the following properties are satisfied:*

(i) $\&$ *and* $\swarrow$ *are order-preserving in the first argument.*
(ii) $\perp_1 \, \& \, y = \perp_3$, $\top_3 \swarrow y = \top_1$, *for all* $y \in P_2$, *when* $(P_1, \leq_1, \perp_1, \top_1)$ *and* $(P_3, \leq_3, \perp_3, \top_3)$ *are bounded posets.*
(iii) $z \swarrow y = \max\{x \in P_1 \mid x \, \& \, y \leq_3 z\}$, *for all* $y \in P_2$ *and* $z \in P_3$.
(iv) $x \, \& \, y = \min\{z \in P_3 \mid x \leq_1 z \swarrow y\}$, *for all* $x \in P_1$ *and* $y \in P_2$.

Another interesting property which relates right adjoint pairs with Galois connection is given below.

Proposition 3. *Let* $(P_1, \leq_{P_1})$, $(P_2, \leq_{P_2})$, $(P_3, \leq_{P_3})$ *be posets and* $\&_y \colon P_1 \to P_3$, $\swarrow^y \colon P_3 \to P_1$ *be mappings defined as* $\&_y(x) = x \, \& \, y$, $\swarrow^y (z) = z \swarrow y$, *respectively, for all* $x \in P_1, z \in P_3$. *The pair* $(\&, \swarrow)$ *is a right adjoint pair with respect to* P_1, P_2, P_3 *if and only if* $(\&_y, \swarrow^y)$ *is an isotone Galois connection, for each* $y \in P_2$.

In this paper, we are interested in relating right adjoint algebras to operator left residuated posets. For that reason, we will recall next the formal definition of this structure, which was introduced by Chajda and Länger in [9].

Definition 5 ([9]). *An operator left residuated poset is an ordered seventuple* $(P, \leq, ', M, R, 0, 1)$ *composed of a bounded poset with a unary operation* $(P, \leq, '$ $0, 1)$ *and two mappings* $M \colon P \times P \to 2^P$ *and* $R \colon P \times P \to 2^P$ *satisfying the following conditions, for all* $x, y, z \in P$:

(i) $M(x,1) = M(1,x) = L(x)$
(ii) $R(x,y) = P$ if and only if $x \leq y$
(iii) $R(x,0) = L(x')$

and the equivalence:

$$M(x,y) \subseteq L(z) \quad \text{if and only if} \quad L(x) \subseteq R(y,z) \tag{1}$$

In [9], the operators M and R were extended from P to 2^P as follows:

$$M(X,Y) = \bigcup_{(x,y) \in X \times Y} M(x,y)$$

$$R(Y,Z) = \bigcup_{(y,z) \in Y \times Z} R(y,z)$$

for all $X, Y, Z \in 2^P$. We will show in this paper that these extensions are not suitable to satisfy Equivalence (1).

After introducing the basic concepts associated with the Dedekind-MacNeille completion and residuated algebraic structures, we can establish a comparative study between right adjoint algebras and operator left residuated posets.

3   Adjoint Property in Operator Left Residuated Posets

This section is devoted to comparing operator left residuated posets with right adjoint algebras. The theoretical development will be split into three parts. The first part introduces proper extensions of the mappings M and R, from P to 2^P, in order to satisfy a generalization of Equivalence (1) in Definition 5. The second part identifies the aforementioned extensions of M and R with the operators involved in a right adjoint algebra, for studying the requirements must be verified to guarantee that these extensions semantically generalize the fuzzy modus ponens. Taking into account the obtained results, the third part presents another different extension of the mappings M and R, which semantically generalizes the fuzzy modus ponens.

First of all, we include the following monotonicity properties which are deduced from the definition of an operator left residuated poset.

Proposition 4. *Given an operator left residuated poset* $(P, \leq, ', M, R, 0, 1)$, *given* $x_1, x_2, y_1, y_2 \in P$, *we have that*

(i) *If* $x_1 \leq x_2$, *then* $M(x_1, 1) \subseteq M(x_2, 1)$.
(ii) *If* $y_1 \leq y_2$, *then* $R(y_2, 0) \subseteq R(y_1, 0)$.
(iii) $R(y, 0) \subseteq R(y, z)$, *for all* $z \in P$.

Proof. (i) Given $x_1, x_2 \in P$, with $x_1 \leq x_2$, we trivially have that $M(x_2, 1) \subseteq M(x_2, 1) = L(x_2)$. By the monotonicity of L and Equivalence (1), we obtain that

$$L(x_1) \subseteq L(x_2) \subseteq R(1, x_2)$$

that is, $L(x_1) \subseteq R(1, x_2)$, and considering again Equivalence (1), we obtain the inclusion $M(x_1, 1) \subseteq L(x_2) = M(x_2, 1)$.

Properties (ii) and (iii) hold similarly from the properties of R and Equivalence (1).

Notice that, the usual monotonic properties of right adjoint operators are not satisfied in general. Moreover, M and R are not defined in suitable domains for satisfying adjoint property. Specifically, from the definition of isotone Galois connection (Definition 2), the operators $\downarrow: P \to Q$ and $\uparrow: Q \to P$ are defined between P and Q. However, given $y \in P$, the operators $M(\_, y): P \to 2^P$ and $R(y, \_): P \to 2^P$ are defined in the same domain and codomain and, for instance, they cannot be composed. Therefore, M and R cannot properly generalize the definition of right adjoint pair.

3.1   Extension of M and R to 2^P

The two most simple options in order to modify the domain and codomains of M and R preserving Equivalence (1) are explained below. The first option is to restrict the powerset 2^P to P, but this does not work because $L(x)$ and $L(z)$ are in Equivalence (1). The second option is to consider the extensions presented in [9] and recalled in Sect. 2.2, but this other possibility does not work either because the extension of R to 2^P does not satisfy Equivalence (1) in general. For example, if we assume that $L(x) \subseteq R(y, Z)$, with $x, y \in P$ and $Z \in 2^P$, then $L(x) \subseteq \bigcup_{z \in Z} R(y, z)$, and this inclusion and Equivalence (1) do not imply that $M(x, y) \subseteq L(z)$, for all $z \in Z$, in general.

Next, proper extensions of M and R from P to 2^P are defined, with the main goal of satisfying a generalization of Equivalence (1).

Theorem 1. *The mappings $\bar{M}: 2^P \times 2^P \to 2^P$ and $\bar{R}: 2^P \times 2^P \to 2^P$, defined as*

$$\bar{M}(X, Y) = \bigcup_{(x,y) \in X \times Y} M(x, y)$$

$$\bar{R}(Y, Z) = \bigcap_{(y,z) \in Y \times Z} R(y, z)$$

for all $X, Y, Z \in 2^P$, satisfy the property

$$\bar{M}(X, Y) \subseteq L(Z) \quad iff \quad LU(X) \subseteq \bar{R}(Y, Z) \tag{2}$$

Proof. Given $X, Y, Z \in 2^P$, if we assume $\bar{M}(X, Y) \subseteq L(Z)$, then by the definition of $\bar{M}$ and Proposition 1 we have that

$$\bar{M}(X, Y) = \bigcup_{(x,y) \in X \times Y} M(x, y) \subseteq \bigwedge_{z \in Z} L(z)$$

for all $x \in X$, $y \in Y$, and $z \in Z$. Hence, given $y \in Y$, and $z \in Z$, we have that

$$M(x, y) \subseteq L(z)$$

for all $x \in X$, and applying Equivalence (1) we obtain

$$L(x) \subseteq R(y, z)$$

for all $x \in X$, and by the supremum property we obtain that $\bigvee_{x \in X} L(x) \subseteq R(y, z)$. Therefore, by Proposition 1, we have $LU(X) \subseteq R(y, z)$, for all $y \in Y$, and $z \in Z$. Hence, by the intersection property, we obtain the required inclusion

$$LU(X) \subseteq \bigcap_{(y,z) \in Y \times Z} R(y, z) = \bar{R}(Y, Z)$$

The other implication follows similarly.

Notice that Equivalence (2) really generalizes Equivalence (1), because $LU(x) = L(x)$, for all $x \in P$.

Once the extensions of M and R to 2^P have been defined, we will study if they represent the fuzzy modus ponens in the semantical way.

3.2   Requirements for a Proper Fuzzy Modus Ponens

Petr Hájek highlighted in [27] that an implication operator $\rightarrow$ semantically extends the modus ponens to the fuzzy case if it has an associated conjunctor $\otimes$ satisfying that $y \rightarrow z$ is the maximum element verifying the inequality $(y \rightarrow z) \otimes y \leq z$. For example, right adjoint pairs satisfy this property. By Proposition 2, given a right adjoint pair $(\&, \swarrow)$ with respect to a poset $(P, \leq)$, we have that the implication can be defined as follows:

$$z \swarrow y = \max\{x \in P \mid x \& y \leq z\} \tag{3}$$

for all $y, z \in P$.

This section focuses on studying the requirements that the mappings $\bar{M} \colon 2^P \times 2^P \rightarrow 2^P$ and $\bar{R} \colon 2^P \times 2^P \rightarrow 2^P$, defined in Theorem 1, should satisfy in order to Equivalence (2) semantically represents the fuzzy modus ponens.

Therefore, identifying the mappings $\bar{M}$ and $\bar{R}$ with $\&$ and $\swarrow$, and considering Equivalence (2), we should prove that

$$\bar{M}(\bar{R}(Y, Z), Y) \subseteq L(Z) \tag{4}$$

for all $X, Y, Z \in 2^P$. By Equivalence (1), Inclusion (4) is equivalent to

$$LU(\bar{R}(Y, Z)) \subseteq \bar{R}(Y, Z)$$

Since LU is a closure operator, the element $\bar{R}(Y, Z)$ must satisfy that $LU(\bar{R}(Y, Z)) = \bar{R}(Y, Z)$, but this equality clearly only holds when $\bar{R}(Y, Z)$ is a closure element of LU. Therefore, $\bar{R}(Y, Z) \notin \{X \in 2^P \mid \bar{M}(X, Y) \subseteq L(Z)\}$ in general. As a consequence, $\bar{R}(Y, Z)$ cannot be the maximum of $\{X \in 2^P \mid \bar{M}(X, Y) \subseteq L(Z)\}$, as it happens to operators involved in right adjoint pairs. Thus,

$$\bar{R}(Y, Z) \neq \max\{X \in 2^P \mid \bar{M}(X, Y) \subseteq L(Z)\}$$

which is fundamental in many applications in which a fuzzy modus ponens is required.

Taking into account the previous reasoning, we need to define a new extension of the operators M and R which is carried out below.

3.3   Extension of Operator Left Residuated Posets

This section considers an operator left residuated poset $(P, \leq,', M, R, 0, 1)$, where $R(y, z)$ is an element of $\mathsf{DM}(P)$, and so satisfies Inclusion (4) for singletons in P. On this framework, the following extension of M and R are defined.

Theorem 2. *The mappings $\tilde{M} \colon \mathsf{DM}(P) \times 2^P \to 2^P$ and $\tilde{R} \colon 2^P \times 2^P \to \mathsf{DM}(P)$, defined as*

$$\tilde{M}(X, Y) = \bigcup_{(x,y) \in X \times Y} M(x, y)$$

$$\tilde{R}(Y, Z) = \bigwedge_{(y,z) \in Y \times Z} R(y, z)$$

for all $X, Y, Z \in 2^P$ and where $\bigwedge$ is the infimum operator in $\mathsf{DM}(P)$, satisfy the equivalence

$$\tilde{M}(X, Y) \subseteq L(Z) \quad \text{iff} \quad LU(X) \subseteq \tilde{R}(Y, Z) \tag{5}$$

In addition, we have that

$$\tilde{R}(Y, Z) = \max\{X \in DM(P) \mid \tilde{M}(X, Y) \subseteq L(Z)\}$$

for all $X \in DM(P)$ and $Y, Z \in 2^P$.

Proof. The proof straightforwardly follows from the proof of Theorem 1, the fact that the elements in $\mathsf{DM}(P)$ are closure elements of LU, and the last comments in the previous section.

Notice that, since by hypothesis $R(y, z) \in \mathsf{DM}(P)$, for all $y, z \in P$, we have that

$$\bar{R}(Y, Z) = \bigcap_{(y,z) \in Y \times Z} R(y, z) = \bigwedge_{(y,z) \in Y \times Z} R(y, z) = \tilde{R}(Y, Z)$$

Therefore, the only difference between $\bar{M}$, $\bar{R}$ and $\tilde{M}$, $\tilde{R}$ is the considered domains.

Taking into account this last comment, we can consider the extensions $\bar{M}$ and $\bar{R}$ in the following reasoning. Notice that, if they are identified again with the operators of a right adjoint pair $(\&, \swarrow)$ defined on a poset $(P, \leq)$, then Property (iv) of Proposition 2 is not satisfied. Specifically, if the conjunctor is written from the implication as follows:

$$x \& y = \min\{z \in P \mid x \leq y \swarrow z\}$$

for all $x, y \in P$, considering Equivalence (2) this equality clearly implies the inclusion

$$L(x) \subseteq \bar{R}(y, \bar{M}(x, y))$$

holds for all $x, y, z \in P$, which is equivalent by Equivalence (1) to

$$\bar{M}(x, y) \subseteq L(\bar{M}(x, y))$$

However, this inclusion is only true when $\bar{M}(x, y)$ is a singleton, which is a very restrictive property to $\bar{M}$ by its definition. Since assuming that $\bar{M}(X, Y)$ is a singleton for all $X, Y \in 2^P$ is equivalent to say that $M(x, y)$ is constant for all $x, y \in P$. As a consequence of all previous comments, the extension of M and R to 2^P does not satisfy the usual properties of right adjoint pairs, in general. Thus, Equivalence (1) could not be considered as a generalization of the adjoint property, but as a variant.

The following section will study the operator left residuated poset on the Dedekind-MacNeille completion framework.

4   Operator Left Residuated Posets from a Dedekind-MacNeille Completion

This section will consider Equivalence (1) in the framework of the Dedekind-MacNeille completion of the poset $(P, \leq)$. The first result considers the left side of Equivalence (1).

Proposition 5. *Given an operator left residuated poset $(P, \leq,', M, R, 0, 1)$, the following equivalence holds:*

$$M(x, y) \subseteq L(z) \quad \text{iff} \quad LU(M(x, y)) \subseteq L(z)$$

for all $x, y, z \in P$.

Proof. The proof straightforwardly follows from the properties of operators L and U.

Since both elements $LU(M(x, y))$ and $L(z)$ belong to $\mathsf{DM}(P)$, the inclusion $LU(M(x, y)) \subseteq L(z)$ can be considered on $\mathsf{DM}(P)$. Moreover, if we assume as in the previous section that $R(y, z)$ is an element of $\mathsf{DM}(P)$, for all $y, z \in P$, then the right side of Equivalence (1) is also considered on $\mathsf{DM}(P)$. However, the second argument of M and the first argument of R are not taken into account in the equivalence and so, these arguments could not be extended to any powerset. These reflexions point out the study of the following result.

Theorem 3. *Let $(P, \leq)$ be a poset, $M \colon P \times P \to 2^P$ and $R \colon P \times P \to 2^P$ be two mappings such that M and R are order-preserving in both second arguments. If the mappings $\widehat{M} \colon \mathsf{DM}(P) \times P \to \mathsf{DM}(P)$ and $\widehat{R} \colon P \times \mathsf{DM}(P) \to \mathsf{DM}(P)$, defined as*

$$\widehat{M}(X, y) = \bigcup_{x \in X} M(x, y)$$

$$\widehat{R}(y, Z) = \bigwedge_{z \in Z} R(y, z)$$

for all $X, Z \in \mathsf{DM}(P)$ and $y \in P$,   satisfy the adjoint property, that is,

$$\widehat{M}(X, y) \subseteq Z \quad \text{if and only if} \quad X \subseteq \widehat{R}(y, Z) \tag{6}$$

then M and R satisfy Equivalence (1).

Proof. Given $x, y, z \in P$, by Equivalence (6), we have that

$$\widehat{M}(L(x), L(y)) \subseteq L(z) \quad \text{iff} \quad L(x) \subseteq \widehat{R}(L(y), L(z))$$

which is equivalent to

$$\bigcup_{x' \in L(x)} M(x', y) \subseteq L(z) \quad \text{iff} \quad L(x) \subseteq \bigwedge_{z' \in L(z)} R(y, z')$$

If $M(x, y) \subseteq L(z)$, by the monotonicity of M, we have that

$$\bigcup_{x' \in L(x)} M(x', y) \subseteq M(x, y) \subseteq L(z)$$

and by the equivalence above and the monotonicity of R we obtain that

$$L(x) \subseteq \bigwedge_{z' \in L(z)} R(y, z') \subseteq R(y, z)$$

Since the other implication arises similarly, we have that M and R satisfy Equivalence (1).

As a consequence of the previous result, we can assert that Equivalence (1) can be considered as a weaker property of the adjoint property of operators on the Dedekind-MacNeille completion of a poset P.

5   Conclusions and Future Work

This paper establishes a comparative study among the operators involved in right adjoint algebras and operator left residuated posets. Two different extensions of the operators M and R are introduced, one of these in order to show that it satisfies a generalization of the fuzzy modus ponens in a semantical way. In addition, a detailed reasoning is included to ensure that the equivalence satisfied by operator left residuated posets cannot be considered as a generalization of the usual adjoint property and that they do not satisfy other interesting properties, such as the monotonicity properties. Finally, operator left residuated posets are studied in the framework of the Dedekind-MacNeille completion of a poset. We obtain that the "residuation" property satisfied by the operators of a given operator left residuated poset is a restriction of the adjoint property, of the operators on the Dedekind-MacNeille completion of the poset.

As a consequence, the operators satisfying this restriction could not verify important properties, such as the monotonicity properties, and they could not satisfy the natural semantics fuzzy extension of the modus ponens. Hence, the useful of this restriction in applications will be studied in the future. Moreover, more properties and possible modifications and adaptions will be studied in order to be considered in real cases, such as the ones related to digital forensic and taken into consideration in the COST Action DIGital FORensics: evidence Analysis via intelligent Systems and Practices (DigForASP).

References

1. Alfonso-Robaina, D., Díaz-Moreno, J.C., Malleuve-Martinez, A., Medina, J., Rubio-Manzano, C.: Modeling enterprise architecture and strategic management from fuzzy decision rules. Stud. Comput. Intell. **819**, 139–147 (2020)
2. Antoni, L., Cornejo, M.E., Medina, J., Ramirez-Poussa, E.: Attribute classification and reduct computation in multi-adjoint concept lattices. IEEE Trans. Fuzzy Syst. **29**(5), 1121–1132 (2020)
3. Antoni, L., Krajči, S., Krídlo, O.: Constraint heterogeneous concept lattices and concept lattices with heterogeneous hedges. Fuzzy Sets Syst. **303**, 21–37 (2016)
4. Antoni, L., Krajči, S., Krídlo, O.: On stability of fuzzy formal concepts over randomized one-sided formal context. Fuzzy Sets Syst. **333**, 36–53 (2018)
5. Bělohlávek, R.: Lattices of fixed points of fuzzy Galois connections. Math. Log. Q. **47**(1), 111–116 (2001)
6. Bělohlávek, R.: Concept lattices and order in fuzzy logic. Ann. Pure Appl. Log. **128**, 277–298 (2004)
7. Benítez-Caballero, M.J., Medina, J., Ramírez-Poussa, E., Ślęzak, D.: Rough-set-driven approach for attribute reduction in fuzzy formal concept analysis. Fuzzy Sets Syst. **319**, 117–138 (2020)
8. Benítez-Caballero, M.J., Medina, J., Ramírez-Poussa, E., Ślęzak, D.: Bireducts with tolerance relations. Inf. Sci. **435**, 26–39 (2018)
9. Chajda, I.: A representation of residuated lattices satisfying the double negation law. Soft. Comput. **22**(6), 1773–1776 (2017). https://doi.org/10.1007/s00500-017-2673-9
10. Chajda, I., Länger, H.: Left residuated operators induced by posets with a unary operation. Soft. Comput. **23**(22), 11351–11356 (2019). https://doi.org/10.1007/s00500-019-04028-w
11. Chajda, I., Länger, H.: Residuation in modular lattices and posets. Asian-Eur. J. Math. **12**(02), 1950092 (2019)
12. Cornejo, M.E., Díaz-Moreno, J.C., Medina, J.: Multi-adjoint relation equations: a decision support system for fuzzy logic. Int. J. Intell. Syst. **32**(8), 778–800 (2017)
13. Cornejo, M.E., Lobo, D., Medina, J.: Solving generalized equations with bounded variables and multiple residuated operators. Math. **8**(11), 1992 (2020). 1–22
14. Cornejo, M.E., Lobo, D., Medina, J.: Extended multi-adjoint logic programming. Fuzzy Sets Syst. **388**, 124–145 (2020)
15. Cornejo, M.E., Medina, J., Ramírez-Poussa, E.: A comparative study of adjoint triples. Fuzzy Sets Syst. **211**, 1–14 (2013)
16. Cornejo, M.E., Medina, J., Ramírez-Poussa, E.: Multi-adjoint algebras versus extended-order algebras. Appl. Math. Inf. Sci. **9**(2L), 365–372 (2015)
17. Cornejo, M.E., Medina, J., Ramírez-Poussa, E.: Multi-adjoint algebras versus non-commutative residuated structures. Int. J. Approx. Reason. **66**, 119–138 (2015)

18. Cornejo, M.E., Medina, J., Ramírez-Poussa, E.: Algebraic structure and characterization of adjoint triples, Fuzzy Sets and Systems, 9 February 2021 (In Press)
19. Cornejo, M.E., Medina, J., Ramírez-Poussa, E.: Implication operators generating pairs of weak negations and their algebraic structure. Fuzzy Sets Syst. **405**, 18–39 (2021)
20. Cornelis, C., Medina, J., Verbiest, N.: Multi-adjoint fuzzy rough sets: definition, properties and attribute selection. Int. J. Approx. Reason. **55**, 412–426 (2014)
21. Damásio, C., Medina, J., Ojeda-Aciego, M.: Termination of logic programs with imperfect information: applications and query procedure. J. Appl. Log. **5**, 435–458 (2007)
22. Damásio, C.V., Pereira, L.M.: Monotonic and residuated logic programs. In: Benferhat, S., Besnard, P. (eds.) ECSQARU 2001. LNCS (LNAI), vol. 2143, pp. 748–759. Springer, Heidelberg (2001). https://doi.org/10.1007/3-540-44652-4_66
23. Davey, B., Priestley, H.: Introduction to Lattices and Order, 2nd Edition, Cambridge University Press, Cambridge (2002)
24. Díaz-Moreno, J.C., Medina, J.: Multi-adjoint relation equations: definition, properties and solutions using concept lattices. Inf. Sci. **253**, 100–109 (2013)
25. Díaz-Moreno, J.C., Medina, J., Turunen, E.: Minimal solutions of general fuzzy relation equations on linear carriers. an algebraic characterization. Fuzzy Sets Syst. **311**, 112–123 (2017)
26. Dilworth, R.P., Ward, M.: Residuated lattices. Trans. Am. Math. Soc. **45**, 335–354 (1939)
27. Hájek, P.: Metamathematics of Fuzzy Logic, Trends in Logic. Kluwer Academic, Dordrecht (1998)
28. Julián-Iranzo, P., Medina, J., Ojeda-Aciego, M.: On reductants in the framework of multi-adjoint logic programming. Fuzzy Sets Syst. **317**, 27–43 (2017)
29. Julián, P., Moreno, G., Penabad, J.: On fuzzy unfolding: a multi-adjoint approach. Fuzzy Sets Syst. **154**(1), 16–33 (2005)
30. Lobo, D., Cornejo, M.E., Medina, J.: Abductive reasoning in normal residuated logic programming via bipolar max-product fuzzy relation equations. In: Conference of the International Fuzzy Systems Association and the European Society for Fuzzy Logic and Technology (EUSFLAT 2019), vol. 1 of Atlantis Studies in Uncertainty Modelling, pp. 588–594. Atlantis Press (2019)
31. Madrid, N., Ojeda-Aciego, M.: On the measure of incoherent information in extended multi-adjoint logic programs. In: Proceedings of the 2013 IEEE Symposium on Foundations of Computational Intelligence (FOCI), Institute of Electrical and Electronics Engineers (IEEE) (2013)
32. Medina, J.: Multi-adjoint property-oriented and object-oriented concept lattices. Inf. Sci. **190**, 95–106 (2012)
33. Medina, J., Ojeda-Aciego, M., Ruiz-Calviño, J.: Formal concept analysis via multi-adjoint concept lattices. Fuzzy Sets Syst. **160**(2), 130–144 (2009)
34. Medina, J., Ojeda-Aciego, M., Vojtáš, P.: A multi-adjoint logic approach to abductive reasoning. In: Codognet, P. (ed.) ICLP 2001. LNCS, vol. 2237, pp. 269–283. Springer, Heidelberg (2001). https://doi.org/10.1007/3-540-45635-X_26
35. Moreno, G., Penabad, J., Riaza, J.A.: Symbolic Unfolding of Multi-adjoint Logic Programs, pp. 43–51. Springer International Publishing, Cham (2019)
36. Moreno, G., Penabad, J., Vázquez, C.: Beyond multi-adjoint logic programming. Int. J. Comput. Math. **92**(9), 1956–1975 (2015)
37. Rubio-Manzano, C., Díaz-Moreno, J.C., Alfonso-Robaina, D., Malleuve, A., Medina, J.: A novel cause-effect variable analysis in enterprise architecture by fuzzy logic techniques. Int. J. Comput. Intell. Syst. **13**, 511–523 (2020)

Adapting Fuzzy Rough Sets
for Classification with Missing Values

Oliver Urs Lenz[1]([✉])[ID], Daniel Peralta[1,2][ID], and Chris Cornelis[1][ID]

[1] Computational Web Intelligence, Department of Applied Mathematics, Computer Science and Statistics, Ghent University, Ghent, Belgium
{oliver.lenz,chris.cornelis}@ugent.be, daniel.peralta@irc.vib-ugent.be
[2] Data Mining and Modelling for Biomedicine Group, VIB Center for Inflammation Research, Ghent University, Ghent, Belgium
http://www.cwi.ugent.be, https://www.irc.ugent.be

Abstract. We propose an adaptation of fuzzy rough sets to model concepts in datasets with missing values. Upper and lower approximations are replaced by interval-valued fuzzy sets that express the uncertainty caused by incomplete information. Each of these interval-valued fuzzy sets is delineated by a pair of optimistic and pessimistic approximations. We show how this can be used to adapt Fuzzy Rough Nearest Neighbour (FRNN) classification to datasets with missing values. In a small experiment with real-world data, our proposal outperforms simple imputation with the mean and mode on datasets with a low missing value rate.

Keywords: Fuzzy rough sets · Interval-valued fuzzy sets · Machine learning · Missing values

1 Introduction

Fuzzy and rough sets can be used to model different types of uncertainty. Fuzzy sets [3,23] allow us to model partial membership of a concept, while rough sets [20,21] capture the conflicting ways in which a concept may be predicted from a set of independent attributes. The two concepts are unified in the fuzzy rough set [8]. If $X \subset \mathbb{R}^m$ is a dataset, and $C \subseteq X$ a fuzzy subset, then the fuzzy rough set induced by C is the pair of fuzzy sets $(\overline{C}, \underline{C})$. The upper approximation $\overline{C}$ generalises the positive evidence for C in X, whereas the lower approximation $\underline{C}$ generalises the negative evidence for C in X.

In this paper, we consider a third type of uncertainty: incomplete information. There exists a wide range of strategies to deal with missing data [2], including proposals that involve rough or fuzzy rough sets [22]. In particular, fuzzy rough sets have been used for imputation [1], there have been proposals to adapt both crisp and fuzzy decision rules to the presence of missing values [10,12,17], and in the context of classical rough sets, [10] has provided three alternative definitions of upper and lower approximations in datasets with missing values. In contrast,

© Springer Nature Switzerland AG 2021
S. Ramanna et al. (Eds.): IJCRS 2021, LNAI 12872, pp. 192–200, 2021.
https://doi.org/10.1007/978-3-030-87334-9_16

our strategy is to incorporate the uncertainty of incomplete information directly into the representation of concepts, by extending the notion of fuzzy rough set.

We propose to mimic the dual approach of rough sets by modelling an optimistic and a pessimistic scenario when comparing a missing value with another value. The optimistic scenario is that the two values are really identical, while the pessimistic scenario is that they are maximally different. We cannot know what the ground truth is, but we know that it must lie somewhere in between these two extremes. Formally, we can represent this with an interval-valued fuzzy set [9, and references therein]. Since the uncertainty of incomplete information is orthogonal to the uncertainty that arises from positive and negative information, the resulting interval-valued fuzzy rough set is defined by four fuzzy sets: the optimistic and pessimistic upper and lower approximations $\overline{C}^{\min}, \overline{C}^{\max}, \underline{C}_{\min}, \underline{C}_{\max}$.

This work builds on the earlier proposal for interval-valued fuzzy rough sets in the context of feature selection [14], as well as a related proposal of *ill-known* fuzzy rough sets [5] based on twofold fuzzy sets [7], but this approach has otherwise remained relatively underexplored. We present an up-to-date definition in Sect. 2. In Sect. 3, we modify Fuzzy Rough Nearest Neighbour (FRNN) classification to incorporate interval-valued fuzzy rough sets, and evaluate its performance on a number of real-live datasets.

2 Interval-Valued Fuzzy Rough Sets

Recall the formal definitions of the upper and lower approximations with Ordered Weighted Averaging (OWA) operators [4]:

Definition 1 (Soft maxima and minima). *Let w be a weight vector of length k, with values in $[0, 1]$ that sum to 1. The soft maximum $w \downarrow$ and soft minimum $w \uparrow$ induced by w transform a collection Y of values in $\mathbb{R}$ into, respectively, the weighted sums*

$$
\begin{aligned}
w \underset{i \leq k}{\downarrow} Y &= \sum_{i \leq k} w_i \cdot y_{i+}, \\
w \underset{i \leq k}{\uparrow} Y &= \sum_{i \leq k} w_i \cdot y_{i-},
\end{aligned}
\tag{1}
$$

where y_{i+} and y_{i-} are the ith largest and ith smallest elements in Y.

Definition 2 (Upper and lower approximations). *Let $X \subset \mathbb{R}^m$ be a finite multisubset for some $m \in \mathbb{N}$, let R be a tolerance relation on $\mathbb{R}^m$, w a weight vector of some length k, T a t-norm and I a fuzzy implication. Then for any fuzzy submultiset C of X, the upper and lower approximations $\overline{C}$ and $\underline{C}$ are the fuzzy subsets of $\mathbb{R}^m$ defined by:*

$$
\begin{aligned}
\overline{C}(y) &= w \underset{x \in X}{\downarrow} (T(R(y, x), C(x)) \\
\underline{C}(y) &= w \underset{x \in X}{\uparrow} (I(R(y, x), C(x))
\end{aligned}
\tag{2}
$$

While a dataset can contain instances with identical attribute values but different membership degrees in a concept C, the upper and lower approximations of C only depend on attribute values, and so can be defined as fuzzy subsets of the attribute space.

It is convenient to aggregate R from attribute-specific tolerance relations R_i on $\mathbb{R}$ by means of some monotonic function $f : [0,1]^m \longrightarrow [0,1]$. In line with recent works, we let f be the mean, and we write $R_f := f((R_i)_{i \leq m})$.

Next, recall the definition of the interval-valued fuzzy set [9, and references therein]:

Definition 3 (Interval-valued fuzzy set). *Let X be a set. An interval-valued fuzzy set in X is a pair of fuzzy sets (F_1, F_2) in X such that $F_1(x) \leq F_2(x)$ for all $x \in X$.*

Equivalently, an interval-valued fuzzy set in X can also be defined as a function $X \longrightarrow \mathcal{I}([0,1])$, where the range is the set of intervals in $[0,1]$, i.e. the subset of $[0,1] \times [0,1]$ whose values in the first component are always less than or equal to the values in the second component.

We can accommodate the possibility of missing data by adjoining a formal symbol denoting a missing value to each copy of $\mathbb{R}$ to obtain $\mathbb{R}_? := \mathbb{R} \cup \{?\}$, and by letting X be a multisubset of $\mathbb{R}_?^m$. The task then is to extend any choice of R_i to $?$. We define *optimistic* and *pessimistic* per-attribute relations $R_i^{\max}$ and $R_i^{\min}$ by stipulating that for any $a, b \in \mathbb{R}$:

$$
\begin{aligned}
R_i^{\max}(a, b) &= R_i^{\min}(a, b) &&= R_i(a, b) \\
R_i^{\max}(a, ?) &= R_i^{\max}(?, b) &&= R_i^{\max}(?, ?) &&= 1 \\
R_i^{\min}(a, ?) &= R_i^{\min}(?, b) &&= R_i^{\min}(?, ?) &&= 0
\end{aligned}
\tag{3}
$$

Accordingly, we define interval-valued upper and lower approximations through the aggregated relations $R_f^{\max}$ and $R_f^{\min}$:

Definition 4 (Interval-valued upper and lower approximations). *Let $X \subset \mathbb{R}_?^m$ be a finite multisubset for some $m \in \mathbb{N}$, let w be a weight vector of some length k, T a t-norm and I a fuzzy implication, and let $(R_i)_{i \leq m}$ be a series of similarity measures and $f : \mathbb{R}^m \longrightarrow [0,1]$ an aggregation function such that R_f is a tolerance relation. Then for any fuzzy submultiset C of X, the interval-valued upper and lower approximations of C are, respectively, the interval-valued fuzzy sets $(\overline{C}^{\min}, \overline{C}^{\max})$ and $(\underline{C}_{\min}, \underline{C}_{\max})$, defined as:*

$$
\begin{aligned}
\overline{C}^{\min}(y) &= w \!\downarrow_{x \in X} (T(R_f^{\min}(y, x), C(x)) \\
\overline{C}^{\max}(y) &= w \!\downarrow_{x \in X} (T(R_f^{\max}(y, x), C(x)) \\
\underline{C}_{\min}(y) &= w \!\uparrow_{x \in X} (I(R_f^{\max}(y, x), C(x)) \\
\underline{C}_{\max}(y) &= w \!\uparrow_{x \in X} (I(R_f^{\min}(y, x), C(x))
\end{aligned}
\tag{4}
$$

Because t-norms and fuzzy implications are respectively monotonic and anti-monotonic in the first argument, the pessimistic approximations $\overline{C}^{\min}$ and $\underline{C}_{\min}$ encode the minimum membership degrees in the upper and lower approximations, while the optimistic approximations $\overline{C}^{\max}$ and $\underline{C}_{\max}$ encode the maximum membership degrees.

The computational complexity of calculating (4) is in principle the same as that of calculating (2), which is the computational complexity of a k-nearest neighbour query. However it requires the implementation of k-nearest neighbour algorithms with the distance measure corresponding to (3).

Membership in the optimistic and pessimistic approximations—like membership in ordinary upper and lower approximations—is determined purely on the basis of the attribute values of an instance, so it is possible to plot membership degrees across the attribute space. This is illustrated for a toy example in Fig. 1. Here, C is a crisp set containing two elements, one of which has a missing attribute value, which we have represented with a line. We have chosen $R_i(y, x) = 1 - |y_i - x_i|$. For crisp sets, the choice of t-norm becomes void: $T(R(y, x), C(x))$ is equal to $R(y, x)$ if $C(x) = 1$, and equal to 0 otherwise. Similarly, the choice of fuzzy implication resolves to a choice of fuzzy negation; we use the standard negation $z \longmapsto 1 - z$. We set $w = \langle \frac{2}{3}, \frac{1}{3} \rangle$. Darker shades of red indicate higher membership degrees. It can be seen that membership degrees of the optimistic approximations are uniformly higher than membership degrees of the pessimistic approximations.

The treatment in this section is essentially an updated version of [14]. The differences are mainly practical. Firstly, [14] uses a more general setting, where R_i is an interval-valued relation, but this greater generality potentially obscures the fact that this approach can be applied in any context that currently uses ordinary fuzzy rough sets, where R_i is scalar-valued. And secondly, [14] requires the aggregation function f to be a t-norm. As a result, $R_f^{\min}$ will always be 0 if any of the attribute values are missing, and we lose the information encoded by the non-missing attribute values.

3   FRNN with Interval-Valued Approximations

Upper and lower approximations can be used for Fuzzy Rough Nearest Neighbour (FRNN) classification [13], by calculating the membership of an unknown instance in the upper and lower approximations of the crisp decision classes and identifying the class with the highest membership degree. For datasets with missing data, we can instead use the interval-valued upper and lower approximations for classification. We test this with a small experiment. As upper and lower approximations produce equivalent results with two-class datasets, we simplify the experiment by only using the upper approximation.

For crisp sets C, the choice of t-norm in (4) becomes void. In line with previous work [18] we use linearly decreasing weights $\frac{k}{k(k+1)/2}, \frac{k-1}{k(k+1)/2}, \ldots, \frac{1}{k(k+1)/2}$, and set k to 20 or the size of the decision class, whichever is smaller. For the

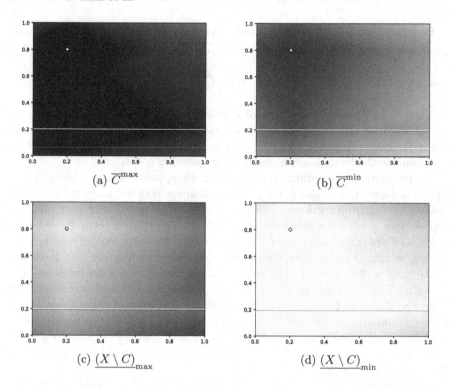

Fig. 1. Toy example with $C = \{(0.2, 0.8), (?, 0.2)\}$. Missing value displayed as a line. Optimistic and pessimistic upper approximations of C and optimistic and pessimistic lower approximations of $X \setminus C$.

tolerance relation, we select $R_i(y, x) = 1 - |y_i - x_i|/r_i$, where r_i is the range of values in the training set.

We evaluate performance with the mean Area Under the Receiver Operator Curve (AUROC) across 5-fold cross-validation. For multi-class datasets, we use the extension of AUROC by Hand & Till (2001) [11]. We apply this to eleven datasets with missing values selected from the UCI archive of machine learning datasets [6]. Where applicable, we remove classes with fewer than five instances, and select a stratified sample of 2000 instances.

We experiment with two strategies: using the mean membership values in the optimistic and pessimistic approximations, and optimising a weighted mean on the basis of training data.

For the second strategy, we use the efficient form of leave-one-out validation detailed in [19]. Briefly, this entails taking a single nearest neighbour query for the entire training set, and correcting it by removing nearest neighbour distances from a training instance to itself. The remaining values can then be used to calculate optimistic and pessimistic approximation memberships $\overline{C \setminus \{x\}}^{\max}(x)$

and $\overline{C \setminus \{x\}}^{\min}(x)$. We parametrise the average of these two values with a value $\lambda \in [0, 1]$ (5).

$$(1 - \lambda) \cdot \overline{C \setminus \{x\}}^{\min}(x) + \lambda \cdot \overline{C \setminus \{x\}}^{\max}(x) \tag{5}$$

We optimise λ by calculating the resulting AUROC and applying Malherbe-Powell optimisation [15,16] with a budget of 20 evaluations.

Note that the computational complexity of this strategy is equal to the computational complexity of a $k + 1$-nearest neighbour query with n query instances and n target instances, where n is the size of the training set. For large n, this can potentially be mitigated by using only a subset of the training set to optimise λ.

Table 1. Datasets with the number of classes, instances and attributes, the rate of missing values, and the AUROC from classification with the mean of optimistic and pessimistic upper approximation memberships, with an optimised ratio of both, and with normal upper approximation memberships after imputation with the mean and mode.

Dataset	c	n	m	#?	Mean	Optimised	Imputation
adult	2	2000	13	0.010	**0.863**	0.863	0.860
aps-failure	2	2000	170	0.083	0.969	0.985	**0.993**
arrhythmia	10	443	279	0.003	0.878	**0.880**	0.877
ckd	2	400	24	0.105	1.000	1.000	**1.000**
exasens	4	399	7	0.428	0.738	**0.748**	0.734
hcc	2	165	49	0.102	0.746	0.741	**0.771**
hepatitis	2	155	19	0.057	0.879	**0.884**	0.877
mammographic-masses	2	961	4	0.042	0.833	**0.834**	0.827
primary-tumor	15	330	17	0.039	**0.779**	0.777	0.775
secom	2	1567	590	0.045	0.678	0.681	**0.689**
soybean	19	683	35	0.098	0.993	0.995	**0.996**
Mean					0.851	0.854	**0.854**

The results are displayed in Table 1. Optimising the weighted mean increases AUROC for 7 datasets and decreases it for 3. Applying a one-sided Wilcoxon signed-rank test, we find that this is weakly significant ($p = 0.057$).

For comparison, we have also included the results obtained from simple imputation with the mean (numerical attributes) or mode (categorical attributes) of the known values in the training data. For 6 datasets, both the mean and optimised weighted mean optimistic and pessimistic approximations achieve a higher AUROC than simple imputation, whereas for 5 datasets, simple imputation achieves a higher AUROC. If we exclude the outlying dataset *exasens*, we see that the optimistic and pessimistic approximations perform better on datasets with a lower missing value rate, and imputation on datasets with a higher missing value rate (Fig. 2). When we fit a logistic regression model, the odds are even at a missing value rate of 0.056.

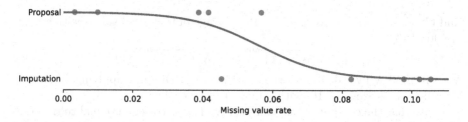

Fig. 2. Distribution of datasets for which imputation or the proposal of this paper achieves higher AUROC as a function of the missing value rate, with logistic regression fit.

4   Conclusion

In this paper we have presented an approach towards datasets with missing values that has received relatively little attention so far. While the existing literature is typically devoted to resolving these missing values in an optimal manner, we have argued that in the context of fuzzy rough sets, which are motivated by a desire to model different kinds of uncertainty, it is worthwhile to also model the uncertainty of incomplete information.

We have recalled the concept of interval-valued fuzzy rough set, which iterates on the dualistic nature of rough sets and replaces the upper and lower approximations by interval-valued fuzzy sets: secondary pairs of approximations, reflecting optimistic and pessimistic assumptions about the values that are missing. These define a bandwidth that contains the true (but unknown) upper and lower approximation memberships. We think that this can offer a valuable perspective for qualitative analyses of datasets with missing values.

We have shown how the interval-valued upper and lower approximations can be used to extend Fuzzy Rough Nearest Neighbour (FRNN) classification to problems with missing values. From an evaluation on several real-world datasets, we found that the best results can be obtained by taking a weighted average of the optimistic and pessimistic approximation memberships, and by optimising the relative weight on the basis of training data. This results in a comparable overall performance to simple imputation with the mean and mode, but is more directly interpretable as it does not involve the insertion of artificial values. Further analysis revealed that our proposal in particular outperforms imputation on datasets with a missing value rate below 0.056.

We leave the integration of interval-valued fuzzy rough sets into other algorithms like fuzzy rule induction for future research.

Acknowledgement. We would like to thank two anonymous reviewers for bringing the similar proposals [14] and [5] to our attention.

The research reported in this paper was conducted with the financial support of the Odysseus programme of the Research Foundation – Flanders (FWO). D. Peralta is a Postdoctoral Fellow of the Research Foundation – Flanders (FWO, 170303/12X1619N).

References

1. Amiri, M., Jensen, R.: Missing data imputation using fuzzy-rough methods. Neurocomputing **205**, 152–164 (2016)
2. Aste, M., Boninsegna, M., Freno, A., Trentin, E.: Techniques for dealing with incomplete data: a tutorial and survey. Pattern Anal. Appl. **18**(1), 1–29 (2014). https://doi.org/10.1007/s10044-014-0411-9
3. Bellman, R., Kalaba, R., Zadeh, L.: Abstraction and pattern classification. Technical report, RM-4307-PR, The Rand Corporation, Santa Monica, California (1964)
4. Cornelis, C., Verbiest, N., Jensen, R.: Ordered weighted average based fuzzy rough sets. In: Yu, J., Greco, S., Lingras, P., Wang, G., Skowron, A. (eds.) RSKT 2010. LNCS (LNAI), vol. 6401, pp. 78–85. Springer, Heidelberg (2010). https://doi.org/10.1007/978-3-642-16248-0_16
5. Couso, I., Dubois, D.: Rough sets, coverings and incomplete information. Fund. Inform. **108**(3–4), 223–247 (2011)
6. Dua, D., Graff, C.: UCI machine learning repository (2019). http://archive.ics.uci.edu/ml
7. Dubois, D., Prade, H.: Twofold fuzzy sets and rough sets – some issues in knowledge representation. Fuzzy Sets Syst. **23**(1), 3–18 (1987)
8. Dubois, D., Prade, H.: Rough fuzzy sets and fuzzy rough sets. Int. J. Gen. Syst. **17**(2–3), 191–209 (1990)
9. Dubois, D., Prade, H.: Interval-valued fuzzy sets, possibility theory and imprecise probability. In: EUSFLAT-LFA 2005: Proceedings of the Joint 4th Conference of the European Society for Fuzzy Logic and Technology and the 11th Rencontres Francophones sur la Logique Floue et ses Applications, pp. 314–319 (2005)
10. Grzymala-Busse, J.W.: Rough set strategies to data with missing attribute values. In: Young, L.T., Ohsuga, S., Liau, C.J., Hu, X. (eds.) Foundations and Novel Approaches in Data Mining. SCI, vol. 9, pp. 197–212. Springer, Heidelberg (2006). https://doi.org/10.1007/11539827_11
11. Hand, D.J., Till, R.J.: A simple generalisation of the area under the ROC curve for multiple class classification problems. Mach. Learn. **45**(2), 171–186 (2001). https://doi.org/10.1023/A:1010920819831
12. Hong, T.P., Tseng, L.H., Chien, B.C.: Mining from incomplete quantitative data by fuzzy rough sets. Expert Syst. Appl. **37**(3), 2644–2653 (2010)
13. Jensen, R., Cornelis, C.: A new approach to fuzzy-rough nearest neighbour classification. In: Chan, C.-C., Grzymala-Busse, J.W., Ziarko, W.P. (eds.) RSCTC 2008. LNCS (LNAI), vol. 5306, pp. 310–319. Springer, Heidelberg (2008). https://doi.org/10.1007/978-3-540-88425-5_32
14. Jensen, R., Shen, Q.: Interval-valued fuzzy-rough feature selection in datasets with missing values. In: FUZZ-IEEE 2009: Proceedings of the 18th International Conference on Fuzzy Systems, pp. 610–615. IEEE (2009)
15. King, D.E.: DLIB-ML: a machine learning toolkit. J. Mach. Learn. Res. **10**(60), 1755–1758 (2009)
16. King, D.E.: A global optimization algorithm worth using (2017). http://blog.dlib.net/2017/12/a-global-optimization-algorithm-worth.html. Accessed 6 Jan 2021
17. Kryszkiewicz, M.: Rough set approach to incomplete information systems. Inf. Sci. **112**(1–4), 39–49 (1998)
18. Lenz, O.U., Peralta, D., Cornelis, C.: Scalable approximate FRNN-OWA classification. IEEE Trans. Fuzzy Syst. **28**(5), 929–938 (2020)

19. Lenz, O.U., Peralta, D., Cornelis, C.: Optimised one-class classification performance (2021). https://arxiv.org/abs/2102.02618
20. Pawlak, Z.: Rough sets. Rep. 431, ICS PAS (1981)
21. Pawlak, Z.: Rough sets. Int. J. Comput. Inf. Sci. **11**(5), 341–356 (1982). https://doi.org/10.1007/BF01001956
22. Thangavel, K., Pethalakshmi, A.: Dimensionality reduction based on rough set theory: a review. Appl. Soft Comput. **9**(1), 1–12 (2009)
23. Zadeh, L.A.: Fuzzy sets. Inf. Control **8**(3), 338–353 (1965)

Areas of Applications

Spark Accelerated Implementation of Parallel Attribute Reduction from Incomplete Data

Qian Cao[1], Chuan Luo[1(✉)], Tianrui Li[2], and Hongmei Chen[2]

[1] College of Computer Science, Sichuan University, Chengdu 610065, China
`caoqian@stu.scu.edu.cn, cluo@scu.edu.cn`
[2] School of Information Science and Technology, Southwest Jiaotong University, Chengdu 610031, China
`{trli,hmchen}@swjtu.edu.cn`

Abstract. Attribute reduction is a significant process of data preprocessing to overcome the challenges posed by multidimensional data analysis. Missing values in the data are usually unavoidable in the real applications, so it is important to select features with high importance efficiently in incomplete data. The theory of rough sets has been widely used in attribute reduction for uncertain data mining. To enable the rough set theory for large-scale incomplete data analysis, this paper develops a novel distributed attribute reduction algorithm based on Apache Spark cluster computing platform. By taking the advantage of positive approximation technique for reducing the data broadcasting gradually while reducing each redundant attribute iteratively, the proposed algorithm can significantly accelerate the attribute reduction in leveraging a computer cluster when processing large-scale incomplete data. Numerical experiments on different UCI data sets evidences the proposed parallel algorithm achieves high performance in terms of extensibility and scalability.

Keywords: Attribute reduction · Incomplete data · Rough sets · Parallel computing · Apache Spark

1 Introduction

Attribute reduction aims at removing redundant attributes in the data, which is useful for reducing the computational burden of learning, while maintaining the accuracy of learning. Rough set theory, originally proposed by Pawlak, has been widely applied in attribute reduction and its related techniques [1]. A great number of attribute reduction algorithms based on rough sets have been designed for classification learning until now.

The presence of missing value is a significant issue in data mining and knowledge discovery, since the valued information available is incomplete and less reliable. It is a challenge to deal with missing values for attribute reduction in

© Springer Nature Switzerland AG 2021
S. Ramanna et al. (Eds.): IJCRS 2021, LNAI 12872, pp. 203–217, 2021.
https://doi.org/10.1007/978-3-030-87334-9_17

incomplete data. Therefore, many researchers are going into efforts to design novel algorithms for attribute reduction in incomplete data to improve learning performance. Du and Hu expand the similarity dominance relation in incomplete ordered information systems and presented an attribute reduction algorithm based on discernibility matrix and discernibility function [2]. Tan et al. measured the uncertainty of rough approximations in incomplete decision systems using belief and plausibility functions, and constructed an attribute reduction method based on multigranulation rough sets [3]. Shu and Shen designed a positive region based incremental attribute reduction algorithm in incomplete information systems [4]. Qian and Shu introduced boolean reasoning techniques into attribute reduction in consistent and inconsistent incomplete ordered information systems with fuzzy decision [5]. Xie and Qin developed incremental attribute reduction algorithm based on the inconsistency degree with three update strategies of reduction [6]. Li and Wang proposed novel approaches for approximate concept construction and attribute reduction based on three-way decisions in incomplete contexts [7]. Sun et al. designed a novel attribute reduction method by using Lebesgue and entropy measures for uncertainty analysis in incomplete neighborhood decision systems [8]. Qian et al. extended the concept of positive approximation to incomplete information systems, and presented a positive approximation framework for accelerating large-scale attribute reduction problem [9]. Most of the existing attribute reduction algorithms can often show good performance on incomplete information systems with small amounts of data. As the complexity of algorithms are usually not so much high when the amount of data is relatively small, and the data will be loaded into the memory at one time for the computation of optimal attribute subset.

With the gradual popularization of computer networks and the rapid development of information technology, the rapid growth of data has become a key problem that needs to be considered in various research fields. Parallel and distributed computing technique provide a practical pathway to meet the challenge of explosive growth of data by exploiting usage of high performance computing resources. Acceleration of attribute reduction algorithms within a reasonable processing time by the adequacy of parallel computing has been addressed in some previous works. Qian et al. designed parallel algorithms for the computation of equivalence classes and attribute significance in order to speed up the attribute reduction precess based on MapReduce mechanism [10]. They further introduced the concept of hierarchical encoded decision table, and developed novel parallel hierarchical attribute reduction algorithms in data and task parallel [11]. Chen et al. investigated parallel attribute reduction approaches based on dominance-based neighborhood rough sets by considering the partial orders among numerical and categorical attributes [13]. Zhang et al. implemented knowledge acquisition approach based on rough sets on the representative MapReduce runtime systems of Hadoop, Phoenix and Twister [12]. El-Alfy and Alshammari introduced parallel genetic algorithm to approximate the minimum attribute reduct in the framework of decision-theoretic rough sets [14]. Zhang et al. developed boolean matrix based parallel algorithm for computing rough approximations on Multi-GPU environment [15]. Raza and Oamar proposed a novel parallel dependency calculation approach without calculating the positive region itself for attribute reduction [16]. Yin et al. developed fast attribute reduction on Apache Spark based on the redundant attribute batch processing method and the single cache iteration strategy [17]. Kong et al. proposed a dynamic data decomposition algorithm to maintain the global information on each distributed node to accelerate attribute reduction using fuzzy rough sets in distributed processing systems [18]. Dagdia et al. proposed an effective distributed algorithm based on rough set theory for large-scale data pre-processing under the Spark framework without sacrificing performance [21]. However, existing studies on parallel attribute reduction methods are mainly devote to the complete data without missing values. In this paper, we focus on the distributed implementation of attribute reduction for large-scale incomplete data classification. A novel parallelization of attribute reduction algorithm based on tolerance rough sets that exploits hardware resources of distributed computing systems is proposed. It is implemented with the positive approximation technique that use granulation orders of iteratively selected attributes to reduce the size of data gradually, in order to mitigate the problem of data broadcasting burden in the distributed heuristic search process of attribute reduction. Experiments on several large-scale incomplete data sets show the time-efficiency and scalability of the proposed algorithm.

Organization of the rest of this paper is given as follows. In Sect. 2, we review some basic notions of attribute reduction based on tolerance rough sets and Spark parallel programming framework. The details of the proposed parallel algorithm for attribute reduction in incomplete information systems is presented in Sect. 3. In Sect. 4, several performance metrics of parallel computing are used to verify the efficiency of our algorithm. Section 5 concludes our work with future research directions.

2   Preliminaries

In this section, we will review some basic concepts, including incomplete information systems, tolerance rough sets, positive approximation based attribute reduction [1,9].

In rough set theory, an information system is defined as $S = (U, A, V, f)$, where $U = \{x_1, x_2, ..., x_{|U|}\}$ is a non-empty finite set of objects, $A = \{a_1, a_2, ..., a_{|A|}\}$ is a non-empty finite set of attributes, $V = \bigcup_{a \in A} V_a$, where V_a denotes the possible value of all objects under a certain attribute a, and $f : U \times A \longrightarrow V$ is a function that satisfies $\forall a \in A, x \in U, f(x, a) \in V_a$. A decision information system can be denoted as $S = (U, A = C \cup d, V, f)$, where C is a condition attribute set, and d is a attribute feature. If the attribute values of some objects are missing, denoted by "$*$", in a decision information system, S can be regarded as an incomplete decision information system, otherwise S is a complete decision information system, which satisfies $* \notin V$.

Given an incomplete decision information system $S = (U, A = C \cup d, V, f)$, for any $P \subseteq C$, we can define a binary relation on U as follows:

$$T(P) = \{(x, y) \in U \times U | f(x, a) = f(y, a) \text{ or } f(x, a) = * \text{ or } f(y, a) = *\}. \quad (1)$$

For any two objects $x, y \in U$, if $(x, y) \in T(P)$, then we say x and y are tolerant under attribute set P. For any $x \in U$, we defined the tolerance class of x as $T_P(x) = \{y \in U | (x, y) \in T(P)\}$. The set of all tolerance classes on U can be defined as $U/T(P) = \{T_P(x) | x \in U\}$.

Let X be a subset of U, the upper and lower approximation can be constructed to characterize X based on the tolerance relation $T(P)$ as follow:

$$\overline{T_P}X = \{x \in U | T_P(x) \cap X \neq \emptyset\}, \underline{T_P}X = \{x \in U | T_P(x) \subseteq X\}. \quad (2)$$

The upper approximation is a set of the objects that may belong to the subset X, and lower approximation is a set of the objects that must belong to the subset X.

Assume that the objects are partitioned into m mutually exclusive crisp subsets by the decision attribute d, and denoted as $U/d = \{d_1, d_2, ..., d_m\}$, where $f(x, d) = f(y, d)$ for any $x, y \in d_i$, $i \in \{1, 2, ..., m\}$. The positive region of d with respect to tolerance relation $T(P)$ is defined as follows:

$$POS_P(d) = \bigcup_{i=1}^{m} \underline{T_P}(d_i) = \bigcup_{i=1}^{m} \{x \in U | T_P(x) \subseteq d_i\}. \quad (3)$$

Based on the Eq. (3), the dependency between conditional and decision attributes can be expressed as $\gamma_P(d) = \frac{|POS_P(d)|}{|U|}$. For any $a \in P$, the inner and outer significance measures of a in P can be defined respectively as $Sig^{inner}(a) = \gamma_P(d) - \gamma_{P-\{a\}}(d)$ and $Sig^{inner}(a) = \gamma_P(d) - \gamma_{P-\{a\}}(d)$, in order to construct a heuristic attribute reduction algorithm from incomplete data to remove redundant attributes while keeping the ratio of correctly classified objects unchanged.

In order to accelerate the heuristic search process of attribute reduction, a concept of positive approximation was proposed by exploiting the structure of set approximation under a granulation order [9].

Let $\mathbf{P} = \{P_1, P_2, ..., P_i\}$ is a family of attribute sets with $P_j \subseteq C$, $P_j \subseteq P_{j+1}$, $j = 1, 2, ..., i - 1$. For any $\mathbf{P} = \{P_1, P_2, ..., P_i\}$, $X \subseteq U$, the $\mathbf{P}_i$-lower approximation and upper approximation of X can be defined as follow:

$$\overline{\mathbf{P}}_i(X) = \overline{T_{P_i}}X, \underline{\mathbf{P}}_i(X) = \bigcup_{n=1}^{i} \underline{T_{P_n}}X_n. \quad (4)$$

where $X_1 = X$, $X_n = X - \bigcup_{j=1}^{n-1} \underline{T_{P_j}}X_j$, $n = 2, 3, ..., i$.

Suppose $U/d = \{d_1, d_2, ..., d_m\}$ is the decision partition of the incomplete decision information system, then the positive region of d with respect to the granulation order of $\mathbf{P} = \{P_1, P_2, ..., P_i\}$ can be defined as $POS_{\mathbf{P}_i}^U(d) =$

$\bigcup_{n=1}^{m} \underline{\mathbf{P}_i}(d_n)$. It is easy to find that $POS_{\mathbf{P}_{i+1}}^{U}(d) = POS_{\mathbf{P}_i}^{U}(d) \cup POS_{\mathbf{P}_{i+1}}^{U_{i+1}}(d)$ always hold, where $U_1 = U$, and $U_{i+1} = U - POS_{\mathbf{P}_i}^{U}(d)$. Hence, one can design accelerated attribute reduction algorithm based on the positive approximation by using granulation orders on the gradually reduced universe during the iteratively heuristic search process.

2.1   Apache Spark Computing Model

In order to alleviate the problems of storage difficulty and computational burden for large-scale data processing in real-time and validity, cluster computing techniques have been appeared to provide reliable and efficient computation. MapReduce, proposed by Google, is one of the representative distributed computing frameworks, which has been frequently adopted for improving the performance of data mining and machine learning algorithms [19]. However, the computational results in MapReduce framework need to be stored on the Hadoop distributed file system, which costs much time for disk I/O operations, and hence is not suitable for iterative algorithms which need to save or reuse the intermediate computational results.

Apache Spark is a unified analytics engine to develop large-scale distributed programs based on the optimization of Hadoop MapReduce computing model [20]. By introducing a Resilient Distributed Datasets (RDD) model, Spark can store intermediate data in memory to save huge amounts of disk I/O operation time. RDD is a fault-tolerant abstraction to represent a read-only set of objects that is distributed across multiple computing elements. RDD can save the results of intermediate calculations in memory, these data will greatly reduce the time it takes to read data from disk during iterative computing. Moreover, Spark provides the function of broadcasting variables to distribute the data into the nodes to compute, and the worker nodes can write increments to special variables named accumulators, which further reduce the time consumption of iterative computing. Hence, we decompose the attribute reduction algorithms in independent tasks for a Spark implementation that will allow the users to speed up the large-scale data analysis by taking advantage of the distributed computational capabilities.

3   Spark Parallelization of Attribute Reduction from Incomplete Data

Parallel algorithms of attribute reduction are implemented by dividing the incomplete data into multiple subsets as RDD and perform calculations of positive region on multiple computational nodes in parallel. Algorithm 1 begins the attribute reduction process by parallelly calculating the positive region. Since the object to be considered needs to compare with the other objects of the input data. So we persist the input RDD in step 1, and broadcast it to slave nodes in step 2. Persist is a RDD operator which can loads RDD on cache or disk physically. Then we reset the input accumulator in step 3. The accumulator can help us to collet the positive region across the data stored on different slave nodes as partitions. In step 4 to step 16, we traverse each object on partitions. If the object is tolerant with one, which has a different value on decision attribute, the object will not be counted by accumulator. Afterwards, RDD will be unpersisted in step 17, which avoids memory usage, and the result will be returned in step 18.

Algorithm 1. Parallel algorithm for calculating the positive region(countPos)

Input: RDD: The input RDD; //The input RDD will be divided into r partitions, and U^k represents the set of objects on each partition, where $k = 1, 2, ..., r$.

   Q:The set of the attributes to be computed

   acc:An accumulator to collect the number of positive region in each partition

Output: the number of total positive region;

1: RDD.persist();
2: bc←broadCast(RDD);
3: acc.reset;
4: **for** each $x \in U^k$ **do**
5:    $t \leftarrow 1$
6:    **for** each $y \in$ bc **do**
7:       **if** $(x, y) \in T(Q) \wedge f(x, d) \neq f(y, d)$ **then**
8:          $t \leftarrow 0$;
9:          **break**;
10:       **end if**
11:    **end for**
12:    **if** $t = 1$ **then**
13:       acc.add(1);
14:    **end if**
15: **end for**
16: RDD.unpersist();
17: **return** acc.value

Based on the parallelly calculated results of positive region by Algorithm 1, Algorithm 2 and 3 present the parallel processes of obtaining the core attributes based on the inner importance measure and selecting the most important attribute in terms of outer importance measure. Furthermore, by adopting the positive approximation technique, Algorithm 4 is developed to reduce the broadcasted data gradually during the heuristic search process of attribute reduction according to the granulation orders of iteratively selected attributes. The implementation of Algorithm 4 is similar to Algorithm 1. But the difference is that we do not use accumulator to collet the calculational result of positive region in Algorithm 4. Instead, we put the object which belongs to the positive region in terms of the currently selected features into a new RDD in step 14, and return the new RDD in order to update the broadcasted data for the later iterative computations.

Algorithm 5 shows the main function of the proposed parallel feature selection algorithm for incomplete incomplete decision information system based on

---

Algorithm 2. Parallel algorithm for getting the core attributes (getCores)

Input: DataRDD:RDD of an incomplete decision information system $IS = (U, A = C \cup d, V, f)$;

        C:The set of all the conditional attributes;

        acc:An accumulator to collect the number of positive region in each partition

Output: *cores*: All the core attributes;

1: *cores* $\leftarrow \emptyset$; // *cores* is the set which conserves all the core attributes
2: $Q \leftarrow \emptyset$;
3: $pos\_count \leftarrow 0$;
4: **for** each a in C **do**
5:     $Q \leftarrow C - a$
6:     $pos\_count \leftarrow$ countPos(DataRDD, Q, acc);
7:     Calculate $Sig^{inner}(a)$ by pos_count;
8:     **if** $Sig^{inner}(a) > 0$ **then**
9:         put a into *cores*;
10:    **end if**
11: **end for**
12: **return** *cores*

Algorithm 3. Parallel algorithm for getting the most important attribute (getAttribute)

Input: subRDD:the RDD of a subset of the original dataset;

        red:The set of the attributes selected currently;

        acc:An accumulator to collect the number of positive region in each partition

Output: a_{max}: the attribute with the biggest outer significance measure;

1: $a_{max} \leftarrow \emptyset$;
2: $Q \leftarrow red$;
3: $pos\_count \leftarrow 0$;
4: $Sig^{outer}(a_{max}) \leftarrow 0$;
5: **for** each a in $C - red$ **do**
6:     $Q \leftarrow red \cup a$;
7:     $pos\_count \leftarrow$ countPos(subRDD, Q, acc);
8:     Calculate $Sig^{outer}(a)$ by pos_count;
9:     **if** $Sig^{outer}(a) > Sig^{outer}(a_{max})$ **then**
10:      $Sig^{outer}(a_{max}) \leftarrow Sig^{outer}(a)$;
11:      $a_{max} \leftarrow a$;
12:    **end if**
13: **end for**
14: **return** a_{max}

---

the positive approximation. In Algorithm 5, we firstly defines an accumulator called acc in order to collet the partial results of positive region across different slave nodes in step 4, and calculate the positive region on the whole incomplete incomplete decision information system by Algorithm 1 in step 6, which was used to determine whether the heuristic iteration feature selection algorithm meets the stopping criterion. In step 7, Algorithm 2 was executed to obtain the all core

Algorithm 4. Parallel algorithm for reducing universe gradually by positive approximation (posApproximation)

Input: RDD: The input RDD;//The input RDD will be divided into r partitions, and U^k represents the set of objects on each partition, where $k = 1, 2, ..., r$.
    red:The set of the attributes selected currently
Output: newRDD:The RDD of the new subset of the input after positive approximation;

1: newRDD ← ∅;
2: RDD.persist();
3: bc←broadCast(RDD);
4: **for** each $x \in U^k$ **do**
5:     $t \leftarrow 1$
6:     **for** each $y \in$ bc **do**
7:         **if** $(x, y) \in T(Q) \land f(x, d) \neq f(y, d)$ **then**
8:             $t \leftarrow 0$;
9:             **break**;
10:        **end if**
11:    **end for**
12:    **if** $t = 1$ **then**
13:        newRDD ← newRDD $\cup x$;
14:    **end if**
15: **end for**
16: RDD.unpersist();
17: **return** newRDD

---

attributes, which are indispensable in a reduct of the incomplete decision information system. If these are some core attributes found by Algorithm 2, positive approximation is used to reduce the universe by Algorithm 4 in steps 8–13. After that, Algorithm 3 in step 17 was used to select an optimal attribute based on the currently selected features, and positive approximation in step 21 will further speed up the next iteration. When the positive region of the selected attributes is the same as the original one, the reduct will be returned in step 25.

Figure 1 describes the process of the proposed parallel attribute reduction algorithm. Firstly, we parallelly compute the inner significance of all attributes in an incomplete decision information system, then we select the features whose inner significance are bigger than zero as core attributes. In the next steps, the slave nodes parallelly compute the outer significance of the remaining features and add the feature which show best outer significance into the reduct. Meanwhile, positive approximation was used to update the dataset if there are core attributes or selecting an optimal attribute. By this way, we filtered the objects in the dataset and reduce the time consumption for subsequent iterative calculations. The algorithm will perform iterative calculations until the selected features can get the same positive region as the original incomplete decision information system.

Algorithm 5. Parallel incomplete attribute reduction algorithm based on the positive approximation (PIARPA)

Input: DataRDD:RDD of an incomplete decision information system $IS = (U, A = C \cup d, V, f)$;

Output: *red*: A reduct of the selected conditional attributes

1: $red \leftarrow \emptyset$; // *red* is the set which conserves all the selected features
2: Pos $\leftarrow 0$; // Pos stores the number of positive region of the original data set
3: redPos $\leftarrow 0$; // redPos stores the number of positive region under the attributes selected
4: acc $\leftarrow 0$; // acc is a accumulator defined in Spark used to calculate the number of positive region
5: subRDD $\leftarrow \emptyset$;
6: Pos $\leftarrow$ countPos(DataRDD, C, acc);
7: $red \leftarrow$ getCores(DataRDD, C, acc);
8: **if** $red \neq \emptyset$ **then**
9: $\quad red$Pos $\leftarrow$ countPos(DataRDD, red, acc);
10: $\quad POS_{P_1}(d) \leftarrow$ posApproximation(DataRDD, red);
11: $\quad U_1 = U - POS_{P_1}(d)$
12: **else** $U_1 = U$
13: **end if**
14: $i \leftarrow 1, R_1 = red, P_1 = \{R_1\}$
15: **while** Pos $\neq red$Pos **do**
16: $\quad$ subRDD $\leftarrow U_i$;
17: $\quad a_{max} \leftarrow$ getAttribute(subRDD, red, acc);
18: $\quad red \leftarrow red \cup \{a_{max}\}$;
19: $\quad red$Pos $\leftarrow$ countPos(subRDD, red, acc);
20: $\quad i \leftarrow i + 1$;
21: $\quad POS_{P_i}(d) \leftarrow$ posApproximation(subRDD, red);
22: $\quad U_i = U_{i-1} - POS_{P_i}(d)$;
23: $\quad R_i \leftarrow red, P_i \leftarrow \{R_1, R_2, ..., R_i\}$
24: **end while**
25: **return** *red*

---

4   Experimental Evaluation

In this section, we designed several experiments to verify the performance of the parallel algorithm. All the experiments were run on a cluster of eleven nodes, where one node is set as master, and the rest ten nodes are set as slaves. Each node has the following configuration, CPU: Intel Xeon E5-2682v4 @ 2.5 GHz, CPU Cores: 4 vCPU, Total Memory: 8 GiB, Network: 0.8 Gbps, Java version: OpenJDK 1.8, Spark version: 2.4.5. Five data sets were selected from UCI data sets for performance testing, which are all symbolic data with missing values. The detailed information of the data set is shown in Table 1.

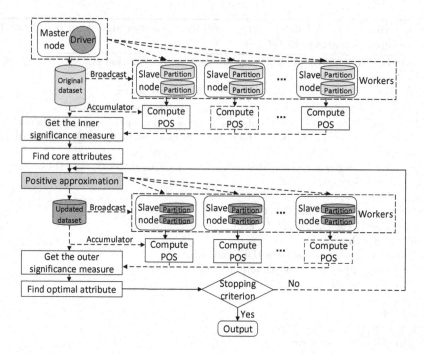

Fig. 1. Abstraction of the process of PIARPA algorithm.

Table 1. Description of data sets.

	Data sets	Cases	Features	Classes
1	Mushroom	8124	22	2
2	Audiology.standardized	200	69	24
3	Breast-cancer-wisconsin	699	9	2
4	Dermatology	366	34	6
5	Soybean-large	307	35	19

4.1 Selection of the Number of Data Partitions

In Spark parallel programming framework, all the tasks will be assigned to each executor to process. Each executor consists of several cores, and each core of one executor can only execute one task at a time. Usually, the number of data partitions will greatly impact on the accelerating performance of the tasks in each stage. So it is important to select an appropriate number of the data partitions, which will directly affect the performance of the proposed parallel algorithms. There are ten slaves of computer in our cluster, and each computer has 4 cores. We use the dataset Mushroom to test the impact of the number of data partitions on performance by increasing the number of data partitions from 1 to 60. The experimental results are shown in Fig. 2.

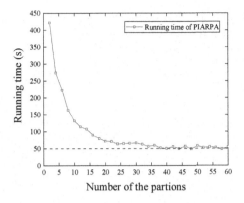

Fig. 2. The relationship between running time and number of data partitions on Mushroom dataset of PIARPA.

From the experimental results, It can be seen that the time consumption of the proposed algorithm is significantly reduced with the number of data partitions increasing from 1 to 40. It is noticed that the total number of cores in our cluster is equals to 40. However, when the number of the data partitions continues to increase above 40, there is not much difference in time consumption. We can conclude that when the number of data partition is less than the total number of the cores in the cluster, increasing the number of data partition will significantly reduce the time consumption of parallel algorithm. But when the number of data partitions is larger than the number of cores, there is no executor to use and the tasks have to wait, so the processing time hardly change. To achieve the maximum parallelism, we select 4× (number of computers) as the number of the data partition in the following experiments.

4.2 Evaluation of the Parallelism Metrics

The parallelism metrics of speedup, scaleup, and sizeup were evaluated in this section to confirm the scalability of algorithm to process massive amount of incomplete data. Speedup describes the changes in the running time of the algorithm when increasing the number of machines in the cluster. It is calculated as $Speedup(m) = T_1/T_m$, where T_1 and T_m are the running time of the same task in a single processor system and a distributed processor system with m computers.

If the speedup can maintain a linear growth, it means that multiple machines can well shorten the time consumption of the parallel algorithm. But it is difficult to achieve the linear speedup, because parallel algorithms are often mixed with other calculations, and data transmission between nodes in the parallel system will also have an impact. To verify the Speedup of PIFSPA, we select the datasets of *Audiology.standardized*, *Breast-cancer-wisconsin*, *Dermatology*, and *Soybean-large*, and the datasets of 10, 20, 30, 40 and 50-times of the original dataset are adopted to evaluate the speedup by increasing the number of computers from 1 to 10. The experimental results are shown in Fig. 3.

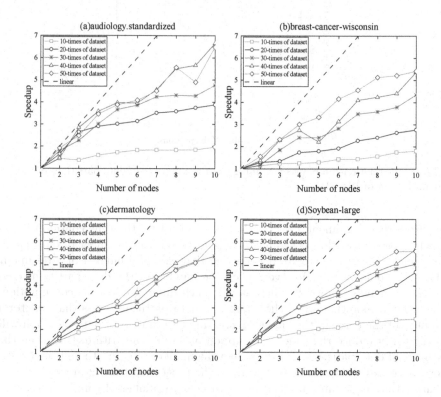

Fig. 3. Speedup of PIARPA on different datasets.

From Fig. 3, we can find that the speedup performance is unsatisfactory when the data sets is small, since the time cost on parallel computing is not dominant. However, the proposed algorithm achieves better speedup performance with the growth of data size. The speedup has shown a linear growth trend when the dataset is 50 times than the original one. Meanwhile, we can analyze that the speedup gradually increases when the number of computers in the cluster increases.

Scaleup is used to verify how the parallel algorithm performs on a larger data set when there are enough computing nodes available. It is specified as $Scaleup(m) = T_1^1/T_m^m$, where T_1^1 is the running time of a task on a single processor system, and T_m^m is the running time to complete m of the same task on a distributed processor system with m computers. Ideally, the scaleup of a good parallel algorithm should be closer to 1. To test the scaleup of our algorithm, we also expand the original datasets by 10, 20, 30, 40 and 50-times, and then tested them on clusters with computing nodes from 1 to 10 respectively. Figure 4 shows the experimental results of scaleup on different datasets.

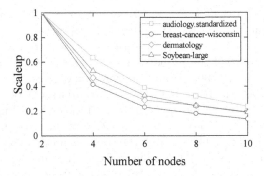

Fig. 4. Scaleup of PIARPA on different datasets.

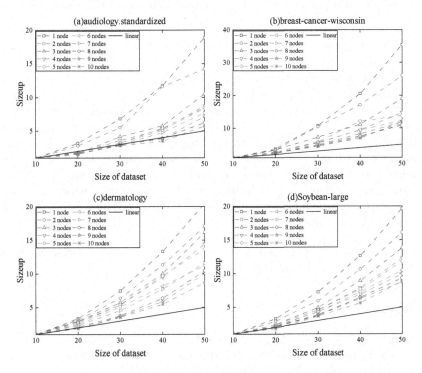

Fig. 5. Sizeup of PIARPA on different datasets.

From Fig. 4, it is clear that the scaleup of the proposed parallel algorithm is difficult to be closer to 1, because the amount of computation is not linear with the growth of the data set. However, the scaleup is gradually flatting out with the increase of the number of nodes, which demonstrates that our algorithm still keeps good parallel performance when the cluster is big enough.

Sizeup measures the computing time one takes on a given system when the dataset is m-times larger than the original dataset. The sizeup is calculated as

$Sizeup(m) = T^m/T^1$, where T^m and T^1 are the running time of a single task and m of the same task on the same distributed processor system. The sizeup represents the variation trend of time consumption with the data set increasing. To measure the sizeup of the proposed parallel algorithm, we also increase the original datasets 10, 20, 30, 40 and 50-times, and select the 10-times of the original datasets as the basic datasets. All the datasets run experiments on clusters with computing nodes from 1 to 10 respectively. Figure 5 shows the experimental results of sizeup with the variation of the data size.

The experimental results show that the sizeup is much larger than the growth multiple when there are less computing nodes in the cluster. As the number of computing nodes gradually increases to 10, the sizeup is closed to m, which is the ideal value, and it is a linear growth. The results of sizeup show that the proposed algorithm can scales well when more computational nodes are added.

5 Conclusions

This work aims at solving the computational difficulties of attribute reduction algorithms with the rapid growth of data volume in incomplete information systems. We proposed an efficient algorithm based on Spark parallel programming framework for large-scale attribute reduction. The concept of positive approximation is incorporated into the proposed algorithm to speed up calculations of importance measure of attribute. Experimental evaluation in terms of several parallelism metrics were carried out extensively. The experimental results show that our parallel algorithm can achieve good parallel performance in terms of the speedup, scaleup and sizeup. Performance tuning will be our future research direction to further optimize the parallel performance.

Acknowledgements. This work was supported by the National Natural Science Foundation of China (Nos. 62076171, 61573292, 61976182).

References

1. Pawlak, Z.: Rough sets. Int. J. Comput. Inf. Sci. **11**, 341–356 (1982)
2. Du, W.S., Hu, B.Q.: Dominance-based rough set approach to incomplete ordered information systems. Inf. Sci. **346–347**, 106–129 (2016)
3. Tan, A.H., Wu, W.Z., Li, J.J., Lin, G.P.: Evidence-theory-based numerical characterization of multigranulation rough sets in incomplete information systems. Fuzzy Sets Syst. **294**, 18–35 (2016)
4. Shu, W.H., Shen, H.: Updating attribute reduction in incomplete decision systems with the variation of attribute set. Int. J. Approximate Reasoning **55**(3), 867–884 (2014)
5. Qian, W.B., Shu, W.H.: Attribute reduction in incomplete ordered information systems with fuzzy decision. Appl. Soft Comput. **73**, 242–253 (2018)
6. Xie, X.J., Qin, X.L.: A novel incremental attribute reduction approach for dynamic incomplete decision systems. Int. J. Approximate Reasoning **93**, 443–462 (2018)

7. Li, M.Z., Wang, G.Y.: Approximate concept construction with three-way decisions and attribute reduction in incomplete contexts. Knowl.-Based Syst. **91**, 165–178 (2016)
8. Sun, L., Wang, L.Y., Ding, W.P., Qian, Y.H., Xu, J.C.: Neighborhood multigranulation rough sets-based attribute reduction using Lebesgue and entropy measures in incomplete neighborhood decision systems. Knowl.-Based Syst. (2020). https://doi.org/10.1016/j.knosys.2019.105373
9. Qian, Y.H., Liang, J.Y., Pedrycz, W., Dang, C.Y.: An efficient accelerator for attribute reduction from incomplete data in rough set framework. Pattern Recogn. **44**(8), 1658–1670 (2011)
10. Qian, J., Miao, D.Q., Zhang, Z.H., Yue, X.D.: Parallel attribute reduction algorithms using MapReduce. Inf. Sci. **279**, 671–690 (2014)
11. Qian, J., Lv, P., Yue, X.D., Liu, C.H., Jing, Z.J.: Hierarchical attribute reduction algorithms for big data using mapreduce. Knowl.-Based Syst. **75**, 18–31 (2015)
12. Zhang, J.B., Wong, J.-S., Li, T.R., Pan, Y.: A comparison of parallel large-scale knowledge acquisition using rough set theory on different MapReduce runtime systems. Int. J. Approximate Reasoning **55**(3), 896–907 (2014)
13. Chen, H., Li, T.R., Cai, Y., Luo, C., Fujita, H.: Parallel attribute reduction in dominance-based neighborhood rough set. Inf. Sci. **373**, 351–368 (2016)
14. El-Alfy, E.-S.M., Alshammari, M.A.: Towards scalable rough set based attribute subset selection for intrusion detection using parallel genetic algorithm in MapReduce. Simul. Model. Pract. Theory **64**, 18–29 (2016)
15. Zhang, J.B., Zhu, Y., Pan, Y., Li, T.R.: Efficient parallel Boolean matrix based algorithms for computing composite rough set approximations. Inf. Sci. **329**, 287–302 (2016)
16. Raza, M.S., Qamar, U.: A parallel rough set based dependency calculation method for efficient feature selection. Appl. Soft Comput. **71**, 1020–1034 (2018)
17. Yin, L.Z., Qin, L.Y., Jiang, Z.H., Xu, X.M.: A fast parallel attribute reduction algorithm using Apache Spark. Knowl.-Based Syst. (2021). https://doi.org/10.1016/j.knosys.2020.106582
18. Kong, L., et al.: Distributed feature selection for big data using fuzzy rough sets. IEEE Trans. Fuzzy Syst. **28**(5), 846–857 (2020)
19. Dean, J., Ghemawat, S.: MapReduce: simplified data processing on large clusters. Commun. ACM **51**(1), 107–113 (2008)
20. Apache Spark Lightning-fast unified analytics engine. https://spark.apache.org/. Accessed Aug 2020
21. Chelly Dagdia, Z., Zarges, C., Beck, G., Lebbah, M.: A scalable and effective rough set theory-based approach for big data pre-processing. Knowl. Inf. Syst. **62**(8), 3321–3386 (2020). https://doi.org/10.1007/s10115-020-01467-y

Attention Enhanced Hierarchical Feature Representation for Three-Way Decision Boundary Processing

Jie Chen[1,2,3] (ID), Yue Chen[1,2,3], Yang Xu[1,2,3], Shu Zhao[1,2,3](✉) (ID),
and Yanping Zhang[1,2,3] (ID)

[1] Key Laboratory of Intelligent Computing and Signal Processing,
Ministry of Education, Hefei 230601, Anhui, People's Republic of China
[2] School of Computer Science and Technology, Anhui University,
Hefei 230601, Anhui, People's Republic of China
[3] Information Materials and Intelligent Sensing Laboratory of Anhui Province,
Hefei 230601, Anhui, People's Republic of China

Abstract. For binary classification, the three-way decision divides samples into positive (POS) region, negative (NEG) region, and boundary region (BND). The correct division of these boundary data is helpful to improve the accuracy of binary classification. However, how to construct the optimal feature representation from certain samples for boundary domain partition is a challenge. In this paper, we propose attention enhanced hierarchical feature representation for three-way decision boundary processing (AHT) to deal with the boundary region. Based on the three-way decision, certain regions (positive, negative) and boundary regions are obtained. Obtaining the hierarchical feature representations on the positive domain and the negative domain respectively. Constructing attention-enhanced fusion feature representation to guide the boundary domain division of the testing set. The experimental results on five UCI datasets show that our algorithm effectively improves binary classification accuracy.

Keywords: Three-way decision · Hierarchical feature representation · Attention

1 Introduction

For binary classification, the three-way decision divides all samples into three possible decisions: positive decision, negative decision, and boundary decision: namely, as a positive region (POS), a negative region (NEG), and a boundary region (BND). The boundary region still needs to be divided into the positive region or the negative region. Certain regions POS/NEG can be classified directly, while these boundary data would be further divided into certain regions (positive region or negative region) when they obtain enough useful information.

S. Ramanna et al. (Eds.): IJCRS 2021, LNAI 12872, pp. 218–224, 2021.
https://doi.org/10.1007/978-3-030-87334-9_18

In order to solve binary classification problems properly, researchers have made many contributions for processing boundary regions based on three-way decision theory, Ma and Yao [12] proposed three types of class-specific attribute reduces in probabilistic rough set models for boundary processing. Li et al. [11] proposed a three-way decision model for dealing with the boundary region to improve the binary text classification performance.

In this paper, we propose an AHT model for binary classification. The hierarchical feature representations on the positive domain and the negative domain are obtained respectively. Then, we use attention-enhanced fusion to obtain the optimal feature representation to process the BND region. The experimental results demonstrate that our algorithm achieves a good classification performance on five UCI datasets.

2   Proposed Method

Figure 1 shows the overall flow of the proposed model. The model consists of two parts: attention enhanced mechanism, attention fusion mechanism. Firstly, we use a three-way decision model (Minimum Covering Algorithm: MinCA) to divide training samples into three regions: POS, NEG, and BND. Based on FQST, m granularities feature representations in the POS regio and n granularities feature representations of the NEG region are obtained. Then, we get the weight of each granularity of feature representation by using the attention enhanced mechanism. In the testing set, the samples in boundary regions are processed by the attention fusion mechanism.

Attention Enhanced Mechanism: In the POS region and the NEG region, the fuzzy equivalence relation is constructed by the variance mutual information. On the basis of fuzzy quotient space theory (FQST), we set a different parameter λ to control the hierarchical structure and obtain hierarchical feature representations. In this way, we can obtain m and n granularities of feature representations from the POS region and the NEG region. Boundary samples in the training dataset are used to get the weight of each granularity of the feature representations which are obtained from the positive region and negative region. The number of samples in the BND region in the training set that are divided correctly by using this layer of feature representation is the weight of this granularity.

$$POS[j] = \sum_{i=1}^{q} sapmle\_sub\_test(BND[i], j, 1) \tag{1}$$

$$NEG[k] = \sum_{i=1}^{q} sapmle\_sub\_test(BND[i], k, 0) \tag{2}$$

$$POS[j] = POS[j]/q \tag{3}$$

$$NEG[k] = NEG[k]/q \tag{4}$$

Where, $POS[j]$ and $NEG[k]$ are the weight of jth granularity of feature representation in the POS region and kth granularity of feature representation in the NEG region, j is the jth granularity of feature representation in POS region, $j \in [1, 2, \cdots, m]$) and k is the kth granularity of feature representation in POS region, $k \in [1, 2, \cdots, n]$); q is the number of samples in the BND region; sample_sub_test $(s, num, class)$ is a class function, s is a sample in BND region, num represents the numth granularity of feature representation, and $class$ means positive or negative domain (0 means the negative region, 1 means the positive region). If the sentiment polarity of comments of sth review obtained by sample_sub_test$(s, num, class)$ is the same as the original sentiment polarity, the return value of the function is 1, otherwise, it returns 0.

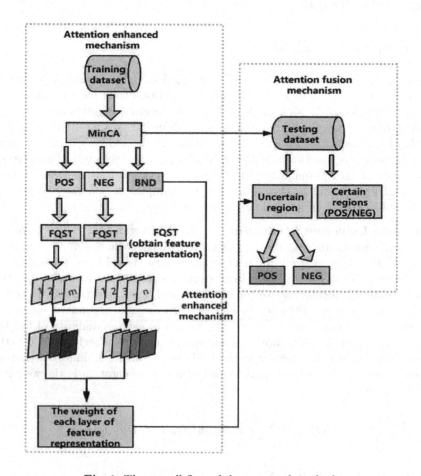

Fig. 1. The overall flow of the proposed method.

Attention Fusion Mechanism: Each boundary sample is predicted by using m granularities of feature representation on the positive domain and the n granularities of feature representation on the negative domain respectively. And then, the prediction result is combined with the attention weight of each granularity of feature representation to obtain the probability of the boundary sample belongs to the positive and negative category respectively. Finally, we can determine which category the boundary sample belongs to.

$$all\_pos = \sum_{j=1}^{m} POS[j] * predictResult[q][j] \qquad (5)$$

$$all\_neg = \sum_{k=1}^{n} NEG[k] * predictResult[q][k] \qquad (6)$$

Where, all_pos and all_neg are the probability of the qth boundary sample that predicted for positive and negative. $predictResult[q][j]$ is the prediction result of the qth boundary sample using the jth granularity feature representation in the POS region, $predictResult[q][k]$ is the prediction result of the qth sample using the kth granularity feature representation in the NEG region. If the predicted result is positive, the value of $predictResult$ is 1, otherwise, the value of $predictResult$ is -1.

During the testing process, the testing set is divided into three parts (positive, negative, boundary) by using the minimum coverage radius obtained by MinCA. The boundary samples are divided into the POS region or the NEG region. If the value of all_pos is higher than the value of all_neg, the qth boundary sample is predicted as positive, otherwise, it is predicted as negative.

3   Performance Evaluation

We use five datasets (Chess, Spambase, WPBC, WDBC, Occupancy) that come from the UCI Machine Learning Repository [1]. Firstly, we compared the methods HFR-TWD [4] and AH3 [5] proposed in the previous work. And then, we compare the latest algorithms for these datasets respectively.

Table 1. Comparative results on five datasets

Data	HFR-TWD	AH3	AHT
Chess	96.9	98.7	**100.0**
Spambase	98.9	98.9	**99.3**
WPBC	**98.0**	**98.0**	89.4
WDBC	99.6	**99.8**	99.6
Occupancy	**99.9**	**99.9**	**99.9**

Table 1 shows the results of boundary processing by three different algorithms on five datasets. It presents that the AHT algorithm has better performance on

Chess and Spambase datasets. For the Occupancy dataset, the classification results of the three methods are the same. It can be seen that our method has certain advantages for the processing of boundary samples.

According to Table 2, we can get the comparison results of the Chess dataset, AHT has the best performance, the classification accuracy is improved. It is clear that our algorithm improves the experimental result on the Spambase dataset, it is up to 99.3%. Although the result of the AHT algorithm on the WPBC dataset is not the best, our algorithm also performs well, the result of our method comes in second. For the WDBC dataset, the result of our method is better than other methods, it is up to 99.6%. Our algorithm also performs well on the Occupancy dataset, the classification accuracy is up to 99.9%, the same as the result of TWKM algorithms. From the above results, our algorithm can get better classification results on these datasets.

Table 2. Comparative results on each dataset

Datasets	Algorithm	Accuracy (%)	Datasets	Algorithm	Accuracy (%)
Chess	ISbFIM [18]	89.0	Spambase	IPSO-J48 [10]	98.3
	WEFPM [6]	93.2		ABBDT [10]	93.7
	AS-KMC [16]	94.8		SVM [8]	96.1
	RfDE [19]	97.1		SVM&K-means [8]	98.0
	AHT	**100.0**		**AHT**	**99.3**
WPBC	PSO-FS [2]	78.2	WPBC	GA-FS [2]	78.1
	ANN-FS [2]	79.2		SMO with Ranker [14]	77.3
	C4.5+DT with Ranker [14]	76.3		NB with Ranker [14]	76.3
	SA-LSTSVM [15]	**97.4**		**AHT**	89.4
WDBC	MBA-FS [9]	96.9	WDBC	GRU-SVM[3]	93.8
	L2-SVM [3]	96.1		PCA-KNN [13]	82.3
	PCA-SVM [13]	86.7		EPCF Rule [13]	93.2
	GNRBA [7]	98.9		PSO-FS [2]	97.2
	GA-FS [2]	96.6		AN-FS [2]	97.3
	TWKM [17]	93.0		**AHT**	**99.6**
Occupancy	k-means [17]	89.4	Occupancy	k-medoids [17]	90.3
	TWKM [17]	**99.9**		**AHT**	**99.9**

4   Conclusion

In this paper, we proposed an AHT method to process boundary samples into a certain region. First of all, all samples are divided into three regions by utilizing MinCA. Then, in POS and NEG regions, we obtain different granularities feature representations respectively based on FQST and use BND samples to get the weight of each granularity of feature representation. Finally, the weight obtained by the attention enhanced mechanism is used to deal with boundary samples in the testing process. Compared with other latest algorithms, the AHT can effectively handle samples from the boundary regions. Therefore, we can conclude that the performance of the AHT algorithm is better.

References

1. http://archive.ics.uci.edu/ml/
2. Aalaei, S., Shahraki, H., Rowhanimanesh, A., Eslami, S.: Feature selection using genetic algorithm for breast cancer diagnosis: experiment on three different datasets. Iran. J. Basic Med. Sci. **19**(15), 476–482 (2016)
3. Agarap, A.F.M.: On breast cancer detection: an application of machine learning algorithms on the wisconsin diagnostic dataset, pp. 5–9. ACM (2018). https://doi. org/10.1145/3184066.3184080
4. Chen, J., et al.: A method for boundary processing in three-way decisions based on hierarchical feature representation. In: Nguyen, H.S., Ha, Q.-T., Li, T., Przybyła-Kasperek, M. (eds.) IJCRS 2018. LNCS (LNAI), vol. 11103, pp. 123–136. Springer, Cham (2018). https://doi.org/10.1007/978-3-319-99368-3_10
5. Chen, J., Xu, Y., Zhao, S., Zhang, Y.: AH3: an adaptive hierarchical feature representation model for three-way decision boundary processing. Int. J. Approx. Reason. **130**, 259–272 (2021). https://doi.org/10.1016/j.ijar.2020.10.009
6. Devi, S.G., Sabrigiriraj, M.: Swarm intelligent based online feature selection (OFS) and weighted entropy frequent pattern mining (WEFPM) algorithm for big data analysis. Clust. Comput. **22**(5), 11791–11803 (2019). https://doi.org/10.1007/ s10586-017-1489-9
7. Dora, L., Agrawal, S., Panda, R., Abraham, A.: Optimal breast cancer classification using gauss-newton representation based algorithm. Expert Syst. Appl. **85**, 134–145 (2017). https://doi.org/10.1016/j.eswa.2017.05.035
8. Elssied, N.O.F., Ibrahim, O., Osman, A.H.: Enhancement of spam detection mechanism based on hybrid k-mean clustering and support vector machine. Soft Comput. **19**(11), 3237–3248 (2015). https://doi.org/10.1007/s00500-014-1479-2
9. Jeyasingh, S., Veluchamy, M.: Modified bat algorithm for feature selection with the wisconsin diagnosis breast cancer (WDBC) dataset. Asian Pac. J. Cancer Prev. **18**(5), 1257–1264 (2017). https://doi.org/10.22034/apjcp.2017.18.5.1257 https:// doi.org/10.22034/apjcp.2017.18.5.1257
10. Lee, Z., Lu, T., Huang, H.: A novel algorithm applied to filter spam e-mails for iPhone. Vietnam. J. Comput. Sci. **2**(3), 143–148 (2015). https://doi.org/10.1007/ s40595-015-0039-8
11. Li, Y., Zhang, L., Xu, Y., Yao, Y., Lau, R.Y., Wu, Y.: Enhancing binary classification by modeling uncertain boundary in three-way decisions. IEEE Trans. Knowl. Data Eng. **29**(7), 1438–1451 (2017). https://doi.org/10.1109/TKDE.2017.2681671
12. Ma, X., Yao, Y.: Three-way decision perspectives on class-specific attribute reducts. Inf. Sci. **450**, 227–245 (2018). https://doi.org/10.1016/j.ins.2018.03.049
13. Nilashi, M., Ibrahim, O., Ahmadi, H., Shahmoradi, L.: A knowledge-based system for breast cancer classification using fuzzy logic method. Telemat. Inform. **34**(4), 133–144 (2017). https://doi.org/10.1016/j.tele.2017.01.007
14. Pritom, A.I., Munshi, R., Sabab, S.A., Shihab, S.: Predicting breast cancer recurrence using effective classification and feature selection technique (2016)
15. Sartakhti, J.S., Afrabandpey, H., Saraee, M.: Simulated annealing least squares twin support vector machine (SA-LSTSVM) for pattern classification. Soft Comput. **21**(15), 4361–4373 (2017). https://doi.org/10.1007/s00500-016-2067-4
16. Surya Narayana, G., Vasumathi, D.: An attributes similarity-based k-medoids clustering technique in data mining. Arab. J. Sci. Eng. **43**(8), 3979–3992 (2018). https://doi.org/10.1007/s13369-017-2761-2

17. Wang, P., Shi, H., Yang, X., Mi, J.: Three-way k-means: integrating k-means and three-way decision. Int. J. Mach. Learn. Cybern. **10**(10), 2767–2777 (2019). https://doi.org/10.1007/s13042-018-0901-y

18. Wu, X., Fan, W., Peng, J., Zhang, K., Yu, Y.: Iterative sampling based frequent itemset mining for big data. Int. J. Mach. Learn. Cybern. **6**(6), 875–882 (2015). https://doi.org/10.1007/s13042-015-0345-6

19. Zainudin, M., et al.: Feature selection optimization using hybrid relief-f with self-adaptive differential evolution (2017). https://doi.org/10.22266/ijies2017.0430.03

An Opinion Summarization-Evaluation System Based on Pre-trained Models

Han Jiang, Yubin Wang, Songhao Lv, and Zhihua Wei$^{(\boxtimes)}$

Department of Computer Science and Technology, College of Electronics and Information Engineering, Tongji University, Shanghai, China
{1853290,1851731,1852635,zhihua_wei}@tongji.edu.cn

Abstract. As social media appeal more frequently used, the task of extracting the mainstream opinions of the discussions arising from the media, i. e. opinion summarization, has drawn considerable attention. This paper proposes an opinion summarization-evaluation system containing a pipeline and an evaluation module for the task. In our algorithm, the state-of-the-art pre-trained model BERT is fine-tuned for the subjectivity analysis, and the advanced pre-trained models are combined with traditional data mining algorithms to gain the mainstreams. For evaluation, a set of hierarchical metrics is also stated. Experiment result shows that our algorithm produces concise and major opinions. An ablation study is also conducted to prove that each part of the pipeline takes effect significantly.

Keywords: Opinion summarization · Subjectivity analysis · Pre-trained model · Evaluation · Hierarchical metrics

1 Introduction

In the post-pandemic era, social media like webinars, message boards, micro blogs, etc., have been increasingly spotlighted and used. Consequently, a special class of data, discussion, is mushrooming all over the Internet. Compared with other textual data, discussion has features as follows: (1) Single topic & multiple opinions; (2) Numerous participants & big volume; (3) Short lifespan; (4) Low structuredness; (5) Multiform expression. The data shows considerable potential for data mining and natural language processing, especially when real-time public sentiment is in demand.

Given the properties above, we place the emphasis on the angles and sentiments of the opinions in discussion. Hence the general process of opinion summarization is to filter the possible opinions out of a discussion, then refine the opinions in terms of their angles and sentiments to obtain the mainstreams.

The reason why opinion summarization requires a two-stage procedure is that a discussion is too extensive to be processed in one go. Meanwhile, speeches in a discussion vary a lot in length, compromising the traditional methods of treating every speech as an equal document. Another trouble is that there are always miscellaneous but semantically identical expressions, which is severely detrimental to generalization.

H. Jiang and Y. Wang—Equal Contributions.

© Springer Nature Switzerland AG 2021
S. Ramanna et al. (Eds.): IJCRS 2021, LNAI 12872, pp. 225–230, 2021.
https://doi.org/10.1007/978-3-030-87334-9_19

To address the aforesaid problems, we propose an opinion summarization-evaluation system including a pipeline and a set of evaluation metrics. For the pipeline, we adopt pre-trained language models to analyze the discussion semantically, and utilize data mining algorithms to accomplish following generalization. For evaluation, we state hierarchical metrics to assess the summary from the relevance, the sentiment orientation, and the one-to-one correspondence between the generated and reference opinions.

In order to implement the algorithm, we also construct two Chinese corpora: a subjectivity analysis corpus for fine-tuning BERT [3], and an opinion summarization corpus for evaluation. An ablation study is subsequently performed by setting several variants of our pipeline, and the result substantiates the effectiveness of our methods.

2 Related Works

There has been a long history of the research on extractive summarization, opinion mining and metrics for these NLP tasks. In recent years, the tasks of extractive summarization are usually fulfilled through neural network modeling, network graph method and data mining. Neural network modeling is the focus of the field [9]. A summary-level framework using SBERT with superior performance was proposed based on this method [8]. Network graph method is a mainstream [9] which stems from a research result: Human language is also a complex network with the characteristic of small world and is scale-free [2]. One of its most representative examples is TextRank [6]. Another important method is data mining. A typical application of this method is clustering. Opinion mining can be divided into three main levels: the text document level, the sentence level and the subject-part level [5]. An important problem in sentence level opinion mining is to classify sentences into subjective ones and objective ones.

Automatic evaluation metrics mainly include BLEU [7], ROUGE [4], and METEOR [1]. BLEU is a similarity evaluation method based on accuracy, which excels on sentences that are well-matched on corpus-level. ROUGE is based on recall, which calculates the co-occurrence probability of n-grams in the candidate sentences and the reference sentences to evaluate the adequacy and fidelity [4]. METEOR is based on single-precision weighted harmonic mean and the recall of single word, and solves the problem of low correlation between BLEU [7] and manual evaluation results [1].

3 An Opinion Summarization-Evaluation Algorithm

In this section, we first introduce our algorithm for extracting the mainstream opinions (see Fig. 1) in Sect. 3.1. In Sect. 3.2, we state our hierarchical evaluation metrics for opinion summarization.

3.1 Subjective Analysis and Opinion Mining

The subjectivity analysis is applied to ensure that the candidate sentences for the final summary are qualified for opinions. With a fine-tuned BERT [3] model, the process is formulated as a binary classification task, where most subjective statements are retained for the following steps and others are removed. Since there are usually extensive collo-quial or objective speeches in a discussion, the process alleviates the problem of data overload as well.

To proceed with the pipeline, we choose *distiluse-base-multilingual-cased-v2* [10] instead of BERT to calculate the semantic representations of the subjective sentences, as BERT is not expert in capturing the semantic meaning of the sentences. Next, the sentences are grouped with spectral clustering algorithm. Spectral clustering relies heavily on the similarity matrix, and the encoder above is verified to work well in extracting semantic information, therefore the two methods are complementary to each other. To balance the integrity and conciseness of the generated summary, we recommend the number of clusters between 3 and 6. Within the interval, we refer to silhouette coefficient, a reasonable and reliable measure to select the optimal clustering result. Moreover, it is necessary to abandon some excessively small clusters.

For each cluster, the vector closest to the geometric center is extracted to be the representation of the cluster, and its corresponding sentence will be the candidate opinion for the summary. Since it may appear colloquial, we just simply remove some irrelevant function words from the sentence to get a mainstream opinion. Finally, all the mainstream opinions acquired constitute the generated summary of our algorithm (Fig. 1).

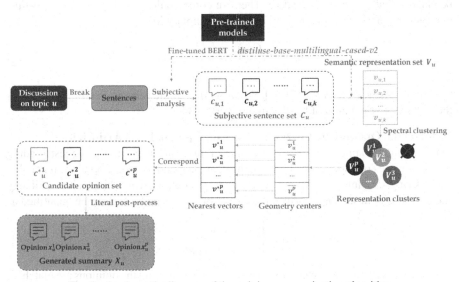

Fig. 1. A schematic diagram of the opinion summarization algorithm

3.2 Hierarchical Metrics

To perfect the opinion summarization algorithm, we state a set of hierarchical metrics, combining automatic and artificial methods to evaluate the generate summary from three aspects progressively.

Evaluate the Relevance on Summary Level Automatically. While assessing a summary, it is most basic to ensure whether it is relevant to the topic, and whether it involves most significant content of the discussion. When design the metric, we refer to the method of collecting the word pairs between the generated and reference summary in ROUGE [4]. For any discussion on topic u, given the generated summary $X_u = \{x_u^1, x_u^2, \ldots, x_u^p\}$ and the reference summary $Y_u = \{y_u^1, y_u^2, \ldots, y_u^q\}$, the relevance between X_u and Y_u can be defined as

$$Relev_u = \frac{1}{q} \sum_{i=1}^{q} Relev\left(X_u, y_u^i\right) \tag{1}$$

$Relev\left(X_u, y_u^i\right)$ denoting the relevance between X_u and opinion y_u^i, is the average value of the cosine similarity between the terms in y_u^i and their most similar terms in X_u. The cosine similarity is computed in the semantic space induced by the model used while clustering. $Relev_u \in [0, 1]$, and the larger $Relev_u$ implies higher relevance.

Evaluate the Sentiment Orientation on Summary Level Automatically. We   take the evaluation a step further by examining how the emotion tendency of the generated summary match expectations. With fine-tuned BERT [3], opinions in the summaries can be classified as positive or negative. Then we compare the proportions of the positive opinions in generated summary and reference summary like

$$Senti_u = 1 - \text{abc}\left(\frac{\sum_i Count_{pos}\left(x_u^i\right)}{p} - \frac{\sum_i Count_{pos}\left(y_u^i\right)}{q}\right) \tag{2}$$

It is knowable that $Senti_u \in [0, 1]$. When $Senti_u = 1$, the generated summary captures the sentiment orientation of the discussion perfectly.

Evaluate the One-to-One Correspondence on Opinion Level Artificially.
Since automatic approaches may be coarse-grained and inexact, we suggest grading the one-to-one correspondence between the generated and reference opinions manually. Considering an opinion x in X_u and y in Y_u, they can compose a matching pair (x, y) if they show similarity in semantics. Thus, the one-to-one correspondence can be quantified as

$$Corre_u = \min\left\{\frac{\theta_m \sum_{x \in X_u, y \in Y_u} Scr_u(x, y)}{\sqrt{pq}}\right\} \tag{3}$$

Here θ_m is a bonus parameter to improve the score when all the opinions are matched. $Scr_u(x, y) \in [0, 1]$ is determined by the graders, and a higher value implies higher similarity.

4   Experiments and Analysis

4.1   Experimental Settings

With discussions from a large-scale Q&A forum named Zhihu, we build two Chinese corpora. To support the subjectivity analysis, we provide a corpus containing 7500 sentences from 15 discussions, annotated by three annotators as subjective or not, for fine-tuning the BERT model. For the sake of evaluation, we generate reference summaries for another 45 discussions to construct an opinion summarization dataset. Considering a discussion appears as one topic or question with numerous answers in Zhihu, each summary is made up of several thesis statements of the most popular answers.

4.2 Experiment Results and Analysis

With the hierarchical evaluation metrics in the proposal, we assess our algorithm on the opinion summarization corpus. An ablation study is performed over our pipeline and its two variants, using the same corpus and metrics. The results listed in Table 1 illustrate how the critical modules mentioned above take effects.

Table 1. Results of ablation study

Pipeline	*Relev*	*Senti*	*Corre*
SA & CE (Ours)	0.715	**0.730**	**0.428**
No SA & CE	**0.718**	0.702	0.257
SA & TextRank	0.685	0.729	0.252

Corresponding to the above two modules of our pipeline, here **SA** represents the subjectivity analysis, and **CE** stands for center extraction, i. e. the method of extracting the mainstream opinions from the centers of the clusters. The results prove that the algorithm brings fantastic sentiment orientation and one-to-one correspondence, also acceptable relevance.

First, we demonstrate the importance of the subjectivity analysis. In Table 1, **No SA & CE** gets a markedly low *Senti* score, which indicates that removing the subjectivity analysis critically hurts performance in capturing the sentiment orientation. Without the subjectivity analysis, the algorithm tends to be misled by salient but overwhelming contents and produce summaries with biased emotional perception.

Second, we observe the necessity of center extraction. As listed, **SA & TextRank** is defeated by our **SA & CE** with especially large drops on the *Relev* and *Corre* score. A noteworthy fact is that the center extraction gets the central sentence of each viewpoint cluster, this way the mainstreams are guaranteed to be juxtaposed, and semantic overlaps between opinions extracted would be minimized.

Besides, note that the *Corre* score of our pipeline is prominently higher than the other two. That is because the two variants can be misled by crucial and overlapping contents easily, and the rule we use to grading the correspondence severely punishes overlaps. Maybe there are still some unknown benefits brought by our algorithm.

5 Conclusion and Future Works

The contributions of our paper are as follows:

First, we observe a class of recently prevalent textual data, namely discussion, analyze its features and value, and conceptualize the task of opinion summarization.

Second, we propose an opinion summarization-evaluation system with two matching Chinese corpora, and accomplish the task well.

Third, we conduct an extra ablation study to substantiate the effectiveness of our peculiar methods, the subjectivity analysis and the center extraction.

Our opinion summarization-evaluation system paves a new way for automatic summarization, while it still requires further research. In our algorithm, a more flexible measure for clustering result shall be introduced to replace the silhouette coefficient, and more semantic information should be taken into account when locate the centers of the clusters. Also, we will try migrate our system to other languages by adjusting the pre-trained model, the corpora, and some strategies accordingly.

Acknowledgement. The work is partially supported by the National Nature Science Foundation of China (Grant No. 61976160, 61906137) and the Technology Research Plan Project of Ministry of Public and Security (Grant No. 2020JSYJD01).

References

1. Banerjee, S., Lavie, A.: METEOR: an automatic metric for MT evaluation with improved correlation with human judgments. In: Proceedings of the acl Workshop on Intrinsic and Extrinsic Evaluation Measures for Machine Translation and/or Summarization, pp. 65–72 (2005)
2. Cancho, R.F., Solé, R.: The small world of human language. In: Proceedings of the Royal Society of London. Series B: Biological Sciences, pp. 2261–2265. The Royal Society (2001)
3. Devlin, J., Chang, M.W., Lee, K., Toutanova, K.: BERT: Pre-training of Deep Bidirectional Transformers for Language Understanding (2018). https://arxiv.org/abs/1810.04805
4. Lin, C.Y.: Rouge: a package for automatic evaluation of summaries. In: Text Summarization Branches out, pp. 74–81 (2004)
5. Liu, B.: Sentiment analysis and opinion mining. Morgan and Claypool Publishers (2012)
6. Mihalcea, R., Tarau, P.: Textrank: bringing order into text. In: Proceedings of the 2004 Conference on Empirical Methods in Natural Language Processing, pp. 404–411 (2004)
7. Papineni, K., Roukos, S., Ward, T., Zhu, W.J.: Bleu: a method for automatic evaluation of machine translation. In: Proceedings of the 40th annual meeting of the Association for Computational Linguistics, pp. 311–318 (2002)
8. Reimers, N., Gurevych, I.: Sentence-BERT: Sentence embeddings using siamese BERT-Networks. In: Proceedings of the 2019 Conference on Empirical Methods in Natural Language Processing (2019)
9. Zhao, J.S., Zhu, Q.M., Zhou, G.D., Zhang, L.: Review of research in automatic keyword extraction. J. Softw. (2017)
10. Zhong, M., Liu, P., Chen, Y., Wang, D., Qiu, X., Huang, X.: Extractive Summarization as text matching. In: ACL 2020 (2020)

Fuzzy-Rough Nearest Neighbour Approaches for Emotion Detection in Tweets

Olha Kaminska[1]([✉])[iD], Chris Cornelis[1][iD], and Veronique Hoste[2][iD]

[1] Computational Web Intelligence Department of Applied Mathematics,
Computer Science and Statistics, Ghent University, Ghent, Belgium
{Olha.Kaminska,Chris.Cornelis,Veronique.Hoste}@UGent.be
[2] LT3 Language and Translation Technology Team, Ghent University,
Ghent, Belgium

Abstract. Social media are an essential source of meaningful data that can be used in different tasks such as sentiment analysis and emotion recognition. Mostly, these tasks are solved with deep learning methods. Due to the fuzzy nature of textual data, we consider using classification methods based on fuzzy rough sets.

Specifically, we develop an approach for the SemEval-2018 emotion detection task, based on the fuzzy rough nearest neighbour (FRNN) classifier enhanced with ordered weighted average (OWA) operators. We use tuned ensembles of FRNN–OWA models based on different text embedding methods. Our results are competitive with the best SemEval solutions based on more complicated deep learning methods.

Keywords: Fuzzy-rough nearest neighbour approach · Emotion detection · Natural language processing

1 Introduction

Over the past decades, the increasing availability of digital text material has allowed the domain of Natural Language Processing (NLP) to make significant headway in a wide number of applications, such as for example in the detection of hate speech [13] or emotion detection [19].

In this paper, we report on our work on emotion detection for the SemEval-2018 Task 1 EI-oc: Affect in Tweets for English[1] [15]. This task represents a classification problem with tweets labeled with emotion intensity scores from 0 to 3 for four different emotions: anger, sadness, joy, and fear.

We explored this task in our previous work [11], using the weighted k Nearest Neighbours classification approach. We chose this method over popular neural

---

[1] https://competitions.codalab.org/competitions/17751.

This work was supported by the Odysseus programme of the Research Foundation–Flanders (FWO).

© Springer Nature Switzerland AG 2021
S. Ramanna et al. (Eds.): IJCRS 2021, LNAI 12872, pp. 231–246, 2021.
https://doi.org/10.1007/978-3-030-87334-9_20

network based solutions because of its explainability. Explainable models allow to investigate the classification progress and discover new patterns.

Our purpose in this paper is to explore the efficiency of the fuzzy-rough nearest neighbour (FRNN) classifier [10] and its extensions based on ordered weighted average (OWA) operators [3,12] for this task. The motivation behind the usage of FRNN is to investigate the potential of relatively simple and transparent instance-based methods for the emotion detection task, in comparison with the black-box solutions offered by deep learning approaches. While the latter can solve sentiment analysis tasks with remarkable accuracy, they provide very little insight about how they reach their conclusions. This does not mean that we dismiss deep learning technology altogether; indeed, to prepare tweets for classification, we represent them by numerical vectors using some of the most popular current neural network based text embedding models [1,2,4,17]. This strategy should allow us to strike the right balance between interpretability and accuracy of the approach.

The remainder of this paper has the following structure: Sect. 2 contains an overview of related work, focusing on the SemEval-2018 Task 1 winning solutions. Section 3 describes the main steps of our proposal, including data preprocessing and tweet representation and classification, and also recalls the competition's evaluation measures. Section 4 reports on our approach's performance for the training data in different setups, while Sect. 5 evaluates the best approach on the test data. Finally, Sect. 6 provides a discussion of the obtained results and some ideas for further research.

The source code of this paper is available online at the GitHub repository[2].

2   Related Work

We start this section by briefly describing the most successful solutions[3] to the SemEval-2018 shared task. The winning approach [5] used ensembles of XGBoost and Random Forest classification models using tweet embedding vectors, while the second place was taken by [6], who used Long Short Term Memory (LSTM) neural nets with transfer learning. The third place contestants [18] presented a complex ensemble of models with Gated-Recurrent-Units (GRU) and a convolutional neural network (CNN) with the role of an attention mechanism.

As is clear, the best approaches all used deep learning technology in one way or another, thus reflecting the current state-of-the-art and trends in automated text analysis (see e.g. [14] for a comprehensive overview). This tendency is further reinforced by the use of the Pearson Correlation Coefficient (see formula (6) in Sect. 3.5) as the sole evaluation measure for the competition, since this measure lends itself well to NN-based optimization.

To gain more insight into how tweets express different emotions and emotion intensities, instance-based methods may be used that discern tweets based on a

---

[2] The source code: https://github.com/olha-kaminska/frnn_emotion_detection.

[3] Competition results: https://competitions.codalab.org/competitions/17751#resu lts.

similarity or distance metric. In particular, we want to explore the use of fuzzy rough set techniques for this purpose. We are not the first to do so: for example, in [21,22], Wang et al. used fuzzy rough set methods to discover emotions and their intensities in multi-label social media textual data.

In this paper, we will use the fuzzy rough nearest neighbour (FRNN) classification algorithm originally proposed in [10], and refined later with Ordered Weighted Average (OWA) operators [3,12].

3 Methodology

In this section, we describe the key ingredients of our methodology. At the data level, we first discuss the data preprocessing steps and then elaborate on the different text embedding methods we implemented. Furthermore, we introduce the similarity relation we used to compare the tweet vectors and discuss the two main setups we used for classification, i.e., FRNN-OWA used as a standalone classifier and within an ensemble. We end the section by a description of the used evaluation method.

The task we consider is the emotion intensity ordinal classification task (EI-oc, [15]) for the emotions anger, fear, joy, and sadness. The aim is to classify an English tweet into one of four ordinal classes. Each class represents a level of emotion intensity: 0 stands for "no emotion can be inferred", 1 corresponds to "low amount of emotion can be inferred", 2 means "moderate amount of emotion can be inferred", and 3 - "high amount of emotion can be inferred". For each emotion, the training, development, and test datasets were provided in the framework of the SemEval-2018 competition. We merge training and development datasets for training our model.

3.1 Data Cleaning

Before the embedding process, we may apply some operations to clean the tweets. In the first, general step, we delete account tags starting with '@', extra white spaces, newline symbols ('\n'), all numbers, and punctuation marks. We do not delete hashtags because they can be a source of useful information [16], so we just remove '#' symbols. Also, we replace '&' with the word 'and' and replace emojis with their textual descriptions. We save emojis as they can be helpful for precision improvement [23]. Emojis are represented either by punctuation marks and/or a combination of letters, or as a small image decoded with Unicode. For the first type, we used their descriptions from the list of emoticons on Wikipedia[4] for replacement. For the second type, we use the Python package "*emoji*"[5] for transformation.

The second step of tweet preprocessing is stop-word removal. For this purpose, the stop-words list from the NLTK package[6] is used.

---

[4] https://en.wikipedia.org/wiki/List_of_emoticons.
[5] https://pypi.org/project/emoji/.
[6] https://gist.github.com/sebleier/554280.

Table 1. Characteristics of the combined train and development data for the four emotion datasets.

Characteristic	Anger	Joy	Sadness	Fear
IR	1.677	1.47	2.2	8.04
Size of the smallest class	376	410	348	217
Number of instances	2,089	1,906	1,930	2,641

Both general preprocessing and stop-word removal are optional for our purposes: during the experimental stage, we will examine whether they improve classification results or not.

We also explored some important characteristics of the datasets and presented them in Table 1. One of characteristics is the class imbalance. It is quantified by the Imbalance Ratio (IR) which is equal to the ratio of the sizes of the largest and the smallest classes in the dataset.

3.2 Tweet Embedding

We represent each tweet as a vector, or set of vectors, to perform classification. For this purpose, we use the following word embedding techniques:

- Gensim pre-trained Word2Vec[7], which contains a vocabulary with 3 million words and phrases and assigns a 300-dimension vector to each of them, obtained by training on a Google News dataset.
- DeepMoji[8] is a state-of-the-art sentiment embedding model. Millions of tweets with emojis were used to train the model to recognize emotions. DeepMoji provides for each sentence (tweet) a vector of size 2,304 dimensions. The model has implementations for several Python packages, and we used the one on PyTorch, made available by Huggingface[9].
- Universal Sentence Encoder (USE) [2] is a sentence-level embedding method, which means it will create vectors for sentences or tweets as a whole. It was developed by the TensorFlow team[10]. USE provides a 512-dimensional vector for a text paragraph (tweet), and was trained on several data sources for different NLP tasks such as text classification, sentence similarity, etc. The model was trained in two ways, using a deep averaging network (DAN) and a Transformer encoder. We chose the second type of USE after basic experiments for our further experiments.
- Bidirectional Encoder Representations from Transformers (BERT), proposed by Devlin et al. [4]. The Google AI Language Team developed a script[11] that

---

[7] https://drive.google.com/file/d/0B7XkCwpI5KDYNlNUTTlSS21pQmM.

[8] https://deepmoji.mit.edu/.

[9] https://github.com/huggingface/torchMoji.

[10] https://www.tensorflow.org/hub/tutorials/semantic_similarity_with_tf_hub_universal_encoder.

[11] https://github.com/dnanhkhoa/pytorch-pretrained-BERT/blob/master/examples/extract_features.py.

we use to assign pre-computed feature vectors with length 768 from a PyTorch BERT model to all the words of a tweet. If the BERT vocabulary does not contain some word, then during the embedding, this word is split into tokens (for example, if the word *"tokens"* is not in the BERT dictionary, then it can be represented as *"tok"*, *"##en"*, *"##s"*), and a vector is created for each token.

- Sentence-BERT (SBERT) is a tuned and modified BERT model developed by Reimers et al. [17]. The model operates on the sentence level and provides vectors with the same size as the original BERT. SBERT is based on siamese (twin) and triplet network structures, which can processes two sentences (tweets) simultaneously in the same way.
- The Twitter-roBERTa-based model for Emotion Recognition presented by Barbieri et al. [1] provides embeddings on word level similar to the original BERT. We consider one of seven fine-tuned roBERTa-based models trained for different tasks with specific data for each of them. The model we chose was trained for the emotion detection task from the same SemEval competition (E-c) using a different set of tweets [15] with emotions such as anger, joy, sadness, and optimism.

All listed sentence-level embeddings methods are applied to the tweets as a whole, while for the word- and token-level approaches, we calculated a tweet vector by taking its words' or tokens' vectors mean. The experiments were performed for all four emotion datasets and the obtained results are provided in Sect. 4.

3.3 Similarity Relation

To be able to compare tweet vectors, we need an adequate similarity relation. We opted for the cosine metric, given by Formula (1): [7].

$$\cos(A, B) = \frac{A \cdot B}{||A|| \times ||B||}, \tag{1}$$

Here, A and B are elements from the same vector space, $A \cdot B$ is their scalar product, and $||x||$ is the vector norm of element x.

As this metric returns values between -1 (perfectly dissimilar vectors), and 1 (perfectly similar vectors), we rescale them to [0,1] using Formula (2) below, which we will use as our primary similarity relation.

$$cos\_similarity(A, B) = \frac{1 + cos(A, B)}{2}. \tag{2}$$

3.4 Classification Methods

In this section, we first recall the OWA-based Fuzzy Rough Nearest Neighbor (FRNN-OWA) classification method and then explain how to construct ensembles with it to solve the emotion detection task.

FRNN-OWA. The fuzzy rough nearest neighbour (FRNN) method [8–10] is an instance-based classifier that uses the lower (L) and upper (U) approximations from fuzzy rough set theory to make classifications. In order to make the method more robust and noise-tolerant, lower and upper approximations are usually calculated with Ordered Weighted Average (OWA) aggregation operators [3]. The OWA aggregation of a set of values V using weight vector $\overrightarrow{W} = \langle w_1, w_2, ..., w_{|V|} \rangle$, with $(\forall i)(w_i \in [0, 1])$ and $\sum_{i=1}^{|V|} w_i = 1$, is given by Formula (3):

$$OWA_{\overrightarrow{W}}(V) = \sum_{i=1}^{|V|} (w_i v_{(i)}), \tag{3}$$

where $v_{(i)}$ is the ith largest element in V.

In this paper, we used the following types of OWA operators[12]:

- Strict weights, which contain only one non-zero position that does not depend on the actual values that are being aggregated:
  $\overrightarrow{W}_L^{strict} = \langle 0, 0, ..., 1 \rangle$ $\overrightarrow{W}_U^{strict} = \langle 1, 0, ..., 0 \rangle$. Strict weights correspond to the original FRNN proposal from [8].
- Exponential weights (Exp), which are drawn from an exponential function with base 2:
  $\overrightarrow{W}_L^{exp} = \langle \frac{1}{2^p-1}, \frac{2}{2^p-1}, ..., \frac{2^{p-2}}{2^p-1}, \frac{2^{p-1}}{2^p-1} \rangle$
  $\overrightarrow{W}_U^{exp} = \langle \frac{2^{p-1}}{2^p-1}, \frac{2^{p-2}}{2^p-1}, ..., \frac{2}{2^p-1}, \frac{1}{2^p-1} \rangle$.
- Additive weights (Add), which model linearly decreasing or increasing weights:
  $\overrightarrow{W}_L^{add} = \langle \frac{2}{p(p+1)}, \frac{4}{p(p+1)}, ..., \frac{2(p-1)}{p(p+1)}, \frac{2}{p+1} \rangle$
  $\overrightarrow{W}_U^{add} = \langle \frac{2}{p+1}, \frac{2(p-1)}{p(p+1)}, ..., \frac{4}{p(p+1)}, \frac{2}{p(p+1)} \rangle$.
- Inverse additive weights (Invadd) are also based on the ratio between consecutive elements in the weight vectors:
  $\overrightarrow{W}_L^{invadd} = \langle \frac{1}{pD_p}, \frac{1}{(p-1)D_p}, ..., \frac{1}{2D_p}, \frac{1}{D_p} \rangle$
  $\overrightarrow{W}_U^{invadd} = \langle \frac{1}{D_p}, \frac{1}{2D_p}, ..., \frac{1}{(p-1)D_p}, \frac{1}{pD_p} \rangle$,
  with $D_p = \sum_{i=1}^{p} \frac{1}{p}$, the pth harmonic number.
- Mean weights, which weight each element equally:
  $\overrightarrow{W}_L^{mean} = \overrightarrow{W}_U^{mean} = \langle \frac{1}{p}, \frac{1}{p}, ..., \frac{1}{p} \rangle$

We used the implementation of the FRNN-OWA classifier [12] provided by the fuzzy-rough-learn package[13]. To classify a test instance y, the method calculates its membership to the lower and upper approximation of each decision class C:

$$\underline{C}(y) = OWA_{\overrightarrow{W}_L} \{1 - R(x, y) \mid x \in X \setminus C\} \tag{4}$$

$$\overline{C}(y) = OWA_{\overrightarrow{W}_U} \{R(x, y) \mid x \in C\} \tag{5}$$

The algorithm then assigns y to the class C for which $\underline{C}(y) + \overline{C}(y)$ is highest.

---

[12] p refers to the number of elements in the OWA weight vector.
[13] https://github.com/oulenz/fuzzy-rough-learn.

Usually, the computation in Formula (4) is restricted to the k nearest neighbours of y from the training data belonging to classes other than C, while in Formula (5) we consider only y's k nearest neighbours from class C. There is no universal rule to determine the value of the parameter k. As a default, we can put $k = \frac{\sqrt{N}}{2}$, where N is the size of the dataset. In order to examine the influence of k on the obtained classification results, we will use different k values for the best-performing approaches in our experiments for each dataset.

We performed experiments for each emotion dataset with different OWA types for lower and upper approximations with various numbers of k.

Classifier Ensembles. We used the FRNN-OWA method both as a standalone method and as part of a classification ensemble. For this purpose, a separate model was trained for every choice of tweet embedding. Each model was based on each dataset's best setup and embedding (choice of tweet preprocessing, OWA types, and the number of neighbours k).

To determine the test label, we use a weighted voting function on the different outputs of our models. As possible voting functions v, we considered average, median, maximum, minimum, and majority. In the voting function the models' outputs receive some weights.

The full architecture of our ensemble approach is presented in Fig. 1. In Sect. 4, we perform several experiments to detect the most accurate ensemble setup, including the best voting function, the most suitable values of weights $\vec{E}$, and the proper combination of models (feature vectors).

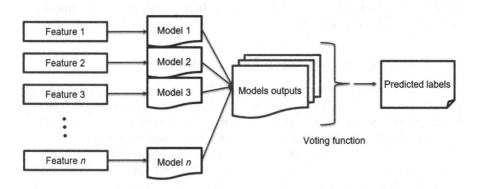

Fig. 1. Scheme of the ensemble architecture.

3.5   Evaluation Method

We used 5-fold cross-validation to evaluate the results of our approaches. As evaluation measure the Pearson Correlation Coefficient (PCC) (6) was chosen, as it was also the evaluation measure used for the competition.

Assuming that y is the vector of predicted values and x is the vector of correct values, we compute

$$PCC = \frac{\sum_i (x_i - \bar{x})(y_i - \bar{y})}{\sqrt{\sum_i (x_i - \bar{x})^2 \sum_i (y_i - \bar{y})^2}}, \tag{6}$$

where x_i and y_i present the ith components of vectors x and y respectively and $\bar{x}$ and $\bar{y}$ are their means.

The PCC measure provides a value between -1, which corresponds to a total negative linear correlation, and 1 - a total positive linear correlation, where 0 represents no linear correlation. Hence, the best classification model should provide the highest PCC.

After submitting the obtained test labels to the competition web page, the PCC scores for each emotion dataset were averaged.

4   Experiments

In this section, we present our results for the classification approaches discussed in the previous section. Initially, we explore the best individual FRNN-OWA setup, including the preprocessing options, the chosen tweet embedding, the OWA types and the number of neighbours k. In a second set of experiments, we evaluate various ensemble approaches.

4.1   Detecting the Best Setup for Embeddings

We performed experiments with different OWA types to detect the best setup for each dataset. We also investigated for each dataset whether it was beneficial to apply tweet preprocessing and stop-words cleaning. Finally, we explored the most suitable k value for each embedding for each emotion dataset.

First, the pipeline was performed for each embedding and emotion dataset after general preprocessing with the same OWA type for upper and lower approximations (strict, additive, exponential, and mean) and a different number of neighbors (from 5 to 23 with step 2). As the results showed, the best results were obtained with the additive ("add") OWA type for most embeddings, so we chose them for the further experiments. The best results for each dataset and each embedding are presented in Table 2.

Next, we calculated the PCC score for all embeddings and datasets with the best add OWA types, while varying the preparation level of the tweets: raw tweets (no preparation at all), standard preprocessing (text transformation steps mentioned in Sect. 3.1, excluding stop-words removal), and stop-words cleaning (the same as above, but including stop-words removal). To examine which setup

Table 2. The best setup for each combination of dataset and embedding.

Setup	Anger	Joy	Sadness	Fear
roBERTa-based				
Tweets preprocessing	Yes	Yes	Yes	Yes
Stop-words cleaning	No	No	No	No
Number of neighbors	19	9	23	9
PCC	0.6779	0.6956	0.7062	0.6031
DeepMoji				
Tweets preprocessing	No	No	No	No
Stop-words cleaning	No	No	No	No
Number of neighbors	23	19	23	21
PCC	0.5853	0.6520	0.6380	0.5745
BERT				
Tweets preprocessing	No	No	No	No
Stop-words cleaning	No	No	No	No
Number of neighbors	19	17	23	7
PCC	0.4492	0.5374	0.4391	0.4500
SBERT				
Tweets preprocessing	Yes	Yes	Yes	Yes
Stop-words cleaning	No	No	No	No
Number of neighbors	19	15	23	11
PCC	0.5016	0.5660	0.5655	0.5192
USE				
Tweets preprocessing	Yes	Yes	Yes	Yes
Stop-words cleaning	No	No	No	No
Number of neighbors	23	23	23	21
PCC	0.5054	0.5693	0.5961	0.5764
Word2Vec				
Tweets preprocessing	Yes	Yes	Yes	Yes
Stop-words cleaning	Yes	Yes	Yes	Yes
The number of neighbors	21	23	23	7
PCC	0.5009	0.5099	0.5048	0.4496

works better, we performed a statistical analysis of results with a two-sided t-test (we assume the statistical significance of the p-value on the 0.05 level). For calculation, the Python's package '*stats*' was used. Results are presented in Table 2. As we can see, some embeddings do not require any preprocessing at all, like DeepMoji and BERT. The standard preprocessing showed an improvement for other methods, and only Word2Vec seems to benefit from an additional stop-words removal step.

For most of the experiments, the obtained p-values are below the chosen threshold of 0.05. For some cases, the p-value was above the threshold, which means no significant difference exists between the compared options. In this situation, for the dataset, we chose the option that works better for other datasets. For example, for the BERT embedding, the joy dataset was the only one with the $p > 0.05$, when for anger, sadness, and fear p is below 0.05 (so cleaned tweets performed better). Hence, we will use cleaned tweets for joy because this is the best setup for the other emotions.

Finally, for each embedding and dataset, we examined the PCC of the best setup (the combination of the best OWA types and the most efficient text preparation) for the different number of neighbours. The highest PCC scores and the proper k values are also listed in Table 2.

The best setup for each combination of embedding and dataset was used in further experiments. We also can draw several intermediate conclusions. Remarkably, the highest PCC scores for all datasets among all embeddings were provided by the roBERTa-based model, which does not come as a surprise,since this model was fine-tuned on similar data and its performance is in line with earlier results for similar classification tasks [1]. The second-best approach was DeepMoji, while BERT and Word2Vec provided the lowest scores. Also, we can see that the PCC scores for the fear dataset are often the lowest among the other emotions, which might probably be due to the fact that the fear dataset is the most unbalanced dataset. Similarly, the joy dataset shows high results, as the most balanced one.

4.2   Ensembles

To improve the PCC scores provided by individual embeddings, we also investigated an ensemble approach. To determine the best setup of the ensemble, we tuned several parameters, i.e., the voting function, the models' weights and the selection of the strongest embedding models.

First, we compared different voting functions for all datasets: majority, mean, rounded mean, median, maximum, minimum. We note that for the majority voting function implementation we use the $mode()$ function from the Python package *stats*. It chooses the most frequent label prediction, and in case of ties, this function returns the lowest value.

Noteworthy is that some voting functions provide a float value between 0 and 3 instead of the required intensity labels 0, 1, 2, or 3. This was not a problem, though, during training because our labels are not different classes, but ordinal intensity labels. At testing time, the obtained values were rounded to submit our predictions. The general setup for comparing the voting functions was based on the six previously discussed models (one for each embedding method) with the parameters determined in Table 2, where each predicted output has the same weight equal to 1. The results are presented in Table 3. As we can see, the mean voting function consistently provided the best results for all datasets, while median performs second best. Although the rounding of the mean's output decreases the PCC results, it remains the best voting function. So, for further experiments, we will use the average as a voting function.

Table 3. Results for ensembles with different voting functions for all datasets.

Voting function	Anger	Joy	Sadness	Fear
Majority	0.6141	0.6669	0.6591	0.5665
Mean	**0.6933**	**0.7501**	**0.7456**	**0.6723**
Rounded mean	0.6485	0.7126	0.7152	0.6448
Median	0.6414	0.7150	0.7079	0.6050
Maximum	0.4856	0.4668	0.5625	0.5640
Minimum	0.5959	0.6411	0.5016	0.3885

Next, we check the use of weights assigned to the models' outputs in the voting function. In particular, we use confidence scores (CS) to give more weight to the better models.

A confidence score is a float value, usually between 0 and 1, provided by a classification model for each prediction class. This value illustrates the accuracy of the model's prediction for a particular class. For FRNN-OWA, the models return four scores (one for each class). They are the mean membership degrees in the upper and lower approximations.

To get confidence scores, we divide each score by the sum of all four class scores. In this way, we obtain the values $C_{i,j}$: four scores (one per class label $i, i = 0, ..., 3$) for every model j ($j = 1, ..., 6$). We use the confidence scores in the following ways:

- Majority voting. The most intuitive approach, where we take as a prediction the label with the highest sum of confidence scores.
- Weighted average (WA). As we saw above, the best voting function is the mean, so we will upgrade it with confidence scores as weights to calculate the prediction label as a weighted average of labels. The output could be a float number, so we also check the rounded option.

Experiments were performed with all six embedding models. Results are provided in the upper half of Table 4.

As we can see, weighted average with confidence scores performed the best. Predictably, rounding decreased the weighted average's score, and it is similar to the results provided by a majority of confidence scores. If we compare them with the values in Table 3, considering the mandatory rounding step, we can conclude that these approaches with confidence scores do not increase PCC scores.

We analyzed the obtained confidence scores and noticed that they are close to each other, approximately, in the range from 0.4 to 0.6. Our hypothesis is that since we have a high dimensional task like ours, the confidence scores will

Table 4. Results for ensembles with different usage of confidence scores for all datasets.

Approach	Anger	Joy	Sadness	Fear
Original confidence scores				
Majority voting	0.6351	0.7082	0.7016	0.5700
Weighted average	0.7025	0.7424	0.7333	0.6044
WA rounded	0.6302	0.6731	0.6962	0.5549
Rescaled confidence scores				
Weighted average	**0.7187**	**0.7781**	**0.7630**	**0.6763**
	(α =0.0420)	(α =0.0360)	(α =0.0400)	(α =0.0460)
WA rounded	0.6432	0.7512	0.7455	0.6430
	(α =0.0420)	(α =0.0320)	(α =0.0320)	(α =0.0460)

be close to 0.5: the upper approximation memberships will be close to 1 and the lower ones to 0, resulting in similar values for each class. In other words, the contribution of such a classifier is low.

To mend this issue, we perform rescaling of the original membership scores in order to increase the differences among them. For this purpose, we subtract the mean 0.5 from each score $C_{i,j}$ and divide the result by a small value α ($0 < \alpha < 1$). Next, for each class i we compute the sum of the scores for each model. Since the obtained values may be negative, we use the softmax transformation to turn them into probabilities. The steps of this rescaling process are summarized in Formula (7):

$$C_i = \frac{\exp(\sum_j (C_{i,j} - 0.5)/\alpha)}{\sum_k \exp(\sum_j (C_{k,j} - 0.5)/\alpha)}, \qquad (7)$$

where α is a parameter to tune. To detect the best value of α for each dataset, we performed a grid search, calculating PCC scores for different α values to choose the one that provides the biggest PCC.

Finally, to calculate the predicted label, we apply the weighted average on classes, where weights are calculated probabilities. Results of this approach with the best α for each dataset are provided in the lower half of Table 4.

Compared with the original confidence scores and values from Table 3, scaled scores performed better for each dataset for both average and rounded average. Hence, we will use scaled confidence scores as models' output weights in the following experiments.

The last step of ensemble tuning is to determine the most accurate set of models in the ensemble. The idea behind this is to see how the PCC score will change depending on the models (embeddings) that we are using in the ensemble to answer the question: is it possible to improve the score by rejecting the weak models' results.

For this purpose, we used grid search, where the PCC score was calculated for each subset of all six models (features) and compared. The predicted label was calculated using a rounded average function with weights equal to the scaled confidence scores. We used a rounded average since it returns integers, so we can use them to submit to the competition web-page. In this way, we detected the best setup for each emotion dataset. The results for cross-validation evaluation are presented in Table 5.

As we can see, all datasets have in common the same features such as roBERTa, DeepMoji, and USE models (we denote them with "r/D/U"). Another one or two features are different for each dataset. We can mainly see that more features provide better results, but the weak models' pruning also takes place.

In the end, we could obtain the best ensemble setup with the required parameters for each emotion dataset.

5   Results on the Test Data

From Sect. 4 we obtained the best setup for each dataset: an ensemble of several models based on different features with proper text preprocessing, k value, and additive lower and upper OWA types for each. The predicted test label is calculated as the mean of the models' outputs with scaled confidence scores as weights.

To measure the best ensemble's effectiveness, we evaluate it on the test data. We calculate PCC values for each emotion dataset and average the results, as was done by the competition organizers. As the output of the ensemble's mean voting function, obtained predictions are in float format, so to satisfy the competition's submitting format, they were rounded to the nearest integer value. The obtained results are presented in Table 5, where we provided results for the combined training and development data to compare them.

Table 5. The best approach on the cross-validation and test data for all datasets.

Dataset	Models	Training and development data	Test data
Anger	r/D/U, Word2Vec, BERT	0.7241	0.6388
Joy	r/D/U, SBERT, BERT	0.7788	0.7115
Sadness	r/D/U, SBERT	0.7719	0.6967
Fear	r/D/U, Word2Vec, SBERT	0.6930	0.5705
Averaged scores		0.7419	0.6544

As we can see from Table 5, results for the test data are predictably worse than those for the combined training and development datasets. The PCC scores for sadness and joy datasets are higher than for anger, and fear, as usual, has lower results.

We submitted the predicted labels for the test data in the required format to the competition webpage[14]. After submission, we took the second place in the competition leader board with PCC = 0.654.

6   Conclusion and Future Work

In this paper, we designed a weighted ensemble of FRNN-OWA classifiers to tackle the emotion detection task. Our approach uses several embeddings, which are mostly sentiment-oriented and applied at sentence-level. We demonstrated that our method, despite its simple design, is competitive to the competition's winning approaches, which are all black-boxes.

As a possible improvement, we may consider additional text preparation steps, for example, bigger weights for hashtags and emojis or exclamation mark usage, before the embedding step.

Finally, we hypothesize that the lower PCC scores for the fear dataset could be related to the dataset's imbalance. As a possible approach to solve this issue, we may use specific classification machine learning methods for imbalanced data. For example, in paper [20], several fuzzy rough set theory methods are described specifically targeting imbalanced data sets.

References

1. Barbieri, F., Camacho-Collados, J., Espinosa Anke, L., Neves, L.: TweetEval: Unified benchmark and comparative evaluation for tweet classification. In: Findings of the Association for Computational Linguistics: EMNLP 2020, pp. 1644–1650. Association for Computational Linguistics, Online (2020). https://doi.org/10.18653/v1/2020.findings-emnlp.148. https://www.aclweb.org/anthology/2020.findings-emnlp.148
2. Cer, D., et al.: Universal sentence encoder for English. In: Proceedings of the 2018 Conference on Empirical Methods in Natural Language Processing: System Demonstrations, pp. 169–174. Association for Computational Linguistics (2018). https://doi.org/10.18653/v1/D18-2029. https://www.aclweb.org/anthology/D18-2029
3. Cornelis, C., Verbiest, N., Jensen, R.: Ordered weighted average based fuzzy rough sets. In: Yu, J., Greco, S., Lingras, P., Wang, G., Skowron, A. (eds.) RSKT 2010. LNCS (LNAI), vol. 6401, pp. 78–85. Springer, Heidelberg (2010). https://doi.org/10.1007/978-3-642-16248-0_16
4. Devlin, J., Chang, M.W., Lee, K., Toutanova, K.: Bert: pre-training of deep bidirectional transformers for language understanding. In: Proceedings of the 2019 Conference of the North American Chapter of the Association for Computational Linguistics: Human Language Technologies, vol. 1 (Long and Short Papers), pp. 4171–4186 (2019)

---

[14] https://competitions.codalab.org/competitions/17751#learn_the_details-evaluation.

5. Duppada, V., Jain, R., Hiray, S.: SeerNet at SemEval-2018 task 1: domain adaptation for affect in tweets. In: Proceedings of The 12th International Workshop on Semantic Evaluation, pp. 18–23. Association for Computational Linguistics, New Orleans (2018). https://doi.org/10.18653/v1/S18-1002. https://www.aclweb.org/anthology/S18-1002

6. Gee, G., Wang, E.: psyML at semeval-2018 task 1: transfer learning for sentiment and emotion analysis. In: Proceedings of The 12th International Workshop on Semantic Evaluation, pp. 369–376 (2018)

7. Huang, A.: Similarity measures for text document clustering. In: Proceedings of the Sixth New Zealand Computer Science Research Student Conference (NZCSRSC2008), Christchurch, New Zealand, vol. 4, pp. 9–56 (2008)

8. Jensen, R., Cornelis, C.: A new approach to fuzzy-rough nearest neighbour classification. In: Chan, C.-C., Grzymala-Busse, J.W., Ziarko, W.P. (eds.) RSCTC 2008. LNCS (LNAI), vol. 5306, pp. 310–319. Springer, Heidelberg (2008). https://doi.org/10.1007/978-3-540-88425-5_32

9. Jensen, R., Cornelis, C.: Fuzzy-rough nearest neighbour classification. In: Peters, J.F., Skowron, A., Chan, C.-C., Grzymala-Busse, J.W., Ziarko, W.P. (eds.) Transactions on Rough Sets XIII. LNCS, vol. 6499, pp. 56–72. Springer, Heidelberg (2011). https://doi.org/10.1007/978-3-642-18302-7_4

10. Jensen, R., Cornelis, C.: Fuzzy-rough nearest neighbour classification and prediction. Theoret. Comput. Sci. **412**(42), 5871–5884 (2011)

11. Kaminska, O., Cornelis, C., Hoste, V.: Nearest neighbour approaches for emotion detection in tweets. In: Proceedings of the Eleventh Workshop on Computational Approaches to Subjectivity, Sentiment and Social Media Analysis, pp. 203–212. Association for Computational Linguistics, Online (2021). https://www.aclweb.org/anthology/2021.wassa-1.22

12. Lenz, O.U., Peralta, D., Cornelis, C.: Scalable approximate FRNN-OWA classification. IEEE Trans. Fuzzy Syst. **28**(5), 929–938 (2019)

13. MacAvaney, S., Yao, H.R., Yang, E., Russell, K., Goharian, N., Frieder, O.: Hate speech detection: challenges and solutions. PLoS ONE **14**(8), e0221152 (2019)

14. Minaee, S., Kalchbrenner, N., Cambria, E., Nikzad, N., Chenaghlu, M., Gao, J.: Deep learning based text classification: a comprehensive review. arXiv e-prints, pp. arXiv-2004 (2020)

15. Mohammad, S.M., Bravo-Marquez, F., Salameh, M., Kiritchenko, S.: Semeval-2018 Task 1: Affect in tweets. In: Proceedings of International Workshop on Semantic Evaluation (SemEval-2018), New Orleans, LA, USA (2018)

16. Mohammad, S.M., Kiritchenko, S.: Using hashtags to capture fine emotion categories from tweets. Comput. Intell. **31**(2), 301–326 (2015)

17. Reimers, N., Gurevych, I.: Sentence-BERT: sentence embeddings using Siamese BERT-networks. In: Proceedings of the 2019 Conference on Empirical Methods in Natural Language Processing and the 9th International Joint Conference on Natural Language Processing (EMNLP-IJCNLP), pp. 3982–3992. Association for Computational Linguistics, Hong Kong (2019). https://doi.org/10.18653/v1/D19-1410. https://www.aclweb.org/anthology/D19-1410

18. Rozental, A., Fleischer, D.: Amobee at SemEval-2018 task 1: GRU neural network with a CNN attention mechanism for sentiment classification. In: Proceedings of The 12th International Workshop on Semantic Evaluation. pp. 218–225. Association for Computational Linguistics, New Orleans (2018). https://doi.org/10.18653/v1/S18-1033. https://www.aclweb.org/anthology/S18-1033

19. Sailunaz, K., Dhaliwal, M., Rokne, J., Alhajj, R.: Emotion detection from text and speech: a survey. Soc. Netw. Anal. Min. **8**(1), 1–26 (2018). https://doi.org/10.1007/s13278-018-0505-2

20. Vluymans, S.: Dealing with Imbalanced and Weakly Labelled Data in Machine Learning Using Fuzzy and Rough Set Methods. Springer, Heidelberg (2019). https://doi.org/10.1007/978-3-030-04663-7

21. Wang, C., Feng, S., Wang, D., Zhang, Y.: Fuzzy-rough set based multi-labeled emotion intensity analysis for sentence, paragraph and document. In: Li, J., Ji, H., Zhao, D., Feng, Y. (eds.) NLPCC -2015. LNCS (LNAI), vol. 9362, pp. 444–452. Springer, Cham (2015). https://doi.org/10.1007/978-3-319-25207-0_41

22. Wang, C., Wang, D., Feng, S., Zhang, Y.: An approach of fuzzy relation equation and fuzzy-rough set for multi-label emotion intensity analysis. In: Gao, H., Kim, J., Sakurai, Y. (eds.) DASFAA 2016. LNCS, vol. 9645, pp. 65–80. Springer, Cham (2016). https://doi.org/10.1007/978-3-319-32055-7_6

23. Wolny, W.: Emotion analysis of twitter data that use emoticons and emoji ideograms (2016)

Three-Way Decisions Based RNN Models for Sentiment Classification

Yan Ma[1,2], Jingying Yu[2], Bojing Ji[3], Jie Chen[2(✉)] ⓘ, Shu Zhao[2] ⓘ, and Jiajun Chen[1(✉)]

[1] School of Electronic and Information Engineering, West Anhui University,
Lu'an 237012, China
chenjj@wxc.edu.cn
[2] School of Computer Science and Technology, Anhui University, Hefei 230601, China
[3] School of Big Data and Artificial Intelligence, Chizhou University,
Chizhou 247000, China

Abstract. Recurrent neural networks (RNN) has been widely used in sentiment classification. RNN can memorize the previous information and is applied to calculate the current output. For sentiment binary classification, RNN calculates the probabilities and then performs binary classification according to the probability values, and some emotions near the median are forcibly divided. But, it does not consider the existence of some samples that are not very clearly polarized in sentiment binary classification. Three-way decisions theory divides the dataset into three regions, positive region, negative region, boundary region. In the process of training classification, the probabilities of some samples belonging to different categories are very close, and three-way decisions can divide them into the boundary region by setting thresholds. Reasonable processing of the boundary region can get better results for binary classification by adjusting the probability of samples in the boundary region. Therefore, in this paper, we propose three-way decisions based RNN models for sentiment classification. Firstly, we use basic RNN models to classify the data. Secondly, we apply three-way decisions theory to set the thresholds, divide the boundary region based on probability. Finally, the probabilities of samples in the boundary region are adjusted and applied in the next round of training. Experiments on four real datasets show that our proposed models are better than corresponding basic RNN models in terms of classification accuracy.

Keywords: Three-way decisions · Sentiment classification · Recurrent neural networks

1 Introduction

Sentiment classification is an important basic research direction in sentiment analysis [1], which aims to analyze peoples opinions in textual data (such as product reviews, movie reviews, and tweets), and extract their polarity and

S. Ramanna et al. (Eds.): IJCRS 2021, LNAI 12872, pp. 247–258, 2021.
https://doi.org/10.1007/978-3-030-87334-9_21

viewpoint. Sentiment classification research is an important aspect of sentiment analysis research which is a hot topic in the natural language processing field [2]. These methods for sentiment classification are mainly divided into two categories: traditional feature-based methods and machine learning methods [3]. Deep learning methods in machine learning are used to extract reasonable features from text and input them into a classifier model to predict sentiment categories [4].

Sentiment classification can be either a binary (such as positive and negative) or a multi-class (such as positive, angry, sad, fearful, surprised, emotionless and ironical [5]) problem. There are many deep learning frameworks proposed for binary sentiment classification problems, such as feed-forward networks [6], RNNs [7], CNNs [8], capsule networks [9], the attention mechanism [10], transformers [11]. RNN is one of the most classic deep learning models. RNN models view a text as a sequence of words and are intended to capture word dependencies and text structures for sentiment classification [12]. To process sequence data, the output of the nodes in the layer will be re-entered into this layer to realize the learning history and predict the future. The two main improvements of RNN are LSTM [13] (Long Short-Term Memory Network) and GRU [14] (Gated Cycling Unit), both of which add additional function gates to the basic neural unit, to better achieve the processing of long-term memory. In order to make up for the lack of information that can not be encoded from back to front, models based on bidirectional RNN (BRNN) [15], BiLSTM [16] and BiGRU [17,18] are proposed.

Although the RNN models are very suitable for processing context-sensitive serial data [19], they are not good at processing boundary data with insignificant polarity. The task of binary sentiment analysis is to divide the texts into two polarities (such as positive and negative). In the process of training classification, some samples are not very clearly polarized. The probability of some samples belonging to different categories is very close, therefore, how to increase the difference of these samples belonging to different categories is a challenge of the current RNN models.

Three-way decisions can divide samples into three regions when the sample information is fuzzy or insufficient, and we can put it in the boundary region for reprocessing. It is an effective way to solve uncertain decision-making problems [20]. Reasonable processing of the boundary region can get better results for binary classification in the boundary region. In this paper, we propose three-way decisions based RNN models for sentiment classification. Firstly, we use basic RNN models to classify the data. And we apply three-way decisions theory to divide the boundary region based on probability. Finally, we adjust the probabilities of samples in the boundary region and apply the adjusted probabilities in the next round of training. Experiments on four real datasets show that our proposed models are better than corresponding basic RNN models in terms of classification accuracy. Our contributions are as follows:

1. We propose three-way decisions based RNN models for sentiment classification. We use the idea of three-way decisions for dealing with the boundary

region to get better handling of fuzzy data and obtain better binary classification results.

2. We divide the boundary region into the positive region and negative region, according to adjusting the probabilities of the samples in the boundary region. We use two adjustment strategies. The first strategy is to move the probabilities closer to two polarities. The second strategy is to use parameters to control the degree of its approach based on the first strategy. We adjust the probability values in the output of samples in the boundary region to enhance polarity. The adjusted data will help us train the data better.

3. Experiments on four real data sets show that our proposed models are better than corresponding basic RNN models in terms of classification accuracy.

2 Related Work

2.1 RNN Models

Models based on RNN (such as long short term memory [21], etc.), which view a text as a sequence of words and are intended to capture word dependencies and text structures [22], have been used in many fields of natural language processing, such as language models and syntax analysis [23], semantic role labeling [24], semantic representation [25], dialogue [26], machine translation [27] and other tasks. They are all excellent and even become the best methods at present. The basic idea of the bidirectional recurrent neural networks (BRNN) is to propose that each training sequence is two recurrent neural networks (RNN) forward and backward, and these two are connected to an output layer. Cheng et al. [28] augment the LSTM architecture with a memory network in place of a single memory cell to model long-span word relations for machine reading. Zhang et al. [29] proposed a Coordinated CNN-LSTM-Attention (CCLA) model to capture the intrinsic semantic, emotional dependence information and the key part of the emotional expression of the text. Feng et al. [30] developed a Context Attention based Long Short-Term Memory (CA-LSTM) network to incorporate preceding tweets for context-aware sentiment classification. Liu et al. [16] proposed an architecture that contains a bidirectional LSTM (BiLSTM), attention mechanism, and the convolutional layer. Feng et al. [31] established a fine-grained feature extraction model based on BiGRU and attention.

2.2 Three-Way Decisions

Three-way decisions theory [32] divides the problem domain into positive, boundary, and negative regions. When dealing with these three regions, the strategies of acceptance, deferment (or further investigation), and rejection were adopted respectively. Let U denote a finite nonempty set of objects and C denote the subset of U, $C \subseteq U$. And $Pr(C|[x]_A)$ denotes the conditional probability of an object in C provided that the object is in $[x]_A$. We normally use the following formulas to describe the positive, negative, and boundary regions respectively: for $0 \leq \beta < \alpha \leq 1$,

$$POS_{(\alpha,\beta)}(C) = \{x \in U \mid Pr(C|[x]_A) \geq \alpha\} \tag{1}$$

$$BND_{(\alpha,\beta)}(C) = \{x \in U \mid \beta < Pr(C|[x]_A) < \alpha\} \tag{2}$$

$$NEG_{(\alpha,\beta)}(C) = \{x \in U \mid Pr(C|[x]_A) \leq \beta\} \tag{3}$$

Based on the three-way decisions theory [32], some researchers have applied to the NLP field and achieved certain results. The practice has proved that the three-way decisions are practical and effective methods for people to deal with uncertain problems. Zhang et al. [33] applied three-way decisions to sentiment classification for solving sentiment uncertainty. Zhou et al. [34] applied the concept of three-way decisions in the classifier that combines lexicon-based methods and supervised learning methods together. Zhang et al. [35] proposed a three-way enhanced convolutional neural network model for sentence-level sentiment classification. Zhang et al. [36] proposed a cost-sensitive combination technique based on sequential three-way decisions in sentiment classification. Zhu et al. [37] presented a model that integrated three-way decision and Bayesian algorithms to distinguish microblog's subjective sentence.

3   The Proposed Method

We propose three-way decisions based RNN models for sentiment classification. Based on the three-way decisions theory, we divide samples into positive, negative, and boundary regions according to the output of training data. And we adjust the probability values in the output of samples in the boundary region to enhance polarity. The adjusted data will help us train the data better.

3.1   Algorithm

Let S denote the set of textual samples, whose samples $x_i \in S$. Suppose that $Pr(x_{ip})$ denotes the positive probability of x_i, and $Pr(x_{in})$ denotes the negative probability. In the process of training data, the sum of the probability values (both positive and negative) that determine the polarity of a sample is 1. Thus the probabilities of every sample satisfy the following equation:

$$Pr(x_{ip}) + Pr(x_{in}) = 1 \tag{4}$$

We use the following formulas to describe the positive, negative, and boundary regions respectively: for $0 \leq \beta < \alpha \leq 1$,

$$POS_{(\alpha,\beta)}(S) = \{x \in S \mid \alpha \leq Pr(x_{ip}) \leq 1\} \tag{5}$$

$$NEG_{(\alpha,\beta)}(S) = \{x \in S \mid 0 \leq Pr(x_{ip}) \leq \beta\} \tag{6}$$

$$BND_{(\alpha,\beta)}(S) = \{x \in S \mid \beta < Pr(x_{ip}) < \alpha\} \tag{7}$$

We view 0.5 as the center and divide the samples with probability values within a certain range into the boundary region. We set a parameter G to control the thresholds α and β to divide three regions by the following formulas. According to the value range of α and β, G has to be less than 0.5.

$$\alpha = 0.5 + G$$

$$\beta = 0.5 - G$$

We adjust the probability value of the boundary region, and the probability value of samples beyond this range remains unchanged. During each iteration of model training, the probability values calculated by the RNN models are processed by the adjustment strategy based on the three-way decisions theory, and the processed result is brought to the next iteration. The entire learning algorithm is summarized as Algorithm 1.

Algorithm 1. M-3Wi

Require:
    dataset S, with m samples $x_i \in S$, dataset V (for validation), dataset T (for testing),
    batch size y, epoch number e, S_k denotes batch k of dataset S, thresholds α and β
Ensure:
    Accuracy, Loss
1: Initialize trainable_weights
2: **for** $j = 1$ to e **do**
3:    **for** $k = 1$ to $\lceil m/y \rceil$ **do**
4:       $[[Pr(x_{ip}) + Pr(x_{in})]]|x_i \in S_k \Leftarrow$ M-training(S_k)
5:       $[[Pr'(x_{ip}) + Pr'(x_{in})]]|x_i \in BND_{(\alpha,\beta)}(S_k) \Leftarrow$ 3Wi(S_k)
6:       Loss, Accuracy$\Leftarrow$ Testing(S_k)
7:       Update trainable_weights
8:    **end for**
9:    Loss, Accuracy$\Leftarrow$ Testing(V)
10: **end for**
11: Loss, Accuracy$\Leftarrow$ Testing(T)

In Algorithm 1, M-training() denotes the training process of model M which is one of the RNN models. Testing()is the process of calculating the accuracy and loss of a dataset. And $3Wi()$ denotes the adjusting operations of our proposed methods based on three-way decisions, in which i means the adjustment strategies i. In order to be different from the benchmark model M, we name our improved models based on three-way decisions as M-3W models. Due to the use of two adjustment strategies, the improved models are named M-3W1 and M-3W2 according to the strategy. These two adjustment strategies will be introduced in Sect. 3.2.

3.2 Probability Adjustment Strategies

After dividing the boundary region, we need to adjust the probability values of the boundary region samples. The following are the two adjustment strategies proposed in this paper.

3.2.1 Probability Adjustment Strategy 1 (M-3W1)

In this strategy, the probability values of the boundary region samples is adjusted by:

$$Pr'(x_{ip}) = 2Pr(x_{ip}) - 0.5 \tag{8}$$

$$Pr'(x_{in}) = 2Pr(x_{in}) - 0.5 \tag{9}$$

where $Pr'(x_{ip})$ denotes the positive probability adjusted by M-W1 of a boundary region sample x_i, and $Pr'(x_{in})$ denotes the negative probability adjusted by M-W1 of the boundary region sample x_i. Since the probability value $0 \le Pr'(x_{ip}) \le 1$, the value of G have to be set between 0 and 0.25. We compare the sample distribution before and after the strategy adjustment, as shown in Fig. 1.

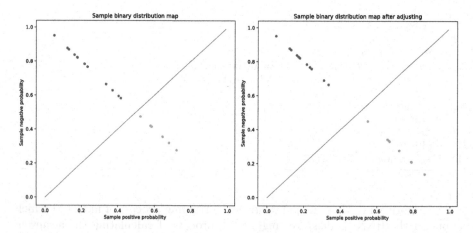

Fig. 1. Comparison chart of sample binary classification before and after adjustment

As can be seen from Fig. 1, the adjusted samples will move towards two poles. In this strategy, this adjustment will lead to the fluctuation of training results in the process of training. In order to make the training results more smooth and stable, we use a coefficient to control the degree of convergence to the two poles. We describe the second strategy in Sect. 3.2.2.

3.2.2   Probability Adjustment Strategy 2 (M-3W2)

In this strategy, the probability values of the boundary region samples is adjusted by:

$$Pr'(x_{ip}) = Pr(x_{ip}) + (Pr(x_{ip}) - 0.5) * w_i \tag{10}$$

$$Pr'(x_{in}) = Pr(x_{in}) + (Pr(x_{in}) - 0.5) * w_i \tag{11}$$

where $Pr'(x_{ip})$ denotes the positive probability adjusted by M-W2 of a boundary region sample x_i, and $Pr'(x_{in})$ denotes the negative probability adjusted by M-W2 of the boundary region sample x_i, and w_i denotes the strategy coefficient of the sample x_i.

The strategy coefficient is calculated by:

$$w_i = G - |Pr(x_{ip}) - 0.5| \tag{12}$$

When $w_i = 1$, the two strategies are the same. But according to formula 12, the value of w_i will be less than 1.

The two above strategies we proposed is to better classify the polarity of the dataset samples. In adjustment strategy 1, we add the difference between probability and center point 0.5 to the positive and negative polar probabilities of a sample in the boundary region, respectively, so that the positive and negative polar probabilities converge towards two poles, or even out of the boundary region and into the positive or negative region, in order to enhance its polarity. To better control the degree of convergence to two poles, we added a coefficient w_i to the adjusted strategy 2.

4   Experiment

In this section, we will construct experiments to evaluate the effectiveness of the proposed methods.

4.1   Datasets and Baseline Methods

We test our model on four binary sentiment classification benchmark datasets. We describe each detailed dataset as follows:

- IMDB: It is a movie review dataset for binary sentiment classification. In the dataset, there are 25,000 training samples and 25,000 test samples, of which 50% are positive samples and negative samples. The average length of each review sample is 255.
- MR (Movie review dataset): The average length of each review is 20 in this review set, in which there are positive samples and negative samples each accounting for 50%.
- CR (customer review dataset): This is an annotated customer review dataset of 14 products from Amazon. Each customer reviews the product and divides the reviews into positive and negative. The average length of each review is 19.

- SUBJ: This is a binary classification movie review dataset that classifies a sentence as subjective or objective. The dataset provides 10,000 samples, with subjective samples and objective samples each accounting for 50%. The average length of each review is 23.

We use five RNN models as the benchmark classifiers, including RNN, LSTM, BiLSTM, CLSTM, and GRU. We compare our methods with the benchmark models and then analyze the accuracy in the results. In addition, we also made a comparative experiment by adjusting the thresholds and adjusting the probability change rate.

- RNN: We use one hidden layer with 32 hidden units, batch size of 512.
- LSTM: We use one hidden layer with 64 hidden units, dropout rate of 0.3, batch size of 512.
- CLSTM: We use one convolution layer and one max-pooling layer, filter windows of 3 with 32 feature maps one hidden layer with 64 hidden units, dropout rate of 0.2, and then use one hidden layer with 64 hidden units, batch size of 128.
- BiLSTM: We use one forward hidden layer and backward hidden layer with 64 hidden units, dropout rate of 0.3, batch size of 128.
- GRU: We use one hidden layer with 128 hidden units, dropout rate of 0.3, batch size of 128.

4.2   Experimental Results

For the RNN models, we compare the performances of vanilla RNN, LSTM, CLSTM, BiLSTM, GRU, and their improved models based on three-way decisions. The experimental results are shown in Table 1 (% is omitted).

We compare the average accuracy of the RNN models and the models improved with the two adjustment strategies and find that the improved models are significantly better than their benchmark models. M-3W1 and M-3W2 give superior performances on the four datasets.

For text classification, the methods for processing the samples in the boundary region can affect classification accuracy. The models improved with the two adjustment strategies M-3W1 and M-3W2 are also different in experimental results. In most cases, the accuracy of M-3W2 is better than the accuracy of M-3W1. The accuracy improvement of the benchmark model RNN is significantly higher than that of other benchmark models.

4.3   Parameter Analysis

In the process of tuning, it is found that different value of parameter G will lead to different experimental results, so we also analyze the classification accuracy under different parameters, and the results are shown in the Fig. 2 and Fig. 3. In the following experiment, we fix the parameter G from 0.1 to 0.24, and the step size used is 0.02. Except for the parameter G, all other parameters are kept

Table 1. Classification accuracy of different models

MODEL	IMDB	MR	CR	SUBJ
RNN	85.94	66.85	70.20	87.96
RNN-3W1	87.54	72.18	75.36	90.65
RNN-3W2	87.90	72.51	74.83	90.65
LSTM	87.32	76.91	79.14	90.50
LSTM-3W1	89.76	78.45	80.47	92.55
LSTM-3W2	89.80	79.21	80.81	92.25
BiLSTM	87.50	76.89	79.47	91.30
BiLSTM-3W1	89.80	78.84	80.74	93.85
BiLSTM-3W2	89.68	79.22	81.60	93.95
CLSTM	87.70	76.93	78.81	90.65
CLSTM-3W1	89.24	78.27	79.27	92.80
CLSTM-3W2	89.20	78.95	80.13	92.90
GRU	86.72	75.43	78.01	90.20
GRU-3W1	87.94	77.12	79.60	92.85
GRU-3W2	88.30	77.25	80.79	92.80

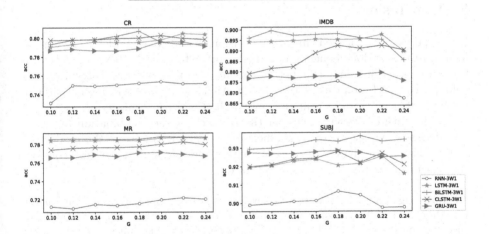

Fig. 2. Classification results of M-3W1 models with different parameter G

unchanged. We plot the two strategies of each dataset with parameter G changes on the two following figures separately for comparison.

It can be seen from the two figures that the fluctuation of some broken lines is not very obvious, which shows that the accuracy is not significantly influenced by the parameter G on some RNN models. However, we can still find that the fluctuation degree of the broken line increases with the increase of the average length of the dataset, and the dataset that varies the most with the parameter is

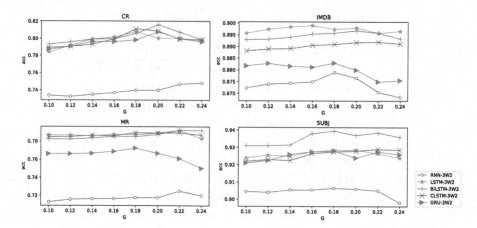

Fig. 3. Classification results of M-3W2 models with different parameter G

IMDB. In addition, the results show that the classification accuracy of proposed models is better when the value of parameter G is one value from 0.18 to 0.22.

5   Conclusion

In this paper, we proposed three-way decisions based RNN models. The experimental results on four benchmark datasets indicate that the RNN models with three-way decision ideas can better classify samples into two categories and enhance the performances of the classification. Compared with some baseline methods, it demonstrates that the new method is more accurate.

In the future, we will focus on the research of the combination of three-way decisions and other text classification models. In addition, the models we proposed is for the binary classification of text analysis, and then we can consider the sentiment analysis of multi-classification.

Acknowledgments. This work was supported by the Universities Natural Science Key Project of Anhui Province (No. KJ2020A0637) and the Universities Natural Science Foundation of Anhui Province (No. KJ2011Z400). The authors of the paper express great acknowledgment of these supports.

References

1. Zhang, Y., Xiang, X., Yin, C., Shang, L.: Parallel sentiment polarity classification method with substring feature reduction. In: Pacific-Asia Conference on Knowledge Discovery and Data Mining, pp. 121–132 (2013)
2. Ju, S., Li, S.: Active learning on sentiment classification by selecting both words and documents. In: Ji, D., Xiao, G. (eds.) CLSW 2012. LNCS (LNAI), vol. 7717, pp. 49–57. Springer, Heidelberg (2013). https://doi.org/10.1007/978-3-642-36337-5_6

3. Hailong, Z., Wenyan, G., Bo, J.: Machine learning and lexicon based methods for sentiment classification: a survey. In: 2014 11th Web Information System and Application Conference, pp. 262–265 (2015)
4. Tang, D., Qin, B., Liu, T.: Deep learning for sentiment analysis: successful approaches and future challenges. Wiley Interdisc. Rev.: Data Min. Knowl. Discov. 5(6), 292–303 (2015)
5. Jia, X., Deng, Z., Min, F., Liu, D.: Three-way decisions based feature fusion for Chinese irony detection. Int. J. Approx. Reason. 113, 324–335 (2019)
6. Raffel, C., Ellis, D.P.: Feed-forward networks with attention can solve some long-term memory problems. arXiv preprint arXiv:1512.08756 (2015)
7. Schmidt, R.M.: Recurrent neural networks (RNNs): a gentle introduction and overview. arXiv preprint arXiv:1912.05911 (2019)
8. Kim, Y.: Convolutional neural networks for sentence classification. arXiv preprint arXiv:1408.5882 (2014)
9. Dong, Y., Fu, Y., Wang, L., Chen, Y., Dong, Y., Li, J.: A sentiment analysis method of capsule network based on biLSTM. IEEE Access 8, 37014–37020 (2020)
10. Tao, H., Tong, S., Zhao, H., Xu, T., Liu, Q.: A radical-aware attention-based model for Chinese text classification. In: Proceedings of the AAAI Conference on Artificial Intelligence, vol. 33, pp. 5125–5132 (2019)
11. Myagmar, B., Li, J., Kimura, S.: Cross-domain sentiment classification with bidirectional contextualized transformer language models. IEEE Access 7, 163219–163230 (2019)
12. Sharfuddin, A.A., Tihami, M.N., Islam, M.S.: A deep recurrent neural network with biLSTM model for sentiment classification. In: 2018 International Conference on Bangla Speech and Language Processing (ICBSLP), pp. 1–4 (2018)
13. Pal, S., Ghosh, S., Nag, A.: Sentiment analysis in the light of LSTM recurrent neural networks. Int. J. Synthetic Emot. 9(1), 33–39 (2018)
14. Yu, H., Ji, Y., Li, Q.: Student sentiment classification model based on GRU neural network and TF-IDF algorithm. J. Intell. Fuzzy Syst. 40(2), 2301–2311 (2021)
15. Tran, N.M.: Aspect based sentiment analysis using neuroner and bidirectional recurrent neural network. In: Proceedings of the Ninth International Symposium on Information and Communication Technology, pp. 1–7 (2018)
16. Liu, G., Guo, J.: Bidirectional LSTM with attention mechanism and convolutional layer for text classification. Neurocomputing 337(APR.14), 325–338 (2019)
17. Han, Y., Liu, M., Jing, W.: Aspect-level drug reviews sentiment analysis based on double BiGRU and knowledge transfer. IEEE Access 8, 21314–21325 (2020)
18. Pan, Y., Liang, M.: Chinese text sentiment analysis based on bi-GRU and self-attention. In: 2020 IEEE 4th Information Technology, Networking, Electronic and Automation Control Conference (ITNEC), pp. 1983–1988 (2020)
19. Arevian, G., Panchev, C.: Optimising the hystereses of a two context layer RNN for text classification. In: International Joint Conference on Neural Networks, pp. 2936–2941. IEEE (2007)
20. Chen, J., Li, Y., Zhao, S., Wang, X., Zhang, Y.: Three-way decisions community detection model based on weighted graph representation. In: Bello, R., Miao, D., Falcon, R., Nakata, M., Rosete, A., Ciucci, D. (eds.) IJCRS 2020. LNCS (LNAI), vol. 12179, pp. 153–165. Springer, Cham (2020). https://doi.org/10.1007/978-3-030-52705-1_11
21. Hochreiter, S., Schmidhuber, J.: Long short-term memory. Neural Comput. 9(8), 1735–1780 (1997)
22. Ranzato, M., Chopra, S., Auli, M., Zaremba, W.: Sequence level training with recurrent neural networks. In: ICLR (2016)

23. Xu, X., Ye, F.: Sentences similarity analysis based on word embedding and syntax analysis. In: 2017 IEEE 17th International Conference on Communication Technology (ICCT), pp. 1896–1900. IEEE (2017)

24. Qian, F., Sha, L., Chang, B., Liu, L.C., Zhang, M.: Syntax aware LSTM model for semantic role labeling. In: Proceedings of the 2nd Workshop on Structured Prediction for Natural Language Processing, pp. 27–32 (2017)

25. Zhu, R., Yang, D., Li, Y.: Learning improved semantic representations with tree-structured LSTM for hashtag recommendation: an experimental study. Information 10, 127 (2019)

26. Skantze, G.: Towards a general, continuous model of turn-taking in spoken dialogue using LSTM recurrent neural networks. In: Sigdial Meeting on Discourse Dialogue, pp. 220–230 (2017)

27. Su, C., Huang, H., Shi, S., Jian, P., Shi, X.: Neural machine translation with Gumbel tree-LSTM based encoder. J. Vis. Commun. Image Represent. 71, 102811 (2020)

28. Cheng, J., Dong, L., Lapata, M.: Long short-term memory-networks for machine reading. arXiv preprint arXiv:1601.06733 (2016)

29. Yangsen, Z., Jia, Z., Yuru, J., Gaijuan, H., Ruoyu, C.: A text sentiment classification modeling method based on coordinated CNN-LSTM-attention model. Chin. J. Electron. 28(001), 120–126 (2019)

30. Feng, S., Wang, Y., Liu, L., Wang, D., Yu, G.: Attention based hierarchical LSTM network for context-aware microblog sentiment classification. World Wide Web 22(1), 59–81 (2019)

31. Feng, X., Liu, X.: Sentiment classification of reviews based on BiGRU neural network and fine-grained attention. In: Journal of Physics: Conference Series. vol. 1229, p. 012064. IOP Publishing (2019)

32. Yao, Y.: Three-way decisions with probabilistic rough sets. Inform. Sci. 180(3), 341–353 (2010)

33. Zhang, Z., Wang, R.: Applying three-way decisions to sentiment classification with sentiment uncertainty. In: Miao, D., Pedrycz, W., Ślęzak, D., Peters, G., Hu, Q., Wang, R. (eds.) RSKT 2014. LNCS (LNAI), vol. 8818, pp. 720–731. Springer, Cham (2014). https://doi.org/10.1007/978-3-319-11740-9_66

34. Zhou, Z., Zhao, W., Shang, L.: Sentiment analysis with automatically constructed lexicon and three-way decision. In: Miao, D., Pedrycz, W., Ślęzak, D., Peters, G., Hu, Q., Wang, R. (eds.) RSKT 2014. LNCS (LNAI), vol. 8818, pp. 777–788. Springer, Cham (2014). https://doi.org/10.1007/978-3-319-11740-9_71

35. Zhang, Y., Zhang, Z., Miao, D., Wang, J.: Three-way enhanced convolutional neural networks for sentence-level sentiment classification. Inform. Sci. 477, 55–64 (2018)

36. Zhang, Y., Miao, D., Wang, J., Zhang, Z.: A cost-sensitive three-way combination technique for ensemble learning in sentiment classification. Int. J. Approx. Reason. 105, 85–97 (2018)

37. Zhu, Y., Tian, H., Ma, J., Liu, J., Liang, T.: An integrated method for micro-blog subjective sentence identification based on three-way decisions and naive bayes. In: Miao, D., Pedrycz, W., Ślęzak, D., Peters, G., Hu, Q., Wang, R. (eds.) RSKT 2014. LNCS (LNAI), vol. 8818, pp. 844–855. Springer, Cham (2014). https://doi.org/10.1007/978-3-319-11740-9_77

Tolerance-Based Short Text Sentiment Classifier

Vrushang Patel and Sheela Ramanna$^{(\boxtimes)}$ iD

Department of Applied Computer Science, University of Winnipeg,
Winnipeg, MB R3B 2E9, Canada
patel-v30@webmail.uwinnipeg.ca, s.ramanna@uwinnipeg.ca

Abstract. Sentiment classification identifies the polarity of text such as positive, negative or neutral based on textual features. A tolerance near set-based text classifier (TSC) is introduced in this paper to classify sentiment polarities of text with vectors from a pre-trained SBERT algorithm. One of the datasets (Covid-Sentiment) was hand-crafted with tweets from Twitter of opinions related to COVID. Experiments demonstrate that TSC outperforms five classical ML algorithms with one dataset, and is comparable with all other datasets using a weighted F1-score.

Keywords: Tolerance near sets · Transformer architecture · Sentiment classification · BERT · SBERT

1 Introduction

Opinion mining and sentiment analysis are two popular research areas where opinions of users are analyzed to detect sentiment polarity [4]. Microblogging platforms such as Twitter with a large user-base, have led to a vast amount of information for sentiment analysis [2]. Sentiment polarity detection is one of the tasks of sentiment analysis. Several classical machine learning approaches such as supervised, semi-supervised and unsupervised methods have been applied to this task. These methods use either sentiment lexicons and/or a variety of hand-crafted textual features for classification. Deep learning approaches that were originally used to learn word embeddings from large amounts of text corpus are now used extensively in sentiment analysis. The Attention-based transformer architecture BERT introduced in [10] as well as several variants like RoBERTa, ELECTRA, DistilBERT, and ALBERT [3] have gained immense popularity due to their high classification accuracy.

In this paper, we introduced a modified form of tolerance-based algorithm (TSC) to classify sentiments in short texts using the vector embeddings generated by the BERT model. The Tolerance near set-based (TNS) classifier was

Vrushang Patel's work was supported by the UW President's Distinguished Graduate Student Scholarship and Sheela Ramanna's work was supported by NSERC Discovery Grant # 194376.

© Springer Nature Switzerland AG 2021
S. Ramanna et al. (Eds.): IJCRS 2021, LNAI 12872, pp. 259–265, 2021.
https://doi.org/10.1007/978-3-030-87334-9_22

first introduced in [7] and applied in [9] to classify images and audio signal data respectively. We experimented with seven datasets and the following algorithms: Stochastic Gradient Descent (SGD), Light Gradient Boosting Machine (LGBM), Random Forest (RF), Maximum Entropy (ME), and Support Vector Machine (SVM). The chosen datasets were a mix of long and short words as well as several sentiment classes. A Covid-Sentiment dataset was hand-crafted by extracting tweets of opinions about COVID. Using a weighted F1-score, we evaluated the performance of all the algorithms. We also experimented with the DistilBERT model which outperformed all algorithms on all datasets. However, our goal was to compare TSC with non-deep learning approaches. The contribution of this work is a new application of tolerance near sets in sentiment polarity classification. This paper is organized as follows: in Sect. 2, we describe the datasets used in our experiments. In Sect. 3, we give the formal model, TSC algorithm, and the transformer models used to derive vectors for our experiments. In Sect. 4, we give analysis of our experimental results.

2   Data Sets

Table 1 shows the datasets used in our experiments. We handcrafted a dataset referred to as **Covid-Sentiment** which is a subset derived from [1] using Tweets ID from 1st April 2020 to 1st May 2020. We extracted 47,386 tweets with the help of Twitter API. The tweets in languages other than English (ex: French, Hindi, Mandarian, Portuguese) as well as duplicate tweets were removed. The collected tweets used hashtags such as #Covid19, #WHO, #Wuhan, #Corona. Extensive pre-processing of 29,981 English language tweets from the original dataset such as removal of HTML tags, @username, hashtags, URLs, and incorrect spellings were also performed. A final set of 8003 tweets were hand-labelled into 3 categories: positive, negative, and neutral sentiments. With other datasets, subsets were created for training and test sets using random and balanced class distribution using the Scikit-learn library[1].

Table 1. Dataset information

Dataset	Type	Sizes	Positive	Negative	Neutral	Irrelevant
Covid–Sentiment	Train,Test	7000,1003	1542,236	2124,374	3334,393	–,–
U.S. Airline Sentiment	Train,Test	12000,1000	2015,130	7322,675	2663,195	–,–
IMDB Movie Review	Train,Test	20000,2000	10055,1007	9945,993	–,–	–,–
SST–2	Train,Test	15000,1500	8306,833	6694,667	–,–	–,–
Sentiment140	Train,Test	15000,1000	7500,500	7500,500	–,–	–,–
SemEval 2017	Train,Test	17001,3546	6915,1473	2551,559	7534,1513	–,–
Sanders corpus	Train,Test	4059,1015	416,100	462,107	1837,484	1344,324

---

[1] https://scikit-learn.org/stable/.

3 Models

In this section, a brief introduction to the main definitions underlying near sets is given. In addition, we give the algorithm used to generate the tolerance classes and representative vectors. We also briefly describe the specific transformer model used in our experiments.

3.1 Tolerance Near Sets

Tolerance near sets [6] provide an intuitive as well as mathematical basis in defining what it means for pairs of objects to be similar. The basic structure which underlies near set theory is a perceptual system which consists of perceptual objects (i.e., objects that have their origin in the physical world [5]).

Definition 1. Perceptual System [5]
A perceptual system is a pair $\langle O, F \rangle$, where O is a nonempty set of perceptual objects and F is a countable set of real-valued probe functions $\phi_i : O \to \mathbb{R}$.

An object description is defined by means of a tuple of probe function values $\Phi(x)$ associated with an object $x \in X$, where $X \subseteq O$ as defined by Eq. 1:

$$\Phi(x) = (\phi_1(x), \phi_2(x) \ldots, \phi_n(x)) \tag{1}$$

where $\phi_i : O \to \mathbb{R}$ is a probe function of a single feature. The probe functions give rise to a number of perceptual relations. This approach is useful when decisions on nearness are made in the context of a perceptual system i.e., a system consisting of objects and our perceptions of what constitutes features that best describe these object. The notion of tolerance is directly related to the idea of closeness between objects, that resemble each other with a tolerable level of difference. A tolerance space $(X, \simeq)$ consists of a set X endowed with a binary relation $\simeq$ (i.e., a subset $\simeq \subset X \times X$) that is reflexive (for all $x \in X$, $x \simeq x$) and symmetric (for all $x, y \in X$, $x \simeq y$ and $y \simeq x$) but transitivity of $\simeq$ is not required.

Definition 2. Perceptual Tolerance Relation [6]
Let $\langle O, F \rangle$ be a perceptual system and let $\mathcal{B} \subseteq F$,

$$\cong_{\mathcal{B},\epsilon} = \{(x, y) \in O \times O : \| \phi(x) - \phi(y) \|_2 \le \varepsilon\} \tag{2}$$

where $\|\cdot\|_2$ denotes the L^2 norm of a vector.

In this paper, given a set of labelled text (objects) T, where $t_i \in T$, $i \in N$ and s_i is a sentiment label for t_i. Each tweet or text t_i can be represented as a k-dimensional word vector $\phi(t_i)$. We define a tolerance class of texts TC using the definition in [6] as follows:

$$\{(t_i, t_j) \in T \times T : \frac{\phi(t_i).\phi(t_j)}{\|\phi(t_i)\| \, \|\phi(t_j)\|} \le \varepsilon\}. \tag{3}$$

where text similarity is measured using the cosine distance measure and ε is a user-defined tolerance level. In other words, our universe of text described by set of vectors ϕ, is spread amongst tolerance classes with a tolerance level ε for semantic textual similarity, where a tolerance class TC is maximal with respect to inclusion. Algorithm 1 gives the method for generating tolerance classes from the training set and Algorithm 2 gives the method for generating tolerance classes from the test set.

Algorithm 1: Training Phase: Generating class representative vectors

---

Input : $\varepsilon > 0$, // Tolerance level
$\quad\quad\quad\quad TR = \{TR_1, \ldots, TR_M\}$, // Training Data Set
Output: $(NC, \{R_1, \ldots, R_{NC}\}, \{TextCat_1, \ldots, TextCat_{NC}\})$

1 **for** $i \leftarrow 1$ **to** M **do**
2 $\quad$ **for** $j \leftarrow i + 1$ **to** M **do**
3 $\quad\quad$ computeCosine($TR_i, TR_j, Cosine_{ij}$);

4 **for** $i \leftarrow 1$ **to** M **do**
5 $\quad$ **for** $j \leftarrow i + 1$ **to** M **do**
6 $\quad\quad$ generateToleranceclass($Cosine_{ij}, \varepsilon$; SetOfPairs);
$\quad\quad\quad$ computeNeighbour(SetOfPairs, $i, TR; N_i$); // Compute the
$\quad\quad\quad$ neighbourhood N_i of ith training data TR_i
7 $\quad\quad$ **for** *all* $x, y \in N_i$ **do**
8 $\quad\quad\quad$ **if** $x, y \notin SetOfPairs$ **then**
9 $\quad\quad\quad\quad$ $C_i \leftarrow N_i$; // Include y from class N_i into C_i

10 $\quad$ $H \leftarrow H \cup \{C_i\}$;
11 $\quad$ // C_i is one tolerance class induced by the tolerance relation
12 $\quad$ computeMajorityCat(C_i; $TextCat_i$); // Determine Category by majority
$\quad\quad$ voting for each C_i
13 $NC \leftarrow |H|$; // Number of classes
14 // End of defineClass
15 defineClassRepresentative(NC, H; $\{R_1, \ldots, R_{NC}\}, \{TextCat_1, \ldots, TextCat_{NC}\}$);
$\quad$ // Based on mean value

3.2 Transformer Model

Transformers overcome the limitation of RNN (sequential processing of text), and CNN (high computational cost of learning representations of long sentences) by applying self-attention to compute in parallel for every word in a sentence or document an "attention score" to model the influence each word has on another [3]. SBERT uses siamese and triplet network structures to derive semantically meaningful sentence embedding and adds a pooling operation to the output of BERT/RoBERTa to derive a fixed sized sentence embedding [8].

Algorithm 2: Learning Phase: Assigning Sentiment Classes

Input : $\varepsilon > 0$, // Tolerance level
$TS = \{TS_1, \ldots, TS_M\}$, // Testing Data Set
$\{R_1, \ldots, R_{NC}\}, \{TextCat_1, \ldots, TextCat_{NC}\}$ // Representative
Class and their associated categories
Output: $(TS' = \{TS'_1, \ldots, TS'_M\})$ // Testing Data Set with assigned
categories

1 **for** $i \leftarrow 1$ **to** M **do**
2   **for** $j \leftarrow i+1$ **to** NC **do**
3   computeCosine(TS_i, RC_j, Cosine$_{ij}$);

4 DetermineTextCat($Cosine_{ij}; TS'$) // Computes min. distance and assigns
category

---

4 Experiments and Analysis of Results

The BERT-base-uncased model provided by Huggingface[2] was used with PyTorch[3] to generate BERT vector embeddings for the machine learning algorithms. The SBERT (bert-base-nli-mean-tokens) transformer model was used to generate vector embeddings of 1x768 dimensional vectors for the TSC algorithm. Prototype vectors or class representatives and their classes shown in Algorithm 1 (line 15) were determined using mean values. Other schemes such as mode and median for prototype vectors were also considered, but rejected due to poor classification performance. The training data was restricted to a maximum of 20,000 examples due to large distance matrix and memory limitations. The weighted F1-score was used since we have imbalanced datasets. Python regex module, NLTK stemming and lemmatization were used in pre-processing prior to generating the transformer vector embeddings.

Table 2 gives F1-scores with the best result shown in bold-face and best value for ε. It can seen that TSC algorithm performs best with the IMDB dataset, second best with three datasets and comparable with the others. The IMBD dataset has an average word length of 230 words. This word length when encoded in a 768 dimensional vector is able to establish good semantic similarity. The Covid-Sentiment dataset contains tweets from all around the globe (in several languages) so there are very few tweets that are similar. Hence it has the lowest score with all algorithms. From the above result, TSC algorithm performs best for binary sentiment classification where the average number of words in a text are larger such as IMDB (230 words) compared to the smaller sizes of 8, 10 or 12 words.

---

[2] https://huggingface.co/.
[3] https://pytorch.org/.

Table 2. F1-score results

Dataset	TSC (best ε value)	RF	ME	SVM	SGD	LGBM
Covid-Sentiment	55 (0.23)	44	57	57	**58**	56
U.S. Airline Sentiment	77 (0.32)	72	**79**	77	75	77
IMDB Movie Review Analysis	**76** (0.26)	69	73	73	72	73
SST-2	85 (0.23)	83	85	**86**	85	85
Sentiment140	70 (0.16)	68	**72**	72	66	70
SemEval 2017	60 (0.26)	54	**64**	63	63	60
Sanders corpus	69 (0.24)	70	**76**	74	76	75

5   Conclusion

In this paper, we implemented a modified form of tolerance-based algorithm (TSC) to classify sentiment polarities of short text. In addition, we also hand-crafted a dataset from Twitter regarding opinions on COVID. We experimented with subsets of seven well-known datasets due to computational challenges. The chosen datasets were a mix of long and short words as well as several sentiment classes. TSC outperforms the reported ML algorithms with one dataset, performs second best with 3 datasets and is comparable with the remaining datasets using a weighted F1-score. Future work will involve more experiments with larger size datasets to investigate the performance of our proposed TSC algorithm.

References

1. Chen, E., Lerman, K., Ferrara, E.: Tracking social media discourse about the COVID-19 pandemic: development of a public coronavirus twitter data set. JMIR Public Health Surveill. **6**(2), e19273 (2020)
2. Giachanou, A., Crestani, F.: Like it or not: a survey of twitter sentiment analysis methods. ACM Comput. Surv. **49**(2), 46 (2016)
3. Minaee, S., Kalchbrenner, N., Cambria, E., Nikzad, N., Chenaghlu, M., Gao, J.: Deep learning based text classification: a comprehensive review. ACM Comput. Surv. (CSUR) **54**(3), 1–40 (2021)
4. Pang, B., Lee, L.: Opinion mining and sentiment analysis. Found. Trends Inf. Retr. **2**(1–2), 1–135 (2008). https://doi.org/10.1561/15000000011
5. Peters, J.: Near sets. special theory about nearness of objects. Fundam. Inform. **75**(1–4), 407–433 (2007)
6. Peters, J.: Tolerance near sets and image correspondence. Int. J. Bio-Inspired Comput. **1**(4), 239–245 (2009)
7. Poli, G., et al.: Solar flare detection system based on tolerance near sets in a GPU-CUDA framework. Knowl.-Based Syst. J. **70**, 345–360 (2014)

8. Reimers, N., Gurevych, I.: Sentence-bert: sentence embeddings using siamese bert-networks. In: Proceedings of the 2019 Conference on Empirical Methods in Natural Language Processing. Association for Computational Linguistics (2019). https://arxiv.org/abs/1908.10084

9. Ulaganathan, A.S., Ramanna, S.: Granular methods in automatic music genre classification: a case study. J. Intell. Inf. Syst. **52**(1), 85–105 (2018). https://doi.org/10.1007/s10844-018-0505-8

10. Vaswani, A., et al.: Attention is all you need. arXiv preprint arXiv:1706.03762 (2017)

Knowledge Graph Representation Learning for Link Prediction with Three-Way Decisions

Zhihan Peng and Hong Yu$^{(\boxtimes)}$ ⓘ

Chongqing Key Laboratory of Computational Intelligence,
Chongqing University of Posts and Telecommunications, Chongqing, China
yuhong@cqupt.edu.cn

Abstract. Relation prediction is one of the important tasks of knowledge graph completion, which aims to predict the missing links between entities. Although many different methods have been proposed, most of them usually follow the closed-world assumption. Specifically, these methods simply treat the unknown triples as errors, which will result in the loss of valuable information contained in the knowledge graphs (KGs). In addition, KGs exist large amounts of long-tail relations, which lack sufficient triples for training, and these relations will seriously affect inference performance. In order to address above-mentioned problems, we propose a novel relation prediction method based on three-way decisions, namely RP-TWD. In this paper, RP-TWD model first obtains the similarity between relations by K-Nearest Neighbors (KNN) to model the semantic associations between them. The semantic association between relations can be considered as supplementary information of long-tail relations, and constrain the learning of KG embeddings. Then, based on the idea of three-way decisions (TWD), the triples of specific relation are further divided into three disjoint regions, namely positive region (POS), boundary region (BND), and negative region (NEG). The introduction of BND intends to represent the uncertainty information contained in unknown triples. The experimental results show that our model has significant advantages in the task of relation prediction compared with baselines.

Keywords: Knowledge graph · Link prediction · Three-way decisions · Semantic association

1 Introduction

The early concept of knowledge graph (KG) originated from the idea of Tim Berners-Lee, i.e. the father of the World Wide Web, about the Semantic Web [13]. It aims to use a graph structure to model and record the knowledge and associations existing in the world, so as to effectively realize a more accurate object-level search. Knowledge graphs represent knowledge in the form of triples, that is, entities (including concepts and attribute values) are represented as nodes on the

© Springer Nature Switzerland AG 2021
S. Ramanna et al. (Eds.): IJCRS 2021, LNAI 12872, pp. 266–278, 2021.
https://doi.org/10.1007/978-3-030-87334-9_23

graph, and the edges between nodes correspond to the relations between entities. In short, KGs utilize network structure to represent obtained knowledge.
KGs have been widely used in many fields such as search engines [19], intelligent
question answering system [7], language understanding [20], recommendation
calculation [17], big data analysis, and decision-making [9].

The objective world has countless entities, and the subjective world of
humans also contains concepts that cannot be measured. Meanwhile, there are a
greater number of complex relations between these entities and concepts. These
problems often lead to the dilemma of missing links for the existing KGs and the
incompleteness of KGs. In practical application, the incompleteness of KGs is
also the primary issue that hinders the development of most downstream tasks.
Thus, knowledge graph completion has also increased a general concern.

Recently, the most popular link prediction methods are mainly based on
knowledge graph embedding models. These methods aim to project the entities
and relations in KGs into a continuous vector space while preserving the inherent
structure and underlying semantic information, such as TransE [4], TransH [18]
and TransR [11]. However, the above methods regard each relation as independent and ignore the interactive information between relations. Lin et al. [11] and
Zhang et al. [24] propose CTransR and HRS by using the semantic information
and structural information between relations. Nevertheless, most work ignores
the rich information contained in the unknown triples in KGs.

To address the above limitations, this paper adopts the idea of three-way
decisions [21] to propose RP-TWD model, a knowledge graph-oriented relation
prediction method based on three-way decisions. Firstly, we employ K-Nearest
Neighbor (KNN) algorithm to get K nearest neighbors of each relation and corresponding entity set. Then, based on the probability that the triples composed
of the relation and entities become correct triple, triples of each relation are
divided into positive region, boundary region, and negative region. Finally, the
representation of KGs is learned according to the three regions, while the boundary region is used to represent uncertain knowledge.

The main contributions of this paper are summarized as follows:

(1) We use KNN algorithm to measure the similarity between relations to obtain
K-nearest neighbors of relations in KGs, and explicitly utilize it to capture
semantic associations between them and constrain the embedding learning
of relations.
(2) We propose to represent the uncertain information hidden in unknown triples
by utilizing three-way decision. Specifically, we divide the set of triples corresponding to each relation in KGs into three disjoint regions based on the
potential correct probability of the triples consisting of relations and entities.
For example, the boundary region for a relation is a set of triples formed by
the relation and the entity set corresponding to its neighbors, and then we
regard boundary region (originally not in KGs) also as training triples.
(3) We evaluate the effectiveness of the proposed model RP-TWD on two benchmark datasets FB15K-237 and WN18RR. Experimental results illustrate

that RP-TWD achieves superior performance on relation prediction task and performs competitively or better than existing state-of-the-art models.

The rest of this paper is organized as follows. First, we briefly introduce the related works of three-way decision theory and knowledge graph relation prediction in Sect. 2. Then, we give the detailed description of the proposed model RP-TWD in Sect. 3. In Sect. 4, experimental results and analysis are reported. Finally, our conclusions and future research directions are given in Sect. 5.

2   Related Work

2.1   Knowledge Graph Embedding Models

In recent years, learning distributed representations for entities and relations in KGs, i.e. KG embedding has received increasing attention and various methods have been designed. Enlightened by the phenomenon of semantic translation that exists between word vectors, TransE [4] embeds both entities and relations in the same vector space. TransE can effectively handle 1-to-1 relations, but has troubles with complex relations like 1-to-N, N-to-1, and N-to-N relations. To cope with these relations, based on TransE, TransH [18] learns an additional mapping vector for each relation, which is used to project entities to relation-specific hyperplanes. TransR [11] learns embeddings of entities and relations in separate entity and relation spaces, respectively. Ji et al. [8] propose TransD, which improving TransR by considering that different entities should be mapped into different semantic spaces and reducing the computational effort. However, KG contains multiple relation types, such as symmetry, inversion and composition types, none of the aforementioned methods can model and infer all of them simultaneously.

Many studies have attempted to represent entities and relations in a complex space. The RotatE model [14] defines each relation as a rotation from head entity to tail entity based on Euler's formula in a complex vector space. The optimization objectives and regularization items of TorusE [5] are similar to those works followed by TransE. In order to avoid the contradiction caused by the regular terms mentioned above, TorusE learns the representation of KGs in a compact space instead of learning features in an open manifold European space. Zhang et al. [23] introduce a more expressive hypercomplex representation for modeling entities and relations for knowledge graph embedding through quaternion embedding. More specifically, quaternion embedding is a hypercomplex-valued embedding with three imaginary components, $Q = a + bi + cj + dk$, where the inner product of the quaternion acts $h \otimes r$ as a composition operator between head entity and relation. However, these methods regard relations as independent and ignore the interactive information between them.

To address this problem, many studies have introduced structural information between relations when learning the representation of KGs. CTransR [11] is a variant model of TransR that uses clustering of entity pairs to represent the semantic similarity between relations. Zhang et al. [24] argue that the relations involved in KGs follow a three-layer hierarchical relation structure (HRS), so through clustering the relations and training them in a combined manner for relations within a cluster, the hierarchical information between the relations can be obtained when learning relation embeddings. However, these works treat all the unobserved triples present in KGs as false, ignoring the uncertain information that participated in them.

2.2 Three-Way Decisions

Three-way decisions is a simple "working with threes" and "simplify the complex" decision theory proposed by Yao at Regina University, Canada, in 2009 [21]. The key point is to divide the domain (a whole) into three subsets or parts through granular computing, and to adopt different decision behaviors or processing strategies for different subsets or parts, and then to evaluate or provide feedback on the corresponding behaviors or strategies.

Three-way decisions is an methodology to make decision under uncertainty or insufficient information. The main idea is to partition the universe of objects into three pair-wise disjoint regions, i.e. the positive, boundary, and negative region. Positive rules acquired from the positive region are used to accept something, negative rules acquired from the negative region are used to deny something, and rules that fall on the boundary region need further observation, which is called delayed decision-making. Zhou et al. [25] apply three-way decisions idea to spam filtering, classifying all messages to be sorted as correct, spam, and suspicious. Yu et al. [22] discuss a cluster analysis model for three-way decision making and an automatic learning algorithm for the number of clusters. Aranda-Corral et al. [1] present a three-way decision model for knowledge harnessing. Gaeta et al. [6] propose a method based on graph theory and three-way decisions to evaluate critical regions in epidemic diffusion. Three-way decisions is consistent with human thinking and cognition, and can better handle the uncertainty that arises in the actual decision-making process.

3   Our Approach

Knowledge graph consists of a collection of triples, which can be represented as $\mathcal{G} = \{(h, r, t)\} \subseteq \mathcal{E} \times \mathcal{R} \times \mathcal{E}$, where $\mathcal{E}$ and $\mathcal{R}$ denote the sets of entities (nodes) and relations (edges) respectively. The overall framework of the RP-TWD model is shown in Fig. 1. First, we employ TransE to initialize the entities and relations embedding in KGs. Then, KNN algorithm is utilized to calculate the nearest neighbors of relations and the corresponding entity sets. And three-way decisions is introduced to divide the triples composed of the relations and different entity sets into positive region, boundary region and negative region. The boundary region is used to express uncertainty contained in KG. Finally, the embeddings of KGs are learned based on the three regions, and the obtained K nearest neighbors are used to constraint relation embedding learning.

Definition 1. *For relation r, $H_r = \{h|h \in (h,r,t) \wedge r = r\}$ represents the set consisting of head entities of r, $T_r = \{t|t \in (h,r,t) \wedge r = r\}$ represents the set consisting of tail entities of r, and $E_r = H_r \cup T_r = \{e|h,t \in (h,r,t) \wedge r = r\}$ represents the entity set of r.*

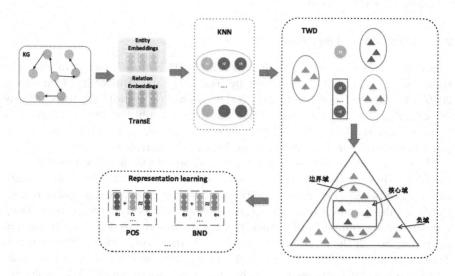

Fig. 1. Architecture of RP-TWD. Triangles of different colors indicate the entitie sets in different regions, where red indicates positive region, orange indicates boundary region, and blue indicates negative region. Green circle indicates the relation r_1. (Color figure online)

Definition 2. *For relation r, $S_r = \{(h,r,t)|r \in (h,r,t) \wedge r = r\}$ denotes the triples set which containing relation r.*

3.1   Relation Neighbor

Previous studies have shown that the embeddings of semantically similar relations are close to each other in the latent space. Therefore, we calculate the K nearest neighbors of each relation by running the KNN algorithm to the relation embeddings learned by TransE [4], where the nearest neighbors of the relation is denoted as N_r. The similarity between two relations r_i and r_j is obtained by $\mathcal{L}_1$ norm,

$$sim_{relation}(r_i, r_j) = ||r_i - r_j||_{\mathcal{L}_1}. \tag{1}$$

It is reasonable to consider that entity set of relations can also reflect the similarity between the relations. For example, the entity sets of relation *producerof* and relation *directorof* have a high overlap rate, which means *producerof* and *directorof* obtaining great semantic similarity. The similarity between the entity sets of relation r_i and its neighbor is calculated according to the following equation:

$$sim_{entity}(E_{r_i}, E_{r_j}) = sim_{Head}(H_{r_i}, H_{r_j}) + sim_{Tail}(T_{r_i}, T_{r_j}), \qquad (2)$$

where $H_{r_i} = \{h\}$ and $T_{r_i} = \{t\}$ denotes the head and tail entity set of relation r_i respectively, and its entity set represented as $E_{r_i} = H_{r_i} \cup T_{r_i}$. $sim_{Head}(H_{r_i}, H_{r_j})$ and $sim_{Tail}(T_{r_i}, T_{r_j})$ severally denotes the similarity of head entity set and tail entity set between relation r_i and r_j. In this paper, we utilize $\mathcal{L}_1$ norm to obtain the similarity between entity pair as follows:

$$sim_{Head}(H_{r_i}, H_{r_j}) = \sum_i \sum_j sim(h_i, h_j), \forall h_i \in H_{r_i}, \forall h_j \in H_{r_j}, \qquad (3)$$

$$sim_{Tail}(T_{r_i}, T_{r_j}) = \sum_i \sum_j sim(t_i, t_j), \forall t_i \in T_{r_i}, \forall t_j \in T_{r_j}, \qquad (4)$$

where $sim(h_i, h_j)$ and $sim(t_i, t_j)$ are the similarity between head entity and tail entity separately, and defined as Eqs. 5, 6:

$$sim(h_i, h_j) = ||h_i - h_j||_{\mathcal{L}_1}, \qquad (5)$$

$$sim(t_i, t_j) = ||t_i - t_j||_{\mathcal{L}_1}. \qquad (6)$$

The final similarity between relation r_i and $r_j \in N_{r_i}$ is defined as:

$$sim(r_i, r_j) = \delta_1 sim_{relation}(r_i, r_j) + \delta_2 sim_{entity}(E_{r_i}, E_{r_j}), \qquad (7)$$

where N_{r_i} represents K neighbors of relation r_i, δ_1 and δ_2 are two hyperparameters respectively weighting the influence of similarity of relations and entity set.

3.2 Knowledge Representation with Three-Way Decisions

In order to effectively handle the uncertain information hidden in the unknown triples in KGs, we make use of the notion of three-way decisions to represent dubious knowledge. RP-TWD adopts the positive region to represent the observed triples of the relation, the set of triples composed of the relation and the entity set of its neighbors are placed in the boundary region, and the negative region consists of triples that neither exists in the positive region nor in the boundary region. For example, due to triple $(Stan\ Lee, place of death, U.S.)$ existing in KG, and relation $place of death$ and relation $place of burial$ obviously have high semantic similarity, it is plausible that triple $(Stan\ Lee, place of burial, U.S.)$ is positive with some probability. Thus, we place triples like $(Stan\ Lee, place of burial, U.S.)$ into boundary region of relation $place of burial$. The positive region, boundary region and negative region of relation r are defined as Eqs. 8, 9, 10 respectively:

$$POS = \{(h, r, t) | (h, r, t) \in S_r\}, \qquad (8)$$

$$BND = \{(h, r, t) | (h, r_i, t) \in S_{N_r} - S_r, r_i \in N_r\}, \tag{9}$$

$$NEG = \mathcal{E} - POS - BND, \tag{10}$$

where S_r denotes the triple set of relation r, N_r denotes the K neighbors belong to relation r, S_{N_r} denotes the triple set of N_r, and is defined as follow:

$$S_{N_r} = S_r \cup S_{r_i}, r_i \in N_r. \tag{11}$$

We give an example below to help the readers understand our work. Suppose there are some triples in KG, $(Ana, place of death, USA)$, $(Bob, place of burial, Japan)$, $(John, place of birth, UK)$, and for relation $place of death$, $place of birth$ and $place of burial$ are its neighbors . So POS of $place of death$ is $(Ana, place of death, USA)$, BND is $(Bob, place of death, Japan)$, $(John, place of death, UK)$ (Fig. 2).

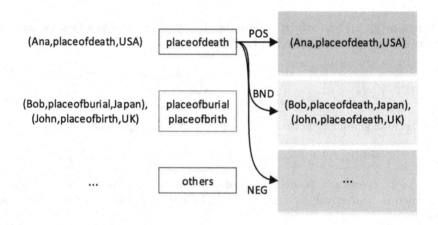

Fig. 2. A example to explain the three regions of relation.

3.3   Loss Function

The objective function for the proposed model RP-TWD consists of three parts, which is formalized as

$$L = L_{POS} + \alpha L_{BND} + \beta L_{neighbor}, \tag{12}$$

where the weight of the positive region is fixed to 1, and α, β are trade-off parameters respectively weighting the influence of boundary region and relation pairs embedding constraint. L_{POS}, L_{BND} and $L_{neighbor}$ are three margin-based loss functions to measure the effectiveness of representation learning in regard to the positive region, boundary region as well as the relation pairs, which are defined as follows:

$$L_{POS} = \sum_{(h,r,t) \in POS} \sum_{(h',r,t') \in NEG} [\gamma_1 + f(h, r, t) - f(h', r, t')]_+, \tag{13}$$

$$L_{BND} = \sum_{(h,r,t)\in BND} \sum_{(h',r,t')\in NEG} [\gamma_2 + f(h,r,t) - f(h',r,t')]_+, \quad (14)$$

$$L_{neighbor} = \sum_{(r)\in S} \sum_{r_{neighbor}\in N_r} \sum_{(r')\in S^-} [\gamma_3 + \lambda f(r,r_{neighbor}) - f(r,r')]_+, \quad (15)$$

where $[x]_+ = max(0,x)$ is defined to obtain the maximum value between 0 and x. γ_1, γ_2 and γ_3 are three positive hyper-parameters denoting each margin of the loss functions in Eqs. 13, 14, 15, respectively, and λ denotes the similarity between r and its neighbors $r_{neighbor}$, $S^- = \{r|r \notin N_r\}$. $f(h,r,t)$ and $f(r,r_{neighbor})$ are score functions as shown in Eqs. 16, 17:

$$f(h,r,t) = ||h + r - t||_{\mathcal{L}_n}, \quad (16)$$

$$f(r,r_{neighbor}) = ||r - r_{neighbor}||_{\mathcal{L}_n}, \quad (17)$$

which can be measured by $\mathcal{L}_1$ or $\mathcal{L}_2$ norm. Positive triples are supposed to have lower scores than negative ones. NEG represents a set that containing the negative triples not in KG.

We utilize stochastic gradient descent (SGD) to optimize the objective function in Eq. 12 and learn parameters of the model.

4   Experiments

4.1   Datasets

In order to evaluate the proposed model RP-TWD, we use two benchmark data sets WN18RR and FB15K-237 as experimental data. They are subsets of the knowledge graph WordNet [12] and Freebase [2] respectively. These datasets are widely employed for relation prediction task of knowledge graph. The statistical details of the datasets are shown in Table 1.

Table 1. Table statistics of datasets.

Dataset	#Rel	#Ent	#Train	#Valid	#Test
FB15K-237	237	14,541	272,115	17,535	20,466
WN18RR	11	40,943	86,835	3,034	3,134

4.2   Baselines and Experiment Setting

To demonstrate the effectiveness of our model, several competitive models are utilized as baselines, including TransE [4], DisMult [3], ComplEx [15], SimplE [10] and RotatE [14].

We evaluate RP-TWD and other baselines on FB15K-237 and WN18RR. Considering the task of relation prediction, i.e., for a given entity pair (h,t) predicting the missing relation r to form an effective triple (h,r,t).

Before training, we adopt TransE to initialize the embeddings of KG. We set the training epochs to 1000, the batch size to 128, the weight coefficients α, β to 0.5 and 0.1 respectively, the embedding dimensions are both set to 100, the margin $\gamma_1 = \gamma_3 = 1$, $\gamma_2 = 2$, λ is the similarity between r and $r_{neighbor}$. Neighbors number of WN18RR is selected to $k = 1$ and 2 to FB15K-237 considering only 11 relations existing in WN18RR. In addition, we employ a grid search to select the other optimal hyper-parameters. We select learning rate lr in $\{0.001, 0.002, 0.005, 0.01\}$, and the weight coefficient δ_1 in $\{0.5, 0.7, 0.8\}$. The final optimal of learning rate and the weight coefficient δ_1 are assigned to: $lr = 0.001$, $\delta_1 = 0.7$ and $\delta_2 = 0.3$.

4.3 Evaluation Metrics

For specific evaluation metrics, we employ three widely used metrics: Mean Rank (MR) which indicates the average rank of correct relations, Mean Reciprocal Rank (MRR) means the mean reciprocal rank of correct relations, and Hit@n represents the proportion of correct relations ranked in the top n. Note that higher MRR or Hit@n signify better performance, while lower value is preferred for MR. Following the standard procedure in prior work, candidate set of relation types is filtered, i.e. the candidate relations for (h, t) do not include any r' where (h, r', t) appears in the training, validation, or test set.

Table 2. Experimental results on FB15K-237 and WN18RR.

Model	FB15K-237				WN18RR			
	MRR	MR	Hits@1	Hits@3	MRR	MR	Hits@1	Hits@3
TransE	0.966	1.352	0.946	0.984	0.784	2.079	0.669	0.870
DisMult	0.924	1.494	0.879	0.970	**0.847**	2.024	0.787	0.891
ComplEx	0.875	1.927	0.806	0.936	0.840	2.053	0.777	0.880
SimplE	0.971	1.407	0.955	**0.987**	0.730	3.259	0.659	0.755
Rotate	0.970	1.315	0.951	0.980	0.799	2.284	0.735	0.823
RP-TWD	**0.976**	**1.25**	**0.962**	0.985	0.830	**1.927**	**0.843**	**0.923**

4.4 Experiment Results

In this section, we evaluate the RP-TWD model through the relation prediction task, and demonstrate the performance of RP-TWD on two datasets. Table 2 shows the experimental results of our proposed RP-TWD model on FB15K-237 and WN18RR. It can be seen from Table 2 that RP-TWD is superior to other baselines in most indicators. Specifically, the Hit@1 of RP-TWD is 0.007% and 0.056% higher than the best baseline on the two datasets, respectively. From experimental results, we can observe that the RP-TWD model outperforms other baselines, which indicates that our model can effectively utilize semantic association between relations to learn the embeddings of relations. The evaluation result of baselines comes from [16].

In addition, in order to further verify the effectiveness of the proposed model RP-TWD in the task of long-tail relation prediction, we conduct tests on two datasets. As shown in Table 3, we select the relations with the frequency of no more than 1000 in the WN18RR dataset, and compare RP-TWD with TransE on the task of predicting these long-tail relations.

Table 3. Long tail relations in WN18RR.

Relation	_similar_to	_member_of_domain_usage	_member_of_domain_region
Frequency	80	629	923

As the experimental results exhibited in Fig. 3, the Hit@1 score of RP-TWD is on average 0.263% higher than TransE. Besides, the correct relation rank (MR) predicted by RP-TWD is 1.92 lower than TransE on average.

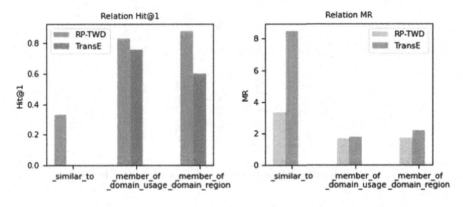

Fig. 3. Hit@1 and MR results of long tail relations prediction on WN18RR

Furthermore, as shown in Table 4, we select the relations that frequency does not exceed 100 from FB15K-237, and the results are shown in Fig. 4 and Fig. 5. In Hit@1 and MRR indicators, RP-TWD outperforms TransE, on average, 0.17% and 0.13% respectively. Moreover, as we can discover from the MR, the correct entity ranking predicted by RP-TWD is 2.54 lower than TransE on average. It can be obtained that RP-TWD is better than TransE in the long-tail relation prediction task. More specifically, the introduction of the boundary domain makes our model can catch uncertainty contained in KG and enhance long-tail relation representation learning.

Table 4. Long tail relations in FB15K-237.

Relation	type_of_appearance	Person	Film	Interests	Family
Frequency	37	90	93	100	100

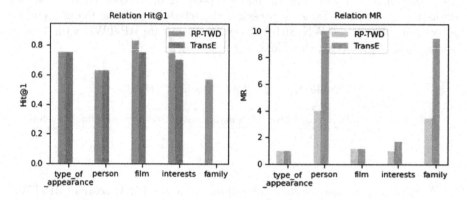

Fig. 4. Hit@1 and MR results of long tail relations prediction on FB15K-237.

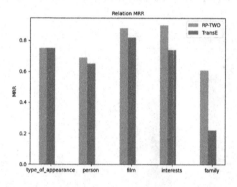

Fig. 5. MRR results of long tail relations prediction on FB15K-237.

5   Conclusion

In this paper, we propose a novel knowledge graph relation prediction model named RP-TWD, which considers uncertain information hidden in KGs. Inspired by three-way decisions theory, we utilize boundary region to represent triples that contained indeterminate knowledge. Furthermore, we employ KNN to explicitly obtain semantic associations between relations and constrain the learning of relation embeddings. Finally, we evaluate the proposed model on two benchmark datasets and experimental results demonstrate the effectiveness of RP-TWD.

In the future work, we will explore to take advantage of some other uncertain knowledge which potentially useful to KG completion. Besides, we plan to use additional information, such as logical rules, relation types, to learn more accurate long-tail relations representation.

Acknowledgements. This work was jointly supported by the National Natural Science Foundation of China (61876027, 61751312, 61533020), and the Natural Science Foundation of Chongqing (cstc2019jcyj-cxttX0002).

References

1. Aranda-Corral, G.A., Borrego-Díaz, J., Galán Páez, J.: A model of three-way decisions for knowledge harnessing. Int. J. Approx. Reason. **120**, 184–202 (2020). https://doi.org/10.1016/j.ijar.2020.02.010
2. Bollacker, K.D., Evans, C., Paritosh, P., Sturge, T., Taylor, J.: Freebase: a collaboratively created graph database for structuring human knowledge. In: Sigmod Conference (2008)
3. Bordes, A., Glorot, X., Weston, J., Bengio, Y.: A semantic matching energy function for learning with multi-relational data. Mach. Learn. **94**(2), 233–259 (2014)
4. Bordes, A., Usunier, N., Garcia-Durán, A., Weston, J., Yakhnenko, O.: Translating embeddings for modeling multi-relational data. In: Proceedings of the 26th International Conference on Neural Information Processing Systems, NIPS 2013, vol. 2, p. 2787–2795. Curran Associates Inc., Red Hook (2013)
5. Ebisu, T., Ichise, R.: Toruse: Knowledge graph embedding on a lie group. arXiv preprint arXiv:1711.05435 (2017)
6. Gaeta, A., Loia, V., Orciuoli, F.: A method based on graph theory and three way decisions to evaluate critical regions in epidemic diffusion. Appl. Intell. **51**(5), 2939–2955 (2021). https://doi.org/10.1007/s10489-020-02173-6
7. Huang, X., Zhang, J., Li, D., Li, P.: Knowledge graph embedding based question answering. In: Proceedings of the Twelfth ACM International Conference on Web Search and Data Mining, pp. 105–113 (2019)
8. Ji, G., He, S., Xu, L., Liu, K., Zhao, J.: Knowledge graph embedding via dynamic mapping matrix. In: Proceedings of the 53rd Annual Meeting of the Association for Computational Linguistics and the 7th International Joint Conference on Natural Language Processing (Volume 1: Long Papers), pp. 687–696 (2015)
9. Kaminski, M., Grau, B.C., Kostylev, E.V., Motik, B., Horrocks, I.: Foundations of declarative data analysis using limit datalog programs. CoRR abs/1705.06927 (2017). http://arxiv.org/abs/1705.06927
10. Kazemi, S.M., Poole, D.: Simple embedding for link prediction in knowledge graphs. In: Advances in Neural Information Processing Systems, pp. 4284–4295 (2018)
11. Lin, Y., Liu, Z., Sun, M., Liu, Y., Zhu, X.: Learning entity and relation embeddings for knowledge graph completion. In: Proceedings of the Twenty-Ninth AAAI Conference on Artificial Intelligence, AAAI 2015, pp. 2181–2187. AAAI Press (2015)
12. Miller, G.A.: WordNet: a lexical database for English. Commun. ACM **38**(11), 39–41 (1995)
13. Shadbolt, N., Hall, W., Berners-Lee, T.: The semantic web revisited. IEEE Intell. Syst. **21**(3), 96–101 (2006)
14. Sun, Z., Deng, Z., Nie, J., Tang, J.: Rotate: knowledge graph embedding by relational rotation in complex space. arXiv preprint arXiv:1902.10197 (2019)

15. Trouillon, T., Welbl, J., Riedel, S., Gaussier, É., Bouchard, G.: Complex embeddings for simple link prediction. In: International Conference on Machine Learning (ICML) (2016)
16. Wang, H., Ren, H., Leskovec, J.: Entity context and relational paths for knowledge graph completion. arXiv preprint arXiv:2002.06757 (2020)
17. Wang, H., Zhao, M., Xie, X., Li, W., Guo, M.: Knowledge graph convolutional networks for recommender systems. In: The World Wide Web Conference, WWW 2019, pp. 3307–3313. Association for Computing Machinery, New York (2019). https://doi.org/10.1145/3308558.3313417
18. Wang, Z., Zhang, J., Feng, J., Chen, Z.: Knowledge graph embedding by translating on hyperplanes. In: AAAI, vol. 14, pp. 1112–1119. Citeseer (2014)
19. Xiong, C., Power, R., Callan, J.: Explicit semantic ranking for academic search via knowledge graph embedding. In: Proceedings of the 26th International Conference on World Wide Web, WWW 2017, pp. 1271–1279. International World Wide Web Conferences Steering Committee, Republic and Canton of Geneva, CHE (2017). https://doi.org/10.1145/3038912.3052558
20. Yang, B., Mitchell, T.: Leveraging knowledge bases in LSTMs for improving machine reading. In: Proceedings of the 55th Annual Meeting of the Association for Computational Linguistics (Volume 1: Long Papers) (2017)
21. Yao, Y.: Three-way decisions with probabilistic rough sets. Inf. Sci. **180**(3), 341–353 (2010)
22. Yu, H., Liu, Z., Wang, G.: An automatic method to determine the number of clusters using decision-theoretic rough set. Int. J. Approx. Reason. **55**(1), 101–115 (2014)
23. Zhang, S., Tay, Y., Yao, L., Liu, Q.: Quaternion knowledge graph embeddings. In: Advances in Neural Information Processing Systems, pp. 2735–2745 (2019)
24. Zhang, Z., Zhuang, F., Qu, M., Lin, F., He, Q.: Knowledge graph embedding with hierarchical relation structure. In: Proceedings of the 2018 Conference on Empirical Methods in Natural Language Processing, pp. 3198–3207 (2018)
25. Zhou, B., Yao, Y., Luo, J.: Cost-sensitive three-way email spam filtering. J. Intell. Inform. Syst. **42**(1), 19–45 (2014)

PNeS in Modelling, Control and Analysis of Concurrent Systems

Zbigniew Suraj$^{(\boxtimes)}$ and Piotr Grochowalski

Institute of Computer Science, Rzeszów University, Rzeszów, Poland
{zbigniew.suraj,piotrg}@ur.edu.pl

Abstract. The paper describes the extended and improved version of the Petri Net System (PNeS) compared to the version published in 2017. PNeS is an integrated graphical computer tool for building, modifying, analyzing Petri nets, as well as controlling a mobile robot. It runs on any computer under any operating system. PNeS can be useful for researchers, educators and practitioners, from both academia and industry, who are actively involved in the work of modelling and analyzing concurrent systems, and for those who have the potential to be involved in these areas.

Keywords: Petri net tool · Computer based tool · Petri net · Concurrent system · Computer modelling · Control

1 Introduction

Petri nets are widely used in both theoretical analysis and practical modelling of concurrent systems. Their graphical aspect allows representation of various interactions between discrete events more easily. However, the mathematical aspect allows formal modelling of these interactions and analysis of the modeled system properties.

Petri nets were proposed by Carl A. Petri as a net-like mathematical tool for the study of communication with automata [27]. Their further development was facilitated by the fact that they present two interesting characteristics. Firstly, they make it possible to model and visualize types of behaviour having parallelism, concurrency, synchronization and resource sharing. These properties characterize concurrent systems. Secondly, the theoretic results are plentiful; the properties of these nets have been and still are extensively studied. There exists a large number of books, articles, and proceedings papers devoted to the theory and applications of Petri nets, see e.g. [5–7, 13, 21, 26, 28, 41, 42].

The practical use of Petri nets is strongly dependent upon the existence of adequate computer tools - helping the user to handle all the details of a large and complex description. For Petri nets one needs at least graphical editor, analyzer,

This work was partially supported by the Center for Innovation and Transfer of Natural Sciences and Engineering Knowledge at Rzeszów University.

S. Ramanna et al. (Eds.): IJCRS 2021, LNAI 12872, pp. 279–293, 2021.
https://doi.org/10.1007/978-3-030-87334-9_24

and simulator programs. The graphical editor gives an opportunity for loading, saving, constructing, and editing Petri nets. The analyzer allows to perform a formal check of the properties related to the behavior of the underlying system e.g. concurrent operations, appropriate synchronization, freedom from deadlock, repetitive activities, and mutual exclusion of shared resources, etc. The simulator allows us to perform a Petri net game, i.e., the flow of tokens in the places of the net through transitions. Simulation gives a vivid graphic description of a system's operation to aid in model design and debugging. Simulation becomes necessary when performance cannot be predicted by the system performance evaluator.

The papers [8] and [14] provide good overviews of typical Petri net tools. Some of these tools and their applications are discussed in details in [6,21,42]. A very good resource of information about Petri net theory and its applications is [37]. The tools such as Charlie [40], PEP [9], Petruchio [20], QPME [17] and Snoopy [11] deserve special emphasis due to the scope of their applicability and the functions they provide (cf. also [14]).

The objective of this paper is to present the basic information about the extended and improved version of PNeS compared to the version released in 2017 [33]. The main changes in the system concern extending the scope of the editor and simulator, improving the methods of analyzing net properties and adding the possibility of controlling mobile robots from the system simulator level. For the editor, the ability to build a hierarchical net has been added in two ways: from detail to general (*bootom - up* method) and from general to detail (*top - down* method). As for the simulator, it can work with non-hierarchical nets in single firings mode, in simple and generalized steps mode, as well as with hierarchical nets by selecting the simulation at any acceptable hierarchy level. Additionally, the mobile robot can be controlled from the simulator level. PNeS allows us to work with six classes of Petri nets, i.e., place/ transition nets, nets with inhibitor arcs, self-modifying nets, place transactor nets, priority nets and FIFO-nets. The paper focuses mainly on place/transition nets, although other types of Petri nets are also discussed in the context of the functional capabilities provided by PNeS. It is also worth emphasizing that PNeS can work with systems such as ROSECON and Rosetta. The first is a proprietary software system supporting automated solving of the problem of synthesis and analysis of concurrent systems using the philosophy of rough sets [23]. The second is for knowledge discovery and data mining, also based on the rough set methodology [22]. Moreover, PNeS also has a help module that not only facilitates the use of the system, but also provides basic concepts of Petri nets with examples to illustrate these concepts. The resulting Petri nets are expressed in the PNML formalism, which is a standard format used by many analysis tools for Petri nets [36].

The system runs on any computer under any operating system. It has been implemented in the *Java* programming language to guarantee portability and efficiency on any computers. PNeS is a follow-up of PN-tools which has been designed and implemented at Pedagogical University in Rzeszów in 1986–1996. PN-tools were run on IBM PC computers under DOS operating system [30].

The rest of this paper is organized in the following way. The Sect. 2 focuses on the basic knowledge of Petri nets in the context of PNeS functional capabilities. In Sect. 3, the definitions of the basic properties of Petri nets examined in the system are recalled. Section 4 describes roughly the most fundamental methods of Petri net analysis available in PNeS. A general description of PNeS with examples of its use is provided in Sect. 5. Section 6 compares the system with other systems of similar purpose, recommended in the professional literature as worth using. In addition, this section presents the directions of further research and final comments.

2 Preliminaries

2.1 Petri Net

The structure of a standard Petri net (a Petri net in short) is a directed graph with two kinds of nodes, *places* and *transitions*, interconnected by *arcs* - in such a way that each arc connects two different kinds of nodes (i.e., a place and a transition and vice versa). Graphically, places are represented by circles and transitions as rectangles. A place is an input place to a transition if there exists a directed arc connecting this place to the transition. A place is an output place of a transition if there exists a directed arc connecting the transition to the place. In a Petri net can exist directed (parallel) arcs connecting a place and a transition (a transition and a place). In such a case, in the graphical representation of a Petri net, parallel arcs connecting a place (transition) to a transition (place) are often represented by a single directed arc labeled with its multiplicity. The arc label '1' is omitted. A pair of a place p and a transition t is called a *self-loop* if p is both an input and output place of t. A Petri net is said to be *pure* if it has no self-loops. A Petri net is said to be *ordinary* if all of its arc multiplicities are 1's.

The dynamic behaviour of the modeled system can be described in terms of its states and their changes, each place may potentially hold either none or a positive number of tokens. Pictorially, the tokens are represented by means of gray "dots" together with the suitable positive numbers placed inside the circles corresponding to appropriate places. We assume that if the number of tokens in a place equals 0 then the place is empty. A distribution of tokens on places of a Petri net is called *a marking*. It defines the current state of the system modeled by a Petri net. A marking of a Petri net with n places can be represented by an n-vector M, elements of which, denoted as $M(p)$, are nonnegative integers representing the number of tokens in the corresponding place p. The *initial marking*, denoted as M_0, is the marking determined by the initial state of a system. A Petri net containing tokens is called *a marked Petri net*. In order to simulate the dynamic behaviour of a system, a marking in a Petri net is changed according to the following *firing rule*. A transition t is said to be *enabled* if each input place p of t contains at least the number of tokens equal to the multiplicity of the directed arc connecting p to t. An enabled transition t may or may not fire (depending on whether or not the transition is selected). A firing of an enabled transition t removes from each input place p the number of tokens equal to the multiplicity of the directed arc connecting p to t, and adds to each output place p of t the number of tokens equal to the multiplicity of the directed arc connecting t to p.

In the case of the firing rule above, it is assumed that each place can hold an unlimited number of tokens. Such a Petri net is referred to as *infinite capacity net*. When modeling many physical systems, it is natural to consider an upper limit to the number of tokens they can store in each place. Such a Petri net is referred to as *finite capacity net*. However, in the case of finite capacity net, each place p has an associated capacity $K(p)$, which is the maximum number of tokens that p can hold at any given time. In the case of finite capacity nets, for a transition t to be enabled, there is an additional condition that the number of tokens in each output place p of t cannot exceed its capacity $K(p)$ when fired.

This rule with the capacity constraint is called the *strong transition rule*, whereas the rule without the capacity constraint is called the *(weak) transition rule*. In PNeS, Petri nets with certain capacities are called place/transition nets [28].

2.2   The Ways of Working of a Petri Net

Many possibilities of increasing the usability of a Petri net by introducing different ways of operating it have been explored. Petri nets can work in the so-called *single firings* or *steps*. Working in the mode of single firings consists in selecting, in accordance with the chosen strategy and the transition rule, from among all enabled, i.e. firable, one transition to fire. There are three strategies available in PNeS, marked symbolically MIN, MAX and RND. These strategies mean, respectively, the selection of the transition with the lowest number, the transition with the highest number, or a transition selected randomly by the system. With regard to the transition rules, there are three rules in PNeS: *weak, medium* and *strong*. The difference between the *medium* rule and the *strong* rule is only visible in nets with self-loops. Generalizing the net operation in single firing mode are *steps* (*simple, generalized*). Working in the steps mode can be treated as firing a selected set of enabled transitions simultaneously or as a single (multiple - in the case of *generalized steps*) firing them in any order [29].

2.3   Extensions of Petri Nets

The simplest extension to standard Petri nets is *nets with inhibitor arcs* proposed in [10]. An inhibitor arc leads from a place p to a transition t and inhibits the firing of t if the token load of p is not less than its multiplicity w. If $w > 1$, then, additionally, an ordinary arc from p to t with multiplicity less than w is allowed. Another extension of standard Petri nets are the so-called *self-modifying* Petri net introduced in [39]. Basically self-modifying nets are standard Petri nets in which integers as well as places are associated with the arcs. Considering places as 'variables' and the marking of a place as its 'value', the multiplicity of an arc is then defined as the sum of the integers and the 'value' of the places associated with this particular arc. *Place transactor nets* are self-modifying Petri nets working under the "conserving" firing rule; here, a transition may only fire if the number of tokens in the net is not changed then. Petri nets with *priorities* have been suggested in [10]. Priorities can be associated with the transition such that if t and t' are both enabled, then the transition with the highest priority will fire first. *FIFO*-nets are Petri nets in which places behave as FIFO (First Input First Output) queues rather than counters [19].

2.4 Hierarchical Petri Nets

The basic idea behind hierarchical Petri nets is to allow to the modeller to construct a large model by combining a number of small nets into a larger net. PNeS enables building hierarchical nets in the class of standard Petri nets. Only transitions can be hierarchical in the net. There are two methods for creating such nets in the system. The first is to replace the transitions with pre-prepared subnets that can be stored in files as blocks. The second method is to replace part of the net (subnet) with a single transition. Both methods of building a hierarchical net can be used simultaneously, i.e. some net elements can be replaced with subnets and other parts of the net with single net transitions. Regardless of how the hierarchical net is structured, the user of PNeS can select a subnet model that is associated with the net transitions. Two models are available: BLOCKS-WF (blocks well formed) [38] and D-BLOCKS (deterministic blocks) [3].

3 Properties of Petri Nets

3.1 Behavioural Properties

A marking M is said to be *reachable* from a marking M_0 if there exists a sequence of transitions firings which transforms a marking M_0 to M. The set of all possible markings reachable from M_0 is called the *reachability set*, and is denoted by $R(M_0)$. A Petri net is said to be *k-bounded* or simply *bounded* if the number of tokens in each place does not exceed a finite number k for any marking reachable from the initial marking M_0. A Petri net is said to be *safe* if it is 1-bounded. A Petri net is said to be *stable* if total number of tokens in the net remains constant. A *deadlock* in a Petri net is a marking such that no transition is enabled. A transition t in a Petri net is *live* if for every marking M in $R(M_0)$, t is firable (enabled) at a certain marking M' reachable from M. A Petri net is *live* if every transition is live. A Petri net, for the initial marking M_0, is said to be *reversible* if the initial marking M_0 can be reached from each marking M in $R(M_0)$. Thus, in a reversible Petri net one can always get back to the initial marking. A marking M' is said to be a *home marking* if, for each marking M in $R(M_0)$, M' is reachable from M.

3.2   Structural Properties

Let a Petri net be ordinary. A Petri net is called: (1) a *marked graph* if each place has exactly one input transition and exactly one output transition, (2) a *state machine* if each transition has exactly one input place and exactly one output place. A Petri net is said to be: (1) *free-choice* if each shared place is the only input place of its output transitions, (2) *extended-free-choice* if the output transitions of each shared place have the same input places, (3) *extended-simple* if it holds that if two places have at least one common output transition then the set of all output transitions of one of the places is a subset of the set of all output transitions of the other one, (4) *state machine coverable* if its set of places is a union of components which have, as subnets, the state machine property, where by a component of a Petri net we understand a set of places where each input transition is also an output transition and reversely, (5) *state machine decomposable* if its set of places is a union of components which, on their part, are strongly connected state machines. A nonempty subset of places S in a Petri net is called a *siphon* if every transition having an output place in S has an input place in S. A nonempty subset of places Q in a Petri net is called a *trap* if every transition having an input place in Q has an output place in Q. A siphon (trap) is said to be *minimal* if it does not contain any other siphon (trap). A trap is said to be *maximal* if it is not included in any other trap. A net has the *siphon-trap-property* if the maximal trap of each minimal siphon is marked.

4   Analysis Methods

The most straightforward kind of analysis is *simulation* - which is very useful for the understanding and debugging of a system, in particular in the design phase. Simulation can be supported by a computer tool or it can be totally manually. Simulation can reveal errors, but in practice never be sufficient to prove the correctness of a system.

The basic idea behind *reachability graphs* is to construct a graph which contains a node for each reachable state and an arc for each possible change of state. A reachability graph can be used to prove properties of the modelled system. For bounded systems a large number of questions can be answered. Deadlocks, reachability and existence of marking bounds can be decided by a simple search through the nodes of the reachability graph, while liveness and home markings can be decided by constructing and inspecting the strongly connected components. The reachability graph method can be totally automated - and this means that the modeler can use the method, and interpret the results, without having much knowledge about the underlying mathematics. For more information see [15].

Invariant analysis allows logical properties of Petri nets to be investigated in a formal way. There are two dual classes of invariants. A *place invariant* (P-invariant) characterizes the conservation of a weighted set of tokens, while a *transition invariant* (T-invariant) characterizes a set of transition sequences having no effect, i.e., with identical start and end markings. The main advantages of invariant analysis are the low computational complexity (in particular, compared to the method of reachability graphs). For more information see [19].

Petri nets can also be analyzed by means of *reductions*. The basic idea behind this method is to modify a Petri net - without changing a selected set of properties, e.g. safeness, boundedness, and liveness. The modification of the net is performed by means of a set of transformations rules and may be carried out manually, automatically or interactively. In the latter case the strategy is decided by a person, while the detailed computations and checks are made by a computer. The purpose of the transformation is to obtain a small and simple net for which it is easy to investigate the given properties. For more information see [1].

For ordinary Petri nets several kinds of analysis methods are known. One of the methods uses structural properties, it means properties, which can be formulated without considering the behaviour (i.e., transition sequences) of Petri net to deduce behavioral properties. For more information see [2].

5   PNeS

5.1   General

PNeS is a fully integrated environment designed to aid engineers in modelling and solving concurrency related problems in parallel and distributed computing systems by using Petri net technology. This system supports the user in constructing of Petri net models as well as in modifying and analyzing. In particular, PNeS may be used to help the system designer in proving the correctness of his or her design.

PNeS consists of three main logical parts as follows: editor, simulator, and analyzer. Editor is a window-based graphical editor for loading, saving, constructing, and editing Petri nets. Simulator is a program for graphical and textual simulation of a Petri net model. The simulation results allow for the detection of design errors that appear at the net design stage, and also result from incorrect modeling of the real system. Analyzer is a set of programs by means of which for standard Petri nets: (1) Basic structural and behavioural properties of Petri nets can be checked. The results of such analysis allow to detect syntactic (sometimes even semantic) design errors. For certain subclasses of standard Petri nets the structural properties can be used to deduce behavioral properties. (2) Simple reductions of the size of a net (and in a consequence of its reachability graph) preserving safeness, boundedness, and liveness can be executed. The use of such reduction methods is necessary if the storage capacity of the given computer system is not sufficient to analyze a given net. (3) Calculation of the invariants and other structural information (e.g. state machine components) from a given net which reflect certain structural properties of the modeled system are possible. Invariant analysis can be done by computing generator sets of all P-/T-invariants and of all nonnegative invariants. Additionally, vectors can be tested for invariant properties.

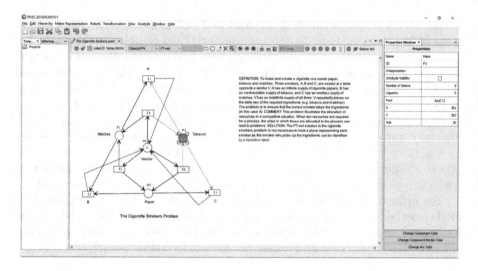

Fig. 1. The main window of PNeS when editing the net.

PNeS has a user-friendly interface. Figure 1 shows its general appearance along with the content in the main editor window.

5.2   Behaviour Analyzers

For testing structural properties from which, in case of ordinary Petri nets, liveness properties follow, the module "Structural Properties..." (available in the option "Analysis" of the Menubar), is used.

If one is given an unknown net, using the module "Basic Properties" one can obtain information on elementary net properties such as: the number of places (transitions), the minimal (maximal) number of the net nodes, as well as the maximal entrance (exit) degree of the net nodes. One can check basic structural properties. Additionally, it is tested, whether the read-in net considered as an undirected graph, is connected. If a net is ordinary one can check whether it is a state machine, a free-choice net, an extended free-choice, or an extended simple net, using the "Structural Properties..." module invoked in the "Analysis" option. These properties are related with the liveness via the siphon-trap-property. Then the minimal siphons are computed [26,29]. If there exists a clean (not marked) siphon, the net is not live. Next, the possible conclusion connected with the liveness of a net follows: (1) If a given net is an extended simple net and the siphon-trap-property holds, then a net is live. (2) If a given net is an extended free-choice and the siphon-trap-property does not hold, then a net is not live. (3) If for a given net the siphon-trap-property holds, then no dead marking can be reached. This module allows also to check whether the net is a state machine decomposable (which implies that it is bounded under any initial marking). In addition, this module decides whether a given net is a state machine coverable, i.e., coverable by components which are state machines.

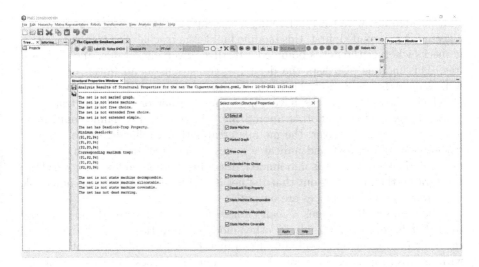

Fig. 2. The results of the analysis of the structural properties of the net from Fig. 1.

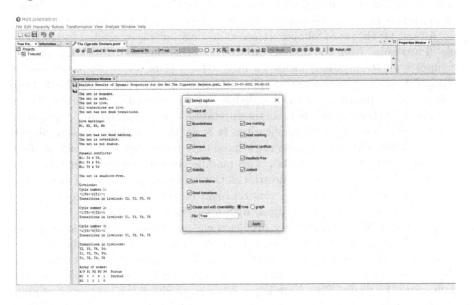

Fig. 3. The results of the analysis of the behavioural properties of the net from Fig. 1 (a fragment).

In the module described above possible conclusions on the behaviour properties are inferred from the structural properties on the basis of an initial marking. If there are no information for such conclusions, we have to investigate the reachability graph. The module "Coverability/Reachability Graphs..." (available in the option "Analysis" of the Menubar) computes the coverability tree/graph [15], which is identical with the reachability tree/graph in the case of a bounded net. If a given net is bounded, first, this module builds the reachability tree/graph, then it checks whether the net is bounded, safe, live, reversible, stable, and deadlock-free. This module also lists the dead/live transitions at an initial marking as well as dead/live markings. Then, the module checks at each reachable marking at which several transitions have concession whether one of these transitions takes the concession of another upon firing. The obtained dynamic conflicts are also written out. Moreover, the module creates the graphical representation of reachability tree/graph and indicates nodes in the tree/graph which are dead, duplicated. If a given net is unbounded, first, this module builds the coverability tree/graph, then it writes out that the net is unsafe, unbounded, and nonstable. Moreover, unbounded places are output.

The module "Reductions..." can be used to reduce the size of the given net, so that it becomes analyzable by other modules from the option "Analysis" of the Menubar on the one hand, and it can be used to find an equivalent small net with known properties, i.e., boundedness, safeness, liveness. In this module has been implemented the most essential local reductions steps known from the literature, e.g. merging of nodes which share all predecessors and successors, fusion of equivalent places, reduction of different kinds of place/transition chains [21].

A basis for the set of all place (transition) invariants is computed by the module "Invariants" from the option "Analysis" of the Menubar. First, this module computes an (transposed) incidence matrix for a given Petri net as a basis for computation of all (nonnegative) place (transition) invariants [28]. Next, from this it derives information on boundedness and reversability properties. Moreover, in general, invariants have an interpretation in terms of the modeled system which can be useful for its verification. It is worth to notice that several conclusions derived from invariants are related only to nonnegative invariants (e.g. net coverability by P-invariants). This module computes a basis for the set of nonnegative place (transition) invariants. Correspondingly, the set of all nonnegative invariants is the set of vectors that can be generated from the computed set by means of linear combinations with nonnegative coefficients.

The reachability/coverability problems can be solved in dialogue using the module "Tests". For this, first, one should use the option "Reachability/ Coverability". Then, the respective marking should be input in the way shown on the screen. The module checks whether this marking is reachable/coverable and announces the result. This module permits also to check the properties of place and transition vectors, respectively. It shows whether the vector investigated is an invariant.

When working with our analyzer we have found it difficult to model complex systems with Petri nets, because the nets are often large and it is not quite easy to see all the interactions between the nodes of the net. Due to that we have found it necessary to use extended nets for our modelling purposes; but then we need an analyzer for those nets. The fastest way to construct an analyzer is to use existing programs as much as possible. Therefore our first step to enable the analysis of extended nets is to translate them into Petri nets and

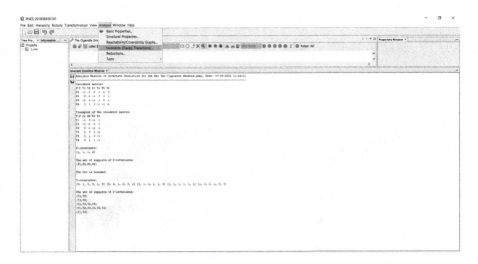

Fig. 4. The results of the analysis of the place/transition invariants of the net from Fig. 1.

to use our analyzer. All we need is an automatic extended Petri net - to the Petri net translator (the translator). The translator is implemented first for: self-modifying nets, nets with inhibitor arcs and some subclass of Petri nets. The translator realizes, among others, the following tasks: (1) It forms a Petri net corresponding to a given self-modifying net (a net with inhibitory arcs). (2) It maps the initial marking of the self-modifying net (the net with inhibitory arcs) into a marking of the Petri net. For more information see [29].

5.3 Controlling Robots

PNeS allows us to control Lego Mindstorms mobile robot using net simulator module. This system acts as a robot controller based on a simulation of the Petri nets running in the system, which is also a control model for the robot. The robot first performs the commands sent from PNeS, and then transmits binary information about the status of its sensors to the system. The entire control process takes place in a feedback loop between PNeS and the robot. The current version of the system allows the robot to perform the following tasks: avoiding the obstacles, reaching a target, following an obstacle, finding the way out of a labyrinth and influencing the environment by selecting the right manipulators. The list of performed tasks depends strictly on the hardware configuration of the available robot (cf. [16]).

Figure 5 shows an example of a Petri net in PNeS controlling the Lego Mindstorms mobile robot performing the task of avoiding the obstacles. In addition, the red color in the drawing shows the options in the system that allow you to prepare a control model for the robot and run it from the system simulator level. Due to the lack of space, a detailed description of the net and its operation was omitted.

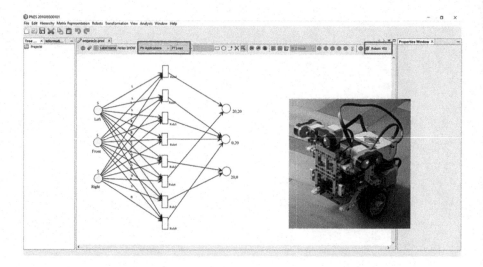

Fig. 5. An example of a Petri net in PNeS controlling a mobile robot.

6   Conclusion and Further Work

In the paper, an overview of the functional capabilities of PNeS, examples of their practical use and differences in relation to the previous version of this system have been presented. By using this system, researchers and practitioners interested in the theory and application of Petri nets obtain a method and a tool with which main design aspects of modeled systems can be analyzed quickly and correctly, as well described in a presentable way. Moreover, in order to allow the Petri nets simulation to interact and synchronize with the real world, a new functionality has been added in PNeS to control the device (robot) via the system net simulator. Analysts, system designers and everyone who, in the framework of project developments, has to describe coherently and vividly complex procedures of system engineering on the basis of a theoretical method, are able to carry out their task in a more economical and time-saving way.

Although in the last three decades, a large number of tools have been reported in the Petri net literature, a majority of these tools are used mostly for research and educational purposes [8,14]. From our observations, in most cases, existing tools concern a concrete class of Petri nets or a particular method for designing and/or analyzing of Petri net model. In a case presented here, the situation is different. We describe in this paper an integrated and modular PNeS which possess features universality in this sense that it can be used for several classes of Petri nets and makes accessible various methods of the designing and analysis of Petri net models.

Considering the need to model and analyze concurrent systems operating in an uncertain environment, for several years we have been conducting intensive research on the development of new fuzzy Petri nets models [31,32,34] and their successive implementation into PNeS. As is commonly known, fuzzy Petri nets are useful, among others, for knowledge representation and approximate reasoning in intelligent decision systems [4,18,25]. We would like to continue this research and the developed methods first to be implemented in PNeS and then thoroughly tested on real data. We hope that one of the examples of meaningful use of PNeS in scientific research may be its application in solving the practical problem of passenger transport logistics modeling based on intelligent computing techniques. Both the problem itself and an example of its solution are described in detail in [35]. In further research related to device control, we plan to control a group of communicating robots performing common tasks and extracting the Petri net control model from real data describing a given type of control.

Table 1. Comparison between the recommended Petri net tools and PNeS

PN Tool	PN Supported						Component												Analysis					Environment					
	High-level Petri nets	Stochastic Petri nets	Petri nets with time	Place/transition nets	Continuous Petri nets	Transfer Petri nets	Graphical editor	State spaces	Condensed state spaces	Token game animation	Fast simulation	Place invariants	Transition invariants	Net reduction	Model checking	Petri net generator	Interchange file format	Petri net translator	Simple performance analysis	Structural analysis	Advanced performance analysis	Reachability graph based analysis	Invariant based analysis	Java	Linux	Sun	HP, HP-UX	Windows	Macintosh
PEP	*						*	*	*	*		*	*	*	*	*	*	*		*				*	*				
SNOOPY		*	*	*	*		*				*	*													*			*	*
CHARLIE		*		*	*		*						*						*		*	*		*	*			*	
PETRUCHIO	*	*	*	*		*	*	*		*	*	*	*			*		*		*				*	*	*	*	*	*
QPME	*	*		*			*					*				*				*				*	*	*	*	*	*
PNeS		*					*	*		*	*	*	*	*		*	*	*		*		*	*	*	*	*	*	*	*

Note: PNeS supports additionally five other types of Petri nets, i.e. self-modifying Petri nets, nets with inhibitory arcs, Petri nets with priorities, place transactor nets, FIFO-nets.

Finishing this section, we would like to compare the current PNeS capabilities with those of other publicly available systems intended for Petri net users. Taking into account five criteria, such as 'Petri Net Supported', 'Component', 'Analysis', 'Environment', 'Free of Charge', the authors of [14] selected five out of 20 assessed tools dedicated to Petri nets: Charlie [40], PEP [9], Petruchio [20], QPME [17], Snoopy [11], which they think are worth recommending. Table 1 compares the five recommended tools with PNeS. Our system seems quite promising in this comparison.

Acknowledgment. We thank the anonymous reviewers for their helpful comments.

References

1. Berthelot, G.: Transformations and decompositions of nets. In: Brauer, W., Reisig, W., Rozenberg, G. (eds.) ACPN 1986. LNCS, vol. 254, pp. 359–376. Springer, Heidelberg (1987). https://doi.org/10.1007/978-3-540-47919-2_13

2. Best, E.: Structure theory of petri nets: the free choice hiatus. In: Brauer, W., Reisig, W., Rozenberg, G. (eds.) ACPN 1986. LNCS, vol. 254, pp. 168–205. Springer, Heidelberg (1987). https://doi.org/10.1007/978-3-540-47919-2_8

3. Bruno, J., Altman, S.M.: A theory of asynchronous control networks, IEEE Trans. Comput. C-20, 629–638 (1971)

4. Chen, S.M., Ke, J.S., Chang, J.F.: Knowledge representation using fuzzy Petri nets. IEEE Trans. Knowl. Data Eng. 2(3), 311–319 (1990)

5. David, R., Alla, H.: Petri Nets and Grafcet: Tools for Modelling Discrete Event Systems. Prentice Hall, New York (1992)

6. Desel, J., Reisig, W., Rozenberg, G.: Lectures on Concurrency and Petri Nets. Springer, Berlin (2004). https://doi.org/10.1007/b98282

7. DiCesare, F., Harhalakis, G., Proth, J.M., Silva, M., Vernadat, F.B.: Practice of Petri nets in Manufacturing. Chapman and Hall, New York (1993)

8. Feldbrugge, F., Jensen, K.: Computer tools for high-level petri nets. In: High-level Petri Nets, pp. 691–717. Theory and Application, Springer, Berlin (1991)

9. Grahlmann, B., Best, E.: PEP - more than a petri net tool. In: Tools and Algorithms for the Construction and Analysis of Systems, pp. 397–401. Springer, Berlin (1996)

10. Hack, M.: Decidability Questions for Petri Nets. Ph.D. Dissertation, Department of Electrical Engineering, MIT, Cambridge (1975)

11. Heiner, M., Herajy, M., Liu, F., Rohr, C., Schwarick, M.: Snoopy - a unifying petri net tool. In: Application and Theory of Petri Nets, pp. 398–407. Springer, Berlin (2012)

12. Jeng, M.D., DiCesare, F.: A review of synthesis techniques for Petri nets with applications to automated manufacturing systems. IEEE Trans. Syst., Man, Cybern. 23(1), 301–312 (1993)

13. Jensen, K., Rozenberg, G. (eds.): High-level Petri Nets. Theory and Application, Springer, Berlin (1991)

14. Jie, T.W., Ameedeen, M.A.: A survey of petri net tools. In: Advanced Computer and Communication Engineering Technology, Lecture Notes in Elect. Eng., vol. 315, pp. 537-551 (2015)

15. Karp, R.M., Miller, R.E.: Parallel program schemata. J. Comput. Syst. Sci. 3, 147–195 (1969)

16. Kindler, E., Nillies, F.: Petri Nets and the Real World, Semantic Scholar, Corpus ID: 1119127 (2006)

17. Kounev, S., Dutz, C.: QPME: a performance modeling tool based on queueing Petri Nets. ACM SIGMETRICS Perform. Eval. Rev. 36(4), 46–51 (2009)

18. Looney, C.G.: Fuzzy petri nets for rule-based decision-making. IEEE Trans. Syst. Man, Cybern. 18–1, 178–183 (1988)

19. Memmi, G., Vautherin, J.: Analysing nets by the invariant method. In: Brauer, W., Reisig, W., Rozenberg, G. (eds.) ACPN 1986. LNCS, vol. 254, pp. 300–336. Springer, Heidelberg (1987). https://doi.org/10.1007/978-3-540-47919-2_11

20. Meyer, R., Strazny, T.: Petruchio: from dynamic networks to nets. In: Computer Aided Verification, pp. 175–179. Springer, Berlin (2010)

21. Murata, T.: Petri nets: properties, analysis and applications. Proc. of the IEEE 77(4), 541–580 (1989)

22. Øhrn, A., Komorowski, J., Skowron, A., Synak, P.: The Rosetta software system. In: Rough Sets in Knowledge Discovery 2, pp. 572–576. Applications, Studies in Fuzziness and Soft Computing, Springer, Berlin (1998)

23. Pancerz, K., Suraj, Z.: Discovering concurrent models from data tables with the ROSECON system. Fundam. Informat. 60(1–4), 251–268 (2004)

24. Pawlak, Z.: Rough sets. Int. J. Comput. & Informat. Sci. **11**, 341–356 (1982)
25. Pedrycz, W., Gomide, F.: A generalized fuzzy Petri net model. IEEE Trans. on Fuzzy Systems **2**–**4**, 295–301 (1994)
26. Peterson, J.L.: Petri Net Theory and the Modeling of Systems. Prentice-Hall Inc, Englewood Cliffs, N.J. (1981)
27. Petri, C.A.: Kommunikation mit Automaten. Bonn: Institut für Instrumentelle Mathematik, Schriften des IIM Nr. 3, 1962 (1966)
28. Reisig, W.: Petri Nets. Springer Publ, Company (1985). https://doi.org/10.1007/978-0-387-09766-4_134
29. Starke, P.H.: On the mutual simulatability of different types of Petri nets. In: Voss, K., Genrich, H.J., Rozenber, G. (eds.) Concurrency and Nets, Springer, Berlin, Heidelberg, pp. 481–495 (1987). https://doi.org/10.1007/978-3-642-72822-8_30
30. Suraj, Z.: PN-tools: environment for the design and analysis of Petri Nets. Control. Cybern. **24**(2), 199–222 (1995)
31. Suraj, Z.: A new class of fuzzy Petri nets for knowledge representation and reasoning. Fundam. Informat. **128**(1–2), 193–207 (2013)
32. Suraj, Z., Bandyopadhyay, S.: Generalized weighted fuzzy petri net in intuitionistic fuzzy environment. In: Proceedings of the IEEE World Congress on Computational Intelligence, 25–29 July, 2016, Vancouver, Canada, pp. 2385–2392 (2016)
33. Suraj, Z., Grochowalski, P.: Petri Nets and PNeS in Modeling and Analysis of Concurrent Systems, pp. 1–12. Proc. CS&P, Warsaw, Poland (2017)
34. Suraj, Z., Hassanien, A.E., Bandyopadhyay, S.: Weighted generalized fuzzy petri nets and rough sets for knowledge representation and reasoning. In: Bello, R., Miao, D., Falcon, R., Nakata, M., Rosete, A., Ciucci, D. (eds.) IJCRS 2020. LNCS (LNAI), vol. 12179, pp. 61–77. Springer, Cham (2020). https://doi.org/10.1007/978-3-030-52705-1_5
35. Suraj, Z., Olar, O., Bloshko, Y.: Modeling of passenger transport logistics based on intelligent computational techniques. Int. J. Comput. Intell. Syst. (accepted) (2021)
36. The Petri Net Markup Language (2017). http://www.pnml.org
37. The Petri Nets World (2017). http://www.informatik.uni-hamburg.de/TGI/PetriNets
38. Valette, R.: Analysis of Petri nets by stepwise refinements. J. Comput. Syst. Sci. **18**(1), 35–46 (1978)
39. Valk, R.: Self-modifying nets: a natural extension of Petri Nets, Lecture Notes in Comput. Sci., vol. 62, Springer, Berlin, pp. 464–476 (1978). https://doi.org/10.1007/3-540-08860-1_35
40. Wegener, J., Schwarick, M., Heiner, M.: A plugin system for charlie. In: Proceedings of the CS&P, pp. 531–554 (2011)
41. Popova-Zeugmann, L.: Time petri nets. In: Time and Petri Nets, pp. 31–137. Springer, Heidelberg (2013). https://doi.org/10.1007/978-3-642-41115-1_3
42. Zurawski, R., MengChu, Z.: Petri nets and industrial applications: a tutorial. IEEE Trans. Ind. Electr. **41**(6), 567–583 (1994)

3RD: A Multi-criteria Decision-Making Method Based on Three-Way Rankings

Yiyu Yao$^{(\boxtimes)}$ and Chengjun Shi$^{(\boxtimes)}$

Department of Computer Science, University of Regina,
Regina, SK S4S 0A2, Canada
{Yiyu.Yao,csn838}@uregina.ca

Abstract. By combining ideas from three-way decision theory, prospect theory, and several families of multi-criteria decision-making (MCDM) methods, including ELECTRE, PROMETHEE, TODIM, and dominance-based rough set analysis (DRSA), we propose a new ranking-based MCDM method called 3RD. With respect to a single criterion, we construct a three-way ranking (i.e., trilevel ranking) of a set of decision alternatives by using an alternative as a reference in the sense of prospect theory and a family of three-way rankings from all alternatives. With respect to a set of criteria, we have multiple families of three-way rankings. By adopting the TODIM procedure, we introduce a ranking function to rank the set of alternatives according to these multiple families of trilevel rankings.

Keywords: Three-way decision · Prospect theory · Three-way ranking · Multi-criteria decision-making

1 Introduction

Multi-criteria decision-making (MCDM) problems involve a set of decision alternatives and another set of possibly conflicting criteria. By assessing and comparing decision alternatives according to the set of criteria, an MCDM method supports a decision maker to search for one optimal alternative, to construct a set of optimal alternatives, or to sort decision alternatives into different categories [8]. For each alternative, the assessment and evaluation of the set decision alternatives can be easily done or are given. A basic task of an MCDM method is essentially an aggregation of the results obtained from multiple criteria [16,21,28].

There are three broad families of MCDM models based on, respectively, the concepts of an outranking relation, a value (utility) function, and a set of decision rules for summarizing and representing preference information about decision alternatives. The group of ranking-based approaches ranks the set of decision alternatives directly or indirectly through pairwise comparison. A ranking

This work was partially supported by a Discovery Grant from NSERC, Canada. The authors thank reviewers for their constructive comments.

S. Ramanna et al. (Eds.): IJCRS 2021, LNAI 12872, pp. 294–309, 2021.
https://doi.org/10.1007/978-3-030-87334-9_25

of decision alternatives captures a decision maker's preference. The group of value-based approaches assigns values to decision alternatives to indicate their performance. Although in general the two groups are not equivalent, they may be transformed into each other. In one direction, it is possible to assign values to decision alternatives to reflect information given by a ranking, which is a topic of utility theory [5]. In the other direction, it is easy to obtain a ranking from the values of decision alternatives, as being done in the ELECTRE family [1,4,19]. To increase the explanation ability of MCDM, rule-based approaches, such as dominance-based rough set models [6,7,20], use easy-to-understand rules.

In a ranking-based aggregation method, for each criterion, a ranking of alternatives is given explicitly or is induced from the performance values of decision alternatives. An MCDM method produces an aggregated ranking by, for example, voting or minimizing the overall distance between a ranking and the set of all rankings induced by the set of criteria [8,21]. Examples of ranking-based methods include the ELECTRE family [1,4,19]. These methods first construct a family of rankings of decision alternatives according to their performance values on individual criteria and then combine the family into an aggregated ranking. In a value-based aggregation method, performance values of decision alternatives on the set of criteria are aggregated to rank the set of decision alternatives. Examples of value-based aggregation MCDM methods include TOPSIS family [12], TODIM family [9], and PROMETHEE family [2]. They normally use arithmetic operations on the performance values. Ranking-based methods are appropriate for criteria with nominal values or linguistic descriptions that only give rankings of decision alternatives and may not allow arithmetic operations. For criteria with numerical values, a transformation from performance values to qualitative rankings may lose some useful quantitative information. Value-based methods are appropriate for criteria with numerical values, where arithmetic operations can be easily applied. For criteria with nominal values, it is necessary to change nominal values into numerical values. A difficulty with such a transformation is the choice of numerical values.

Several studies attempted to combine three-way decision with MCDM [13,23, 28]. The main objective of this paper is to propose a new three-way ranking-based MCDM method, called 3RD, by taking also advantages of value-based methods. Guided by the philosophy of three-way decision as thinking in threes [24–27], 3RD is built based on ideas of reference points and prospect values of gain and loss from the prospect theory [14,22], the notions of pair-wise comparision of decision alternatives from dominance-based rough sets [6,7,20], the ELECTRE family [1,4,19], and the PROMETHEE family [2,3], and the aggregation procedure from the TODIM family [9–11,15,17]. The main features and ingredients of 3RD are summarized as follows:

- Following the idea of reference point of prospect theory [14,22], we treat each decision alternative a as a reference point to compare with other alternatives. Inspired by the principle of three-way decision, with respect to a criterion, we trisect the set of decision alternatives into three subsets (i.e., a trisection): 1) the subset of alternatives that are observably better than a, 2) the subset of alternatives that are observably worse than a, and 3) the subset of alternatives that are approximately the same as a. The first two subsets correspond to the notions of the dominating set and the dominated set of a in dominance-based rough set approaches.

- We design and interpret a constructive way to build a trisection of the set of decision alternatives based on ideas from the prospect theory and the ELECTRE, the PROMETHEE, and the TODIM families of MCDM methods.
- A trisection of the set of decision alternatives may be viewed as a three-way ranking. In this way, we have multiple rankings, rather than a single ranking, with respect to a single criterion. For this reason, instead of rank aggregation, we regenerate values for decision alternatives from multiple trisections with respect to a single criterion and aggregate the values from multiple criteria according to the TODIM procedure.

In summary, 3RD may be viewed as a hybrid method that adopts ideas from ranking-based and value-based methods. Moreover, 3RD only conceptually uses the notions from existing studies and, at the same time, assigns new meanings to these notions. A very preliminary study of 3RD suggests that we may be in a promising territory for exploring new MCDM methods.

2   Three-Way Ranking Based Multi-criteria Decision-Making

We propose a new method for multi-criteria decision-making based on three-way rankings (i.e., trilevel structures or trisections) of the set of decision alternatives.

2.1   Trisecting the Set of Decision Alternatives

The information and knowledge for multi-criteria decision-making may be conveniently represented in a tabular form called a multi-criteria decision-making table (MCDMT).

Definition 1. *A multi-criteria decision-making table (MCDMT) is a triplet* (A, C, p), *where* $A = \{a_1, \ldots, a_n\}$ *is a finite and non-empty set of n alternatives,* $C = \{c_1, \ldots, c_m\}$ *is a finite and non-empty set of m criteria, and* $p : A \times C \longrightarrow V$ *is a mapping function that maps a pair of a decision alternative* a_i *and a criterion* c_j *into a value* $p(a_i, c_j) = p_j(a_i) \in V$.

We divide the set of criteria into two disjoint subsets, namely, a subset of qualitative (ordinal) criteria and another subset of quantitative (cardinal) criteria [18]. An ordinal criterion may use nominal values or linguistic labels, where only qualitative ordering information is meaningful and no arithmetic operations can be used. A cardinal criterion may use real numbers, where, in addition to the ordering real numbers, some arithmetic operations are allowed.

For simplicity, we use the same symbol $>$ (or $<$) for both orderings of ordinal and cardinal criteria and assume that $>$ (or $<$) is asymmetric, transitive, and any two distinct values are comparable. The symbol $\leq$ (or $\geq$) stands for the complement of $>$ (or $<$). An assumption is made that a larger value reflects a better performance. That is, for two alternatives a and b, "a is better than b" (or "a is not worse than b") with respect to criterion c_j if and only if $p_j(a) > p_j(b)$ (or $p_j(a) \geq p_j(b)$).

Consider an alternative $a_r \in A$. With respect to a criterion $c_j \in C$, we can easily divide the set of alternatives into three disjoint subsets: 1) a subset $D_j^{\gg}(a_r)$ consisting of alternatives that are better than a_r, 2) a subset $D_j^{\ll}(a_r)$ consisting of alternatives that are worse than a_r, and 3) a subset $D_j^{\approx}(a_r)$ consisting of alternatives that are the same as a_r. Such a trisection of A in fact gives rise to a three-way (i.e., trilevel) ranking of A. We formally define the trisection for ordinal and cardinal criteria, respectively.

Definition 2. *In an MCDMT, by treating an alternative $a_r \in A$ as a reference alternative, with respect to an ordinal criterion $c_j \in C$ we trisect the set of alternatives A as follows:*

$$D_j^{\gg}(a_r) = \{a_k \in A \mid p_j(a_k) > p_j(a_r)\},$$
$$D_j^{\ll}(a_r) = \{a_k \in A \mid p_j(a_k) < p_j(a_r)\},$$
$$D_j^{\approx}(a_r) = A - (D_j^{\gg}(a_r) \cup D_j^{\ll}(a_r)). \tag{1}$$

In the definition, we assume that distinct values of an ordinal criterion are sufficiently different for us to rank decision alternatives. On the other hand, this assumption may not be reasonable for a cardinal criterion. A pair of alternatives with very close values may be considered to be practically the same. We need to have an observable difference between two quantitative values in order to have a clear ranking. This idea has been considered in several MCDM methods, for example, the ELECTRE [19] and the PROMETHEE families [2].

Consider an alternative $a_r \in A$ and a cardinal criterion c_j. We may use a_r as a reference alternative to evaluate another alternative a_k. When $p_j(a_k) \geq p_j(a_r)$, we would have a gain of $p_j(a_k) - p_j(a_r)$ if we used a_k to substitute a_r. On the other hand, when $p_j(a_k) < p_j(a_r)$, we would have a loss of $p_j(a_k) - p_j(a_r)$ if we used a_k to substitute a_r. Prospect theory investigates the actual behavior of individuals when faced with decisions of potential gains and losses. A fundamental result is that people's perceptions of, and reactions to, gains and losses are different. People prefer risk-averse decisions towards gains and risk-seeking decisions towards losses. This requires an asymmetric S-shaped prospect function, as shown Fig. 1a, to reflect the actual values of gains and losses. A commonly used prospect function in the cumulative prospect theory is given by a two-part formula [22]:

$$v(x) = \begin{cases} x^{\alpha}, & \text{if } x \geq 0, \\ (-\lambda)(-x)^{\beta}, & \text{if } x < 0, \end{cases} \tag{2}$$

where λ, α, and β are three parameters. By following the ideas of the TODIM family [9–11,15,17], we substitute the variable x in the prospect function (2) with the difference $p_j(a_k) - p_j(a_r)$. The result is given by Fig. 1b, in which $p_j(a_r)$ serves as the reference point. In order to capture the notions of "observably better" and "observably worse", we introduce a threshold t on the prospect function. By summarizing these results, we propose a definition of trisection of A with respect to a cardinal criterion.

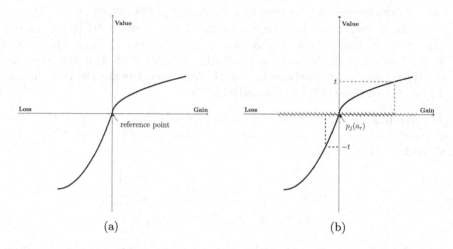

Fig. 1. The prospect theory value function

Definition 3. *In an MCDMT, by considering an alternative $a_r \in A$ as the reference alternative, with respect to a cardinal criterion $c_j \in C$ and a given threshold t, we trisect the set of alternatives as follows:*

$$D_j^{\gg}(a_r) = \{a_k \in A \mid v(p_j(a_k) - p_j(a_r)) > t\},$$
$$D_j^{\ll}(a_r) = \{a_k \in A \mid v(p_j(a_k) - p_j(a_r)) < -t\},$$
$$D_j^{\approx}(a_r) = A - (D_j^{\gg}(a_r) \cup D_j^{\ll}(a_r)). \tag{3}$$

The threshold t may be interpreted as an observable level of the degree of dominance. If $v(p_j(a_k) - p_j(a_r)) > t$, then a_k is observably better than a_r; if $v(p_j(a_k) - p_j(a_r)) < -t$, then a_k is observably worse than a_r; otherwise, a_k is approximately the same as a_r. We have a three-way ranking of the set of alternatives.

In order to have a full understanding of the proposed definitions of trisections, we show some connections to existing studies. In the dominance-based rough set analysis (DRSA) [6,7,20], with respect to a criterion c_j and a reference alternative a_r, a pair of dominating and dominated sets of alternatives of a_r is defined as follows:

$$D_j^+(a_r) = \{a_k \in A \mid p_j(a_k) \geq p_j(a_r)\},$$
$$D_j^-(a_r) = \{a_k \in A \mid p_j(a_k) \leq p_j(a_r)\}. \tag{4}$$

The two subsets are not disjoint, because $a_r \in D_j^+(a_r)$ and $a_r \in D_j^-(a_r)$. On the other hand, a trisection introduced in this paper consists of pairwise disjoint subsets. For an ordinal criterion c_j, it can be easily verified that $D^+ = D^{\gg} \cup D^{\approx}$ and $D^- = D^{\ll} \cup D^{\approx}$. For a cardinal criterion, these relationships hold only under the condition $t = 0$. By introducing the third indifference subset of alternatives and a threshold on the prospect function, the notion of trisections seems to be meaningful in practice, as a small difference in values may not necessarily support a dominance relationship.

The dominance-based rough set models basically use the sign of $p_j(a_k) - p_j(a_r)$, which may be interpreted in terms of a threshold of 0. In contrast, the ELECTRE and PROMETHEE families of MCDM methods [1–4, 19] directly use the difference of $p_j(a_k) - g_j(a_r)$ and a pair of thresholds to determine whether a_k dominates or is dominated by a_r. We consider a special case formulated based on a pair of thresholds (t_p, t_o) satisfying the condition $t_p > t_o > 0$. The same argument is easily extended to the more general cases. With respect to a criterion c_j, the pair of thresholds is used to define a pair of a strict preference relation P_j and an outranking relation S_j as follows:

$$
\begin{aligned}
a_k P_j a_r &\iff p_j(a_k) - p_j(a_r) > t_p, \\
a_k S_j a_r &\iff p_j(a_k) - p_j(a_r) \geq -t_o.
\end{aligned}
\tag{5}
$$

It follows that for any $a_k, a_r \in A$, $a_k P_j a_r \implies a_k S_j a_r$. Based on the two relations, for an alternative $a_r \in A$, we can immediately define the following three-way ranking:

$$
\begin{aligned}
D_j^b(a_r) &= \{a_k \in A \mid a_k P_j a_r\} = \{a_k \in A \mid p_j(a_k) - p_j(a_r) > t_p\}, \\
D_j^w(a_r) &= \{a_k \in A \mid \neg(a_k S_j a_r)\} = \{a_k \in A \mid p_j(a_k) - p_j(a_r) < -t_o\}, \\
D_j^s(a_r) &= A - (D_j^b(a_r) \cup D_j^w(a_r)).
\end{aligned}
\tag{6}
$$

By using the prospect function (2), we can easily re-express $D_j^{\gg}(a_r)$ and $D_j^{\ll}(a_r)$ as follows:

$$
\begin{aligned}
a_k \in D_j^{\gg}(a_r) &\iff p_j(a_k) \geq p_j(a_r) \wedge v(p_j(a_k) - p_j(a_r)) > t \\
&\iff p_j(a_k) \geq p_j(a_r) \wedge (p_j(a_k) - p_j(a_r))^{\alpha} > t, \\
&\iff p_j(a_k) - p_j(a_r) > t^{\frac{1}{\alpha}}; \\
a_k \in D_j^{\ll}(a_r) &\iff p_j(a_k) < p_j(a_r) \wedge v(p_j(a_k) - p_j(a_r)) < -t \\
&\iff p_j(a_k) < p_j(a_r) \wedge (-\lambda)(p_j(a_r) - p_j(a_k))^{\beta} < -t \\
&\iff p_j(a_k) - p_j(a_r) < -(\frac{t}{\lambda})^{\frac{1}{\beta}}.
\end{aligned}
\tag{7}
$$

By setting $t_p = t^{\frac{1}{\alpha}}$ and $t_o = (\frac{t}{\lambda})^{\frac{1}{\beta}}$, we obtain $D_j^b(a_r)$ and $D_j^w(a_r)$. Therefore, the notion of a three-way ranking with a single threshold on the prospect function provides an explanation of the use of a pair of thresholds on the difference value $p_j(a_k) - p_j(a_r)$ in the ELECTRE family of MCDM method.

2.2   Computing Dominance Values of Decision Alternatives

With respect to a single criterion, we treat each alternative as the reference to construct a three-way ranking. For a set of n alternatives, we may have a family of n potentially different three-way rankings. From the m criteria, we can construct m families of three-way rankings. We now consider the problem of building a function from these families of three-way rankings to rank the set of decision alternatives.

Comparison of Two Alternatives on a Single Criterion. Given a criterion $c_j \in C$, a pair of alternatives (a_i, a_k) produces two three-way rankings:

$$(D_j^{\gg}(a_i), D_j^{\ll}(a_i), D_j^{\approx}(a_i)),$$
$$(D_j^{\gg}(a_k), D_j^{\ll}(a_k), D_j^{\approx}(a_k)). \tag{8}$$

We compute the value of alternative a_i in comparison with a_k, denoted by $\phi_j(a_i, a_k)$, in the similar way as being done in the TODIM family. The main difference is that we use the two three-way trisections in Eq. (8) and the qualitative ordering of the values of $p_j(a_i)$ and $p_j(a_k)$. Corresponding to the two-part definition of a prospect function, we consider two cases.

Case 1: $p_j(a_i) \geq p_j(a_k)$. It can be seen that $D_j^{\gg}(a_i) \subseteq D_j^{\gg}(a_k)$, that is, the set of alternatives observably better than a_i is a subset of the set of alternatives observably better than a_k. The value $v(p_j(a_i) - p_j(a_k))$ represents the value of gain, if we substitute a_k by a_i. However, this gain does not provide any information about the relative position of a_i among the set of alternative that are observably better that a_k. Intuitively, the gap between a_k and a_i, in terms of the number of alternatives, determines the relative position of a_i among the set of alternative observably better than a_k. The difference set $D_j^{\gg}(a_k) - D_j^{\gg}(a_i)$ consists of alternatives that are observably better than a_k but not observably better than a_i. If we substitute a_k by a_i, compared with a_k, we would increase the performance by advancing to or above the levels of those alternatives in the set $D_j^{\gg}(a_k) - D_j^{\gg}(a_i)$. Thus, the larger is this difference set, the higher the value of a_i relative to a_k. Based on this observation, we may use the cardinality of the difference set, namely, $|D_j^{\gg}(a_k) - D_j^{\gg}(a_i)|$, to measure the value of a_i related to a_k. In general, we may also use any positive monotonic transformation of the cardinality as such a measure. When $p_j(a_k) = p_j(a_i)$, a_k and a_i have same performance on c_j, the set $D_j^{\gg}(a_k) - D_j^{\gg}(a_i)$ is the empty set. We have a value of 0 for a_i relative to a_k.

Case 2: $p_j(a_i) < p_j(a_k)$. It can be seen that $D_j^{\ll}(a_i) \subseteq D_j^{\ll}(a_k)$, that is, the set of alternatives observably worse than a_i is a subset of the set of alternatives observably worse than a_k. If we substitute a_k by a_i, compared with a_k, we would decrease the performance by declining to or below the levels of those alternatives in the set $D_j^{\ll}(a_k) - D_j^{\ll}(a_i)$. In this case, $|D_j^{\ll}(a_k) - D_j^{\ll}(a_i)|$ is the number of alternatives between the two alternatives. The negative value of the cardinality of the difference set, namely, $-|D_j^{\ll}(a_k) - D_j^{\ll}(a_i)|$, may be used to measure the value of a_i relative to a_k.

By summarizing the results of the two cases, with respect to a criterion c_j, we use a two-part function to measure the dominance value of alternative a_i relative to alternative a_k:

$$\phi_j(a_i, a_k) = \begin{cases} |D_j^{\gg}(a_k) - D_j^{\gg}(a_i)|, & \text{if } p_j(a_i) \geq p_j(a_k), \\ -|D_j^{\ll}(a_k) - D_j^{\ll}(a_i)|, & \text{if } p_j(a_i) < p_j(a_k). \end{cases} \tag{9}$$

In this definition, we focus on a_i. In general, $\phi_j(a_i, a_i) = 0$ and $\phi_j(a_i, a_k)$ is not necessarily equal to $\phi_j(a_k, a_i)$.

Comparison of Two Alternatives on All Criteria. For a set of criteria, we combine the dominance values of a_i over a_k defined by individual criteria. As an example, we consider a simple weighted sum method. Let $\mathbf{w} = \{w_1, \ldots, w_m\}$ denote the weight vector, where w_j is the relative importance of the criterion $c_j \in C$. We assume that the weight vector satisfies two requirements: (i) $\forall w_j, w_j > 0$ and (ii) $\sum_{j=1}^{m} w_j = 1$. By following the ideas of TODIM family methods, the dominance value of a_i over a_k on multiple criteria is computed as the weighted summation:

$$\Phi(a_i, a_k) = \sum_{j=1}^{m} w_j \phi_j(a_i, a_k). \tag{10}$$

In general, $\Phi(a_i, a_i) = 0$ and $\Phi(a_i, a_k)$ is not necessarily equal to $\Phi(a_k, a_i)$.

Overall Performance Values of Alternatives. For a pair of alternatives a_i and a_k, the dominance value $\Phi(a_i, a_k)$ may be viewed as the performance of a_i relative to a_k. To obtain the overall value of a_i, we can simply summarize all dominance values of a_i relative to all alternatives in A. By following TODIM, we define the overall performance of a_i by:

$$\Psi(a_i) = \sum_{k=1}^{n} \Phi(a_i, a_k). \tag{11}$$

By inserting Eq. (10), we immediately have:

$$\Psi(a_i) = \sum_{k=1}^{n} \sum_{j=1}^{m} w_j \phi_j(a_i, a_k). \tag{12}$$

It gives an interpretation of the overall performance in terms of individual dominance values $\phi_j(a_i, a_k)$, $j = 1, \ldots, m$, $k = 1, \ldots, n$.

2.3 3RD Method

Based on the results from the last subsection, we propose a new five-step MCDM method called 3RD. As shown in Fig. 2, 3RD shares the same procedural structure of TODIM. The details of the five steps are given as follows.

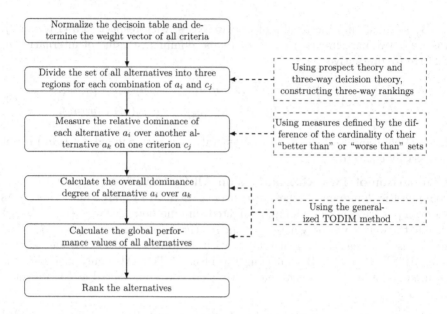

Fig. 2. The workflow of 3RD method

Step 1: For each decision alternative $a_i \in A$ and a criterion $c_j \in C$, divide the set of all alternatives into three regions based on a prospect function v and a threshold $t \geq 0$:

$$(D_j^{\gg}(a_i), D_j^{\ll}(a_i), D_j^{\approx}(a_i)). \tag{13}$$

Step 2: For each pair of alternatives $a_i, a_k \in A$, calculate the dominance degree of alternative a_i over a_k with respect to a criterion $c_j \in C$:

$$\phi_j(a_i, a_k) = \begin{cases} |D_j^{\gg}(a_k) - D_j^{\gg}(a_i)|, & \text{if } p_j(a_i) \geq p_j(a_k), \\ -|D_j^{\ll}(a_k) - D_j^{\ll}(a_i)|, & \text{if } p_j(a_i) < p_j(a_k). \end{cases} \tag{14}$$

Step 3: For each pair of alternatives $a_i, a_k \in A$, calculate the overall dominance degree of alternative a_i over a_k:

$$\Phi(a_i, a_k) = \sum_{j=1}^{m} w_j \phi_j(a_i, a_k). \tag{15}$$

Step 4: For each alternative $a_i \in A$, calculate the overall performance of alternative a_i:

$$\Psi(a_i) = \sum_{k=1}^{n} \Phi(a_i, a_k). \tag{16}$$

Step 5: Rank all of the alternatives according to the their overall performance $\Psi(a_i)$, that is,

$$a_i \succeq a_k \iff \Psi(a_i) \geq \Psi(a_k). \tag{17}$$

In this paper, we use the same prospect function and the same threshold for all criteria. As generalizations, we may consider different prospect functions and different thresholds for different criteria. Like the ELECTRE family methods, it is possible to define thresholds based on the value $p_j(a_i)$. These changes only affect the definitions of the trisections used in **Step 1**.

3   An Illustrative Example

In this section, we illustrate 3RD method by using an example taken from Gomes and Rangel's research [10]. We also analyze the sensitivity of the method with respect to the parameters.

3.1   An Example Solved by the TODIM method

Gomes and Rangel [10] presented an application of TODIM method to order residential properties with different characteristics. Their case study aims to rank the residential properties according to selected evaluation criteria. There are 15 decision alternatives $A = \{a_1, a_2, \ldots, a_{15}\}$ and 8 criteria $C = \{c_1, c_2, \ldots, c_8\}$. The performance function p is described by a normalized matrix in Table 1. The descriptions and weights of criteria are given in Table 2.

Table 1. Normalized decision alternatives' performance taken from [10]

Alternatives	Criteria							
	c_1	c_2	c_3	c_4	c_5	c_6	c_7	c_8
a_1	0.068	0.103	0.100	0.075	0.045	0.069	0.174	0.000
a_2	0.091	0.064	0.067	0.050	0.045	0.046	0.087	0.000
a_3	0.068	0.123	0.033	0.050	0.091	0.057	0.043	0.000
a_4	0.068	0.044	0.067	0.075	0.091	0.057	0.174	0.000
a_5	0.114	0.127	0.100	0.100	0.182	0.103	0.042	0.143
a_6	0.045	0.031	0.067	0.078	0.045	0.057	0.043	0.000
a_7	0.023	0.030	0.033	0.025	0.045	0.046	0.000	0.143
a_8	0.114	0.028	0.067	0.075	0.045	0.069	0.000	0.143
a_9	0.045	0.043	0.067	0.075	0.000	0.069	0.000	0.000
a_{10}	0.045	0.042	0.033	0.075	0.045	0.057	0.043	0.000
a_{11}	0.091	0.099	0.067	0.050	0.091	0.080	0.130	0.143
a_{12}	0.023	0.032	0.033	0.025	0.045	0.057	0.087	0.000
a_{13}	0.045	0.057	0.100	0.075	0.091	0.069	0.043	0.143
a_{14}	0.068	0.113	0.100	0.075	0.091	0.092	0.087	0.143
a_{15}	0.091	0.064	0.067	0.100	0.045	0.069	0.043	0.143

Table 2. Descriptions and weights of criteria taken from [10]

Criterion	Description	Assigned weights	Normalized weights
c_1	Localization	5	0.25
c_2	Constructed area	3	0.15
c_3	Quality of construction	2	0.10
c_4	State of conversation	4	0.20
c_5	Number of garage spaces	1	0.05
c_6	Number of rooms	2	0.10
c_7	Attractions	1	0.05
c_8	Security	2	0.10

In their initial setting, the loss aversion coefficient is $\lambda = 1$ and the parameters for the prospect value function is $\alpha = \beta = 0.5$, which was consistent with the traditional TODIM method. Another set of parameters is given by $\lambda = 2.25$ and $\alpha = \beta = 0.88$, which was suggested in a study by Tversky and Kahneman [22]. Table 3 shows the results under two settings of parameters. The number "1" in the ordering column represents that the corresponding alternative has the highest rank and "15" indicates that the alternative has the lowest rank.

Table 3. TODIM results from different aspects of λ, α, and β

Alternatives	$\lambda = 1$ $\alpha = \beta = 0.5$		$\lambda = 2.25$ $\alpha = \beta = 0.88$	
	Normalized value	Ordering	Normalized value	Ordering
a_1	0.6916	5	0.7500	6
a_2	0.3862	10	0.5196	8
a_3	0.3992	9	0.5076	9
a_4	0.6210	7	0.7527	5
a_5	1.0000	1	0.9696	3
a_6	0.2860	11	0.4044	11
a_7	0.0000	15	0.1870	14
a_8	0.4407	8	0.4670	10
a_9	0.0202	14	0.0000	15
a_{10}	0.2127	12	0.3384	12
a_{11}	0.8576	3	0.9826	2
a_{12}	0.1073	13	0.3242	13
a_{13}	0.7188	4	0.8175	4
a_{14}	0.9372	2	1.0000	1
a_{15}	0.6733	6	0.7213	7

3.2   The Procedure of 3RD Method

We set parameters as $\lambda = 2.25$, $\alpha = \beta = 0.88$, and $t = 0.05$ to explain 3RD method.

Step 1: We consider alternative a_5 as the reference and use Eqs. (3) and (2) to calculate the trisection with respect to each criterion.

For criterion c_1, $p_1(a_5) = 0.114$. $p_1(a_1) = 0.068$. By using Eq. (2), $v(p_1(a_1) - p_1(a_5)) = v(-0.046) = -2.25 * (0.046)^{0.88} = -0.150 < -0.05$. Thus, $a_1 \in D_1^{\lll}(a_5)$. By the same method for all criteria, we arrive at the trisection:

$$D_1^{\lll}(a_5) = \{a_1, a_2, a_3, a_4, a_6, a_7, a_9, a_{10}, a_{11}, a_{12}, a_{13}, a_{14}, a_{15}\},$$
$$D_1^{\ggg}(a_5) = \emptyset,$$
$$D_1^{\approx}(a_5) = \{a_5, a_8\}.$$

By following the same methods, we compute all trisections with respect to the remaining alternatives.

Step 2: We use the pair (a_5, a_4) as an example to show the main ideas. For each criterion $c_j \in C$, we calculate the dominance degree of a_5 over a_4 by Eq. (14).

For criterion c_1, $p_1(a_5) = 0.114 > p_1(a_4) = 0.068$. We have $\phi_1(a_5, a_4) = |D_1^{\ggg}(a_4) - D_1^{\ggg}(a_5)| = |\{a_5, a_8\} - \emptyset| = 2$.

For criterion c_7, We have $\phi_7(a_5, a_4) = -|D_7^{\lll}(a_4) - D_7^{\lll}(a_5)| = -|\{a_2, a_3, a_5, a_6, a_7, a_8, a_9, a_{10}, a_{11}, a_{12}, a_{13}, a_{14}, a_{15}\} - \{a_7, a_8, a_9\}| = -10$.

Similarly, for the rest of criteria, we have $\phi_2(a_5, a_4) = |D_2^{\ggg}(a_4) - D_2^{\ggg}(a_5)| = 5$, $\phi_3(a_5, a_4) = |D_3^{\ggg}(a_4) - D_3^{\ggg}(a_5)| = 0$, $\phi_4(a_5, a_4) = |D_4^{\ggg}(a_4) - D_4^{\ggg}(a_5)| = 0$, $\phi_5(a_5, a_4) = |D_5^{\ggg}(a_4) - D_5^{\ggg}(a_5)| = 1$, $\phi_6(a_5, a_4) = |D_6^{\ggg}(a_4) - D_6^{\ggg}(a_5)| = 2$, and $\phi_8(a_5, a_4) = |D_8^{\ggg}(a_4) - D_8^{\ggg}(a_5)| = 7$.

Step 3: We use the pair (a_5, a_4) to show the computation of the overall dominance degree of a_5 over a_4. According to Eq. (15), $\Phi(a_5, a_4)$ is given as the weighted summation of results from **Step 2**:

$$\Phi(a_5, a_4) = \sum_{j=1}^{m} w_j * \phi_j(a_5, a_4)$$
$$= 0.25 * 2 + 0.15 * 5 + 0.10 * 0 + 0.20 * 0 + 0.05 * 1 + 0.10 * 2$$
$$+ 0.05 * (-10) + 0.10 * 7 = 1.7.$$

Step 4: By Eq. (16), the overall performance of alternative a_5 is calculated by the summation of the dominance of a_5 over the all alternatives in A. That is,

$$\Psi(a_5) = \sum_{k=1}^{n} \Phi(a_5, a_k)$$
$$= 1.1 + 2.15 + 2.95 + 1.7 + 0 + 3.2 + 7.3 + 1.75 + 3.8 + 4.3 + 0$$
$$+ 7 + 2.15 + 0.25 + 1.15 = 38.8.$$

Step 5. After calculating the total performance of all alternatives, we are able to rank the set of alternatives A according to $\Psi(a_r)$, $\forall a_r \in A$. Table 4 gives the total performance values and a ranking of the 15 alternatives.

Table 4. Total performance and ordering of alternatives

Alternatives	$\lambda = 2.25$, $\alpha = \beta = 0.88$, $t = 0.05$	
	Total performance	Ordering
a_1	8.20	5
a_2	−7.30	8
a_3	−13.90	10
a_4	−7.85	9
a_5	38.80	1
a_6	−34.60	12
a_7	−60.60	14
a_8	0.60	6
a_9	−34.45	11
a_{10}	−41.75	13
a_{11}	24.55	3
a_{12}	−65.15	15
a_{13}	−5.85	7
a_{14}	25.75	2
a_{15}	14.45	4

Table 5. Sensitivity analysis of 3RD method with different λ, α, β and t

Approahces	3RD method											
Loss aversion coefficient	$\lambda = 1$						$\lambda = 2.25$					
Parameters	$\alpha = \beta = 0.5$			$\alpha = \beta = 0.88$			$\alpha = \beta = 0.5$			$\alpha = \beta = 0.88$		
Threshold t	0	0.05	0.1	0	0.05	0.1	0	0.05	0.1	0	0.05	0.1
a_1	4	5	5	4	5	8	4	5	5	4	5	5
a_2	9	9	9	9	8	11	9	9	9	9	8	8
a_3	10	10	10	10	10	9	10	10	10	10	10	10
a_4	8	8	8	8	7	10	8	8	8	8	9	9
a_5	1	1	1	1	1	1	1	1	1	1	1	1
a_6	12	12	13	12	11	13	12	12	12	12	12	11
a_7	15	15	14	15	14	7	15	15	15	15	14	12
a_8	6	6	6	6	6	6	6	6	6	6	6	6
a_9	11	11	11	11	12	14	11	11	11	11	11	13
a_{10}	13	13	12	13	13	12	13	13	13	13	13	14
a_{11}	5	4	4	5	2	2	5	4	4	5	3	3
a_{12}	14	14	15	14	15	15	14	14	14	14	15	15
a_{13}	7	7	7	7	9	5	7	7	7	7	7	7
a_{14}	2	2	2	2	3	3	2	2	2	2	2	2
a_{15}	3	3	3	3	4	4	3	3	3	3	4	4

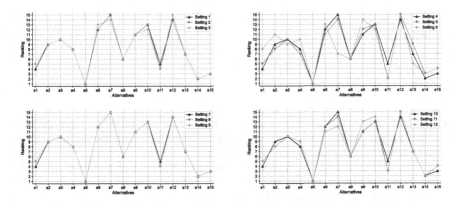

Fig. 3. 3RD rankings under different parameter settings

3.3 Sensitivity Analysis of 3RD

We carried out a sensitivity analysis by altering the parameters and the threshold. We tested λ from $\{1, 2.25\}$, $\alpha = \beta$ from $\{0.5, 0.88\}$, and t from $\{0, 0.05, 0.1\}$. There are a total of twelve settings to verify the stability of 3RD. The results are shown in Table 5 and Fig. 3. It is clear that the introduced threshold t plays an essential role in the pair-wise comparison and the final ordering. The similar trends of rankings under the twelve settings of parameters demonstrated that the proposed 3RD method is not very sensitive to small changes of parameters.

4   Conclusion

Thinking in threes, as the philosophy of three-way decision theory, provides a novel view of MCDM problems. By combining thinking in threes and ideas from existing MCDM methods, we have proposed a new three-way ranking-based approach. Inspired by the prospect theory, with respect to a decision alternative and a criterion, we build a trisection, namely, a three-way ranking, of the alternative set. Two subsets of the trisection correspond to the dominating and dominated sets in the dominance-based rough set approach. The addition of one more subset of approximately the same alternatives takes consideration of the degree of preference, in a similar way as being done in the ELECTRE and the PROMETHEE families methods. Based on trisections of a pair of alternatives, we design measures to quantify dominance of one alternative over another alternative on a single criterion and multiple criteria, respectively. These measures are used to compute the overall performance of alternatives and to rank alternatives, according to the procedure of TODIM method.

Our proposed method is a three-way ranking based method called 3RD. The advantages of 3RD can be seen from two aspects: 1) the prospect theory provides a solid basis for constructing and interpreting a three-way ranking, and 2) there are different ways to determine the threshold for defining an observable degree of dominance. Based on the preliminary results reported in this paper, we may explore further new MCDM methods based on the philosophy of thinking and classification in threes.

References

1. Benayoun, R., Roy, B., Sussman, B.: ELECTRE: une méthode pour guider le choix en présence de points de vue multiples. Note de travail 49, SEMA-METRA International, Direction Scientifique (1966)
2. Brans, J.P., Mareschal, B., Vincke, P.: PROMETHEE: a new family of outranking methods in multicriteria analysis. In: Brans, J.P. (ed.) Operational Research 1984, pp. 477–490. North-Holland, Amsterdam (1984)
3. Brans, J.-P., De Smet, Y.: PROMETHEE methods. In: Greco, S., Ehrgott, M., Figueira, J.R. (eds.) Multiple Criteria Decision Analysis. ISORMS, vol. 233, pp. 187–219. Springer, New York (2016). https://doi.org/10.1007/978-1-4939-3094-4_6
4. Figueira, J.R., Mousseau, V., Roy, B.: ELECTRE methods. In: Greco, S., Ehrgott, M., Figueira, J.R. (eds.) Multiple Criteria Decision Analysis. ISORMS, vol. 233, pp. 155–185. Springer, New York (2016). https://doi.org/10.1007/978-1-4939-3094-4_5
5. Fishburn, P.C.: Nonlinear Preference and Utility Theory. Johns Hopkins University Press, Baltimore (1988)
6. Greco, S., Matarazzo, B., Słowiński, R.: Rough approximation of a preference relation by dominance relations. Eur. J. Oper. Res. **117**, 63–83 (1999)
7. Greco, S., Matarazzo, B., Słowiński, R.: Rough approximation by dominance relations. Int. J. Intell. Syst. **17**, 153–171 (2002)
8. Greco, S., Ehrgott, M., Figueira, J.R.: Multiple Criteria Decision Analysis: State of the Art Surveys, 2nd edn. Springer, New York (2016). https://doi.org/10.1007/978-1-4939-3094-4
9. Gomes, L.F.A.M., Lima, M.M.P.P.: TODIM: basics and application to multicriteria ranking of projects with environmental impacts. Found. Comput. Decis. Sci. **16**, 113–127 (1992)
10. Gomes, L.F.A.M., Rangel, L.A.D.: An application of the TODIM method to the multicriteria rental evaluation of residential properties. Eur. J. Oper. Res. **193**, 204–211 (2009)
11. Gomes, L.F.A.M., González, X.I.: Behavioral multi-criteria decision analysis: further elaborations on the TODIM method. Found. Comput. Decis. Sci. **37**, 3–8 (2012)
12. Hwang, C.L., Yoon, K.: Multiple Attribute Decision Making: Methods and Applications a State-of-the-Art Survey. Springer, Heidelberg (1981). https://doi.org/10.1007/978-3-642-48318-9
13. Jia, F., Liu, P.D.: A novel three-way decision model under multiple-criteria environment. Inf. Sci. **471**, 29–51 (2019)
14. Kahneman, D., Tversky, A.: Prospect theory: an analysis of decision under risk. Econometrica **47**, 263–291 (1979)
15. Lee, Y.S., Shih, H.S.: A study of behavioral considerations based on prospect theory for group decision making. Int. J. Inf. Manage. Sci. **26**, 103–122 (2015)
16. Lin, S.: Rank aggregation methods. Wiley Interdisc. Rev.: Comput. Stat. **2**, 555–570 (2010)
17. Llamazares, B.: An analysis of the generalized TODIM method. Eur. J. Oper. Res. **269**, 1041–1049 (2018)

18. Martel, J.-M., Matarazzo, B.: Other outranking approaches. In: Greco, S., Ehrgott, M., Figueira, J.R. (eds.) Multiple Criteria Decision Analysis. ISORMS, vol. 233, pp. 221–282. Springer, New York (2016). https://doi.org/10.1007/978-1-4939-3094-4_7

19. Roy, B.: The outranking approach and the foundations of ELECTRE methods. In: Bana e Costa, C.A. (ed.) Readings in Multiple Criteria Decision Aid, pp. 155–183. Springer, Heidelberg (1990). https://doi.org/10.1007/978-3-642-75935-2_8

20. Słowiński, R., Greco, S., Matarazzo, B.: Dominance-based rough set approach to reasoning about ordinal data. In: Kryszkiewicz, M., Peters, J.F., Rybinski, H., Skowron, A. (eds.) RSEISP, vol. 4585, pp. 5–11. Springer, Berlin Heidelberg (2007)

21. Triantphyllou, E.: Multi-criteria Decision Making Methods: A Comparative Study. Springer, Boston (2000). https://doi.org/10.1007/978-1-4757-3157-6

22. Tversky, A., Kahneman, D.: Advances in prospect theory: cumulative representations of uncertainty. J. Risk Uncertain. **5**, 297–323 (1992)

23. Wang, J.J., Ma, X.L., Xu, Z.S., Zhan, J.M.: Three-way multi-attribute decision making under hesitant fuzzy environments. Inf. Sci. **552**, 328–351 (2021)

24. Yao, Y.Y.: Three-way decision and granular computing. Int. J. Approx. Reason. **103**, 107–123 (2018)

25. Yao, Y.Y.: Tri-level thinking: models of three-way decision. Int. J. Mach. Learn. Cybern. **11**, 947–959 (2019)

26. Yao, Y.Y.: Set-theoretic models of three-way decision. Granular Comput. **6**, 133–148 (2020)

27. Yao, Y.Y.: The geometry of three-way decision. Appl. Intell. (2021). https://doi.org/10.1007/s10489-020-02142-z

28. Zhan, J.M., Jiang, H.B., Yao, Y.Y.: Three-way multi-attribute decision-making based on outranking relations. IEEE Trans. Fuzzy Syst. (2020). https://doi.org/10.1109/TFUZZ.2020.3007423

Author Index

Printed in the United States
by Baker & Taylor Publisher Services